Artificial Intelligence for Multisource Geospatial Information

Artificial Intelligence for Multisource Geospatial Information

Editors

Gloria Bordogna
Cristiano Fugazza

MDPI • Basel • Beijing • Wuhan • Barcelona • Belgrade • Manchester • Tokyo • Cluj • Tianjin

Editors

Gloria Bordogna
IREA
CNR
Milano
Italy

Cristiano Fugazza
IREA
CNR
Milano
Italy

Editorial Office
MDPI
St. Alban-Anlage 66
4052 Basel, Switzerland

This is a reprint of articles from the Special Issue published online in the open access journal *ISPRS International Journal of Geo-Information* (ISSN 2220-9964) (available at: www.mdpi.com/journal/ijgi/special_issues/Geospatial_Artificial_Intelligence).

For citation purposes, cite each article independently as indicated on the article page online and as indicated below:

LastName, A.A.; LastName, B.B.; LastName, C.C. Article Title. *Journal Name* **Year**, *Volume Number*, Page Range.

ISBN 978-3-0365-6387-9 (Hbk)
ISBN 978-3-0365-6386-2 (PDF)

Contents

About the Editors

Gloria Bordogna

Gloria Bordogna is a Director of research at the National Research Council of Italy within the Institute of Electromagnetic Sensing of the Environment (CNR-IREA). She has been with CNR since 1986. From 2003 to 2010 she was contract professor of Information Retrieval and GIS at the Information Engineering Department of Bergamo University. Her research activity concerns uncertainty and imprecision modeling in information retrieval systems, database and geographic information systems with soft computing. She has defined methods for textual contents' representation, for flexible querying and mining textual information, geographic information and social networks by fuzzy clustering, and for knowledge and data-driven fusion of multidimensional multisource data. Among the main organization activities since 2022 she is in the steering committee of the Int. Conf. on Flexible Query Answering Systems—FQAS, and since 2008 she has organized the special track entitled "Information Access and Retrieval" at the "ACM Symposium on Applied Computing".

She is a member of the editorial board of "MDPI *ISPRS Int. Journal of Geo-information*", of "*Geomatics*", and of the "Newsletter of SIGAPP, ACR, Applied Computing Review". She serves as a reviewer of journals and research agencies (ESF—European Science Foundation, FCT (Portugal), ANR (France); FWO (Belgium), ERCEA, "ERC Starting Grants" program) and participates in the program committee of several international conferences such as ACM SIGIR, ECIR, ACM CIKM, FUZZIEEE, IPMU. As main recognition of her research activity, in 2017 she was elected fellow of the International Fuzzy System Association. Since 2021, she is ranked in the Stanford University list of the top 2% most widely cited researchers in the world.

Cristiano Fugazza

Cristiano Fugazza is a researcher by the Institute for Electromagnetic Sensing of the Environment of the National Research Council of Italy (CNR-IREA), focusing on semantic interoperability, Artificial Intelligence, and distributed metadata management in the context of Spatial Data Infrastructures. He is currently seconded to the Joint Research Centre, European Commission, for the FAIRification of patient registries in the context of the EC/JRC rare disease strategy. He did his post-doc by the Spatial Data Infrastructure Unit of the JRC-IES on the creation, harmonization, and exploitation of multilingual thesauri in the SKOS format for the purpose of annotation and discovery of spatial resources. He supported the development of the Global Earth Observation System of Systems (GEOSS) and the European spatial data infrastructure established by the INSPIRE directive by managing a comprehensive repository of harmonized reference and thematic thesauri. He obtained his Ph.D. in Computer Science by the Dept. of Information Technologies, University of Milan, focusing on Knowledge Management and Semantic Web technologies. He has contributed to, and published in international journals, conferences, and workshops in the domains of spatial data infrastructures, supply chain management, business modeling, digital rights management, personal information protection, and persistent identifiers.

Preface to "Artificial Intelligence for Multisource Geospatial Information"

This volume collects 10 original research contributions published in the Special Issue entitled "Artificial Intelligence for Multisource Geospatial Information" of the *ISPRS International Journal of Geo-Information*. The focus is on different methods of Geospatial Artificial Intelligence (GeoAI) based on deep learning using different network architectures, clustering, soft computing, and semantic approaches. They are proposed to deal with a variety of Geospatial Big Data (GBD), such as georeferenced texts and photos in social networks, remote sensing images, cartographic maps, multidimensional geo databases, metadata in spatial data infrastructures, and for different tasks, such as for multisource georeferenced text integration and geodata flexible querying, for social sensing by applying sentiment analysis, clustering and geo analysis, for segmentation of roads, clouds and snow, and for detection of small targets and people on the streets.

Gloria Bordogna and Cristiano Fugazza
Editors

International Journal of *Geo-Information*

MDPI

Editorial

Artificial Intelligence for Multisource Geospatial Information

Gloria Bordogna * and Cristiano Fugazza

CNR–IREA, Via A. Corti 12, 20133 Milano, Italy
* Correspondence: bordogna.g@irea.cnr.it; Tel.: +39-02-23699-299

1. Introduction

The term Geospatial Artificial Intelligence (GeoAI) is quite cumbersome, and it has no single, shared definition.

An initial, narrow definition characterizes GeoAI as the application of machine learning toolkits to the context of Geographic Information Systems (GISs) in order to simulate future scenarios via data classification and smart predictive analysis with respect to several events and phenomena, such as the occurrence of disasters, human health epidemiology, and the evolution of ecosystems and biodiversity, which, in turn, is undertaken in order to respond to communities and support community resilience by processing traditional kinds of geographic information represented in digital cartography [1].

Another wider definition considers GeoAI as the processing of Geospatial Big Data (GBD) of heterogeneous forms and sources, including both traditional digital cartography managed by GISs, remote-sensing-based multidimensional data including images and image time series, georeferenced unstructured and semi-structured texts, and complex geo databases, with a focus on the geographic dimension [2].

Thus, the application of techniques from AI and data science to GBD, via the exploitation of high-performance-computing platforms, are merged into GeoAI in order to understand natural and social phenomena.

A general definition characterizes GeoAI as the use of artificial intelligence methods, including machine learning and deep learning, to produce knowledge through the analysis of spatial data and imagery [3]. In this sense, GeoAI is regarded as an emergent spatial analytical framework for data-intensive geographic information science, facilitating both environmental sensing and so-called "social sensing" by exploiting both the digital traces people leave behind as they interact with the IoT and the user-generated digital content created on social networks to understand the dynamics related to human mobility patterns and social phenomena.

Moreover, the specificities and importance of the geospatial dimension; its heterogeneity in terms of both conceptualization based on either "place" or "space"; the varied formats of spatial information; the different scales; the need for representing distinct geosemantics, i.e., the semantics of locations; and the different needs of analysis dictated by the goals of the applications, which often necessitate geospatial and temporal reasoning, pose new challenges and opportunities with respect to AI.

The current research topics include multiresolution and multisource GBD fusion; the multiscale geosummarization of information to improve the quality of GBD; multisource, heterogeneous GBD integration for data reuse; and experimentation in deep learning applied to multispectral remote-sensing images, such as CNN, RCNN, LSTM, and GANs generally used for RGB pictures. Finally, GeoAI must bridge the gap between opaque technologies, such as deep learning, which are generally regarded as black-boxes, and more traditional and transparent machine learning approaches to knowledge management, such as decision trees; KNN; clustering algorithms; data mining; soft computing, including genetic algorithms and fuzzy logic; ensemble approaches; and semantic representation and analysis. This can facilitate the advancement of the features of explainable AI, which

Citation: Bordogna, G.; Fugazza, C. Artificial Intelligence for Multisource Geospatial Information. *ISPRS Int. J. Geo-Inf.* **2023**, *12*, 10. https://doi.org/10.3390/ijgi12010010

Received: 22 December 2022
Revised: 23 December 2022
Accepted: 27 December 2022
Published: 30 December 2022

constitutes a mandatory characteristic of software when used for critical tasks impacting people's safety and security, such as in the health and law enforcement domains.

Our motivation to organize this Special Issue stemmed from an observation of the increasing number of academic papers focused on the application of GeoAI and on the evaluation of its potential to analyse natural, environmental, human-driven, and social changes and events.

Nevertheless, the Special Issues published at the launch date of our proposal mostly conceived of GeoAI in the strict sense, and not in the broader view we have already addressed in this Special Issue, wherein we welcomed approaches that merged multisource and heterogeneous GBD.

This Special Issue has received a total of 20 submitted papers; 10 of these papers have been accepted.

The authors' affiliations correspond to the following countries: Italy, Egypt, the United Arab Emirates, South Korea, Turkey, Kazakhstan, China, and the US.

The contributions can be grouped into three main topics:

(1) Social sensing by mining geotagged, user-generated content and traces in the form of either semi-structured textual data or photos;

(2) Environmental monitoring and analysis by employing remote-sensing spatial temporal data;

(3) Methodological approaches to integrating, mining, representing, and interpreting multisource and multidimensional spatial-temporal data.

2. GeoAI for Mining Geotagged, User-Generated Content and Traces

Within this section, we consider the descriptions of original approaches developed to classify and mine geotagged user-generated content and traces, which are created either purposefully or unknowingly by users of social networks:

(i) "*Spatio-Temporal Sentiment Mining of COVID-19 Arabic Social Media*" by Tarek Elsaka et al. [4] is a very interesting paper that mixes several AI techniques for using NLP and GeoAI to mine the available large datasets implicitly geotagged from social media in the Arabic language. This study's goal is to understand people's responses to the COVID-19 pandemic. They first developed a technique for inferring geospatial information from non-geotagged Arabic tweets by performing geo-parsing and geo-coding. Secondly, they designed a sentiment analysis mechanism applied at various location resolutions (regions/countries) and at separate topic abstraction levels (subtopics and main topics). In addition, a correlation-based analysis of Arabic tweets and the official health providers' data was presented. In the conducted experiments, the results were visualized in the combined context of the data from the official health records and lockdown data worldwide, which showed that their method was able to determine the location of tweets so that the total percentage of location-enabled tweets increased from 2% to 46% (about 2.5 M tweets). Furthermore, a positive correlation between the foremost topics such as lockdown and vaccines and new cases of COVID-19 was also reported. In addition, the negative feelings of Arab Twitter users during the pandemic were also analysed, which generally included topics related to lockdowns, closures, and law enforcement, thus demonstrating how social media constitutes a useful and effective means of "social sensing".

(ii) "*Automatic Classification of Photos by Tourist Attractions Using Deep Learning Model and Image Feature Vector Clustering*" by Jiyeon Kim and Youngok Kang [5] is another example of a social-sensing application in which photos created in social media are regarded as representations of tourists' visual preferences for a specific attraction. Thus, the paper proposes a method of automatically classifying tourist photos by tourist attractions. Accordingly, it applies methods of deep learning and image feature vector clustering to identify clusters of photos associated with attractions. The authors conducted experiments by collecting a dataset of photos attached to reviews posted by foreign tourists on TripAdvisor. The advantage of this proposal is that it does not require the creation of a classification category in advance; moreover, it is capable of flexibly extracting categories for each tourist destination and improving classification performance even with rather small data volumes.

(iii) *"Detecting People on the Street and the Streetscape Physical Environment from Baidu Street View Images and Their Effects on Community-Level Street Crime in a Chinese City"* by Han Yue et al. [6] is another example of a social-sensing application, which, in this case, is used to assess street crime via traces unknowingly left by Baidu users. This study is the first to combine Street View images (Baidu Street View), deep learning algorithms, and spatial statistical regression models to retrieve the number of people on a given street and the features of the visual streetscape environment to understand street crime. Finally, this study determines the quantitative measurement of people on a given street and the set of streetscape features that has potential influences on crime by combining the outputs of two deep learning networks. Specifically, they found that the number of people on the street had a significantly positive impact on the total street crime assessment.

3. Remote-Sensing Spatial-Temporal Data for Environmental Monitoring

This section groups three articles that provide novel approaches to the application of GeoAI methods to interpret remote-sensing spatial-temporal data, either acquired from LiDAR or from sensors on satellites. They apply a range of different machine learning and deep learning techniques for distinct environmental applications and tasks, and assess the accuracy of the results by running experiments on real data:

(iv) *"The Use of Machine Learning Algorithms in Urban Tree Species Classification"* by Zehra Cetin and Naci Yastikli [7] analyses LiDAR data with the aim of identifying urban tree species in cities. This is an important objective for planning sustainable smart cities, since knowledge of the locations of tree species in an urban area facilitates the estimation of parameters such as air, water, and land quality; carbon accumulation reduction; the mitigation of urban heat island effects; and the protection of soil and water balance. LiDAR systems are a cost-effective alternative to the traditional methods of identifying tree species based on field surveys and aerial photograph interpretation; thus, their use also constitutes an original application for this kind of GBD. The aim of this work was to assess the usage of machine learning algorithms for classifying the deciduous (broadleaf) and coniferous tree species from 3D raw LiDAR data in the study site, i.e., the Davutpasa Campus of Yildiz Technical University, Istanbul, Turkey. To this end, a total of 25 spatial- and intensity-based features were analysed by three machine learning classifiers—the support vector machine (SVM), random forest (RF), and multi-layer perceptron (MLP)—to discriminate deciduous and coniferous tree species in the study area. The compared evaluation results found that the SVM and RF algorithms generally yielded better classification results than the MLP algorithm for the target task based on the available training data.

(v) *"Multi-Resolution Transformer Network for Building and Road Segmentation of Remote Sensing Image"* by Zhongyu Sun et al. [8] details the authors' extraction of buildings and roads from remote-sensing images for land cover monitoring and soil consumption identification, which are of great help in urban planning. Currently, deep learning algorithms are mainly used for building and road extraction. However, for semantic segmentation, these methods are limited with respect to the receptive field of high-resolution remote-sensing images; thus, the image features need to be compressed by down-sampling so as to determine the loss of detailed information. In order to address this issue while avoiding down sampling, the paper proposes a novel deep learning architecture, the Hybrid Multi-resolution and Transformer semantic extraction Network (HMRT), which stores multiresolution information so as to improve the ability to comprehend a scene. The experiments proved that the proposed method is superior to the existing baseline methods.

(vi) *"RepDarkNet: A Multi-Branched Detector for Small-Target Detection in Remote Sensing Images"* by Liming Zhou et al. [9] addresses the problem of detecting small targets in remote-sensing images, which are often occluded by shadows. To address this shortcoming in the detection of targets, they propose a backbone feature-extraction network called "RepDarkNet" that considerably improves the overall network accuracy, with almost no increase in inference time with respect to the baseline approach. In addition, they propose a multi-scale, cross-layer detector that also significantly improves the network's ability to detect small targets.

(vii) *"Cloud and Snow Segmentation in Satellite Images Using an Encoder–Decoder Deep Convolutional Neural Networks"* by Kai Zheng et al. [10] details the cloud and snow segmentation of satellite images. They propose a cloud-and-snow segmentation method based on a deep convolutional neural network (DCNN) with an enhanced encoder–decoder architecture. Comparative experiments show that the proposed method is superior to the baseline methods. Additionally, a rough-labelled dataset containing more than 20,000 images and fine-labelled data consisting of 310 satellite images are created, with which they studied the relationship between the quality and quantity of the labels of training data and the performance of cloud and snow segmentation. Through experiments on the same network with different datasets, they found that cloud-and-snow segmentation performance is more closely related to the quantity of labels rather than their quality. Namely, under the same labelling consumption, the method performs better when solely using rough-labelled images than when using rough-labelled images plus 10% fine-labelled images.

4. Methodological Approaches to Dealing with Multisource and Multidimensional Numeric and Alphanumeric Spatial-Temporal Data

This section considers approaches whose focus is primarily on the methods for the integration, management, querying, and mining of multisource and multidimensional numeric and alphanumeric spatial-temporal data. Specific attention is paid to the inherent inconsistencies and uncertainty of the spatial temporal information:

(viii) *"Soft Integration of Geo-Tagged Data Sets in J-CO-QL^{+}"* by Paolo Fosci and Giuseppe Psaila [11] tackles the need for the integration of distinct heterogeneous data sets concerning public places located on Earth created by distinct Web applications, social networks, recommendation platforms, and likes using the schemeless JSON format. To exploit complementary and redundant information in such multisource datasets and reuse it for their purposes, analysts usually need to perform complex, long pre-processing tasks including data transformation, homogenisation, and data-cleaning, as well as training activities that require tedious and extensive labelling of data. To perform integration from scratch, by avoiding these burdening activities, this paper proposes a methodology based on a soft integration framework defined within soft-computing and fuzzy sets. The proposed framework, which is a stand-alone tool devised to process JSON datasets stored within distinct JSON document stores, enables researchers to perform flexible querying and transformations of multiple heterogeneous data sets by providing operators with the ability to select, manipulate, and merge JSON objects with distinct structures. The ease of use, effectiveness, and efficiency of this soft integration technique is demonstrated with real data.

(ix) *"Modelling and Querying Fuzzy SOLAP-Based Framework"* by Sinan Keskin and Adnan Yazıcı [12] addresses the need to analyze GBD generated by sensors by considering the uncertainty and fuzziness inherent in spatiotemporal database applications. Spatial Online Analytical Processing (SOLAP) provides appropriate data structures and supports the querying of multidimensional numeric and alphanumeric spatial temporal data. Nevertheless, this technique is limited in terms of its ability to manage uncertainty and fuzziness. Thus, this paper proposes FSOLAP, which is a new framework based on fuzzy logic technologies and SOLAP. The study uses crisp measures as inputs to this framework and applies fuzzy operations to obtain the membership functions and fuzzy classes; then, it generates fuzzy association rules. Therefore, FSOLAP does not require predefined sets of fuzzy inputs. This approach is applied to handle non-spatial and fuzzy spatial queries, as well as spatiotemporal fuzzy query types. Additionally, FSOLAP is not only used to query and analyse historical data but also to handle predictive fuzzy spatiotemporal queries, which typically require an inference mechanism.

(x) "*Implicit, Formal, and Powerful Semantics in Geoinformation*" by the authors herein and our colleagues Paolo Tagliolato Acquaviva D'Aragona and Paola Carrara [13] addresses the need to identify suitable methodologies and frameworks in order to represent and mine GBD depending on their geosemantics—whose classification is often ill-defined. A meta-review of the state of the art in geosemantics is performed to pinpoint relevant "keywords" representing key concepts, challenges, methods, and technologies of the domain. Then, real case studies dealing with geoinformation are first categorized based on three forms of semantics, defined as implicit, formal, and powerful (i.e., soft) depending on the kind of the input data they use; consequently, they are successively associated with the previously identified relevant keywords for the domain of geosemantics. Finally, the similarities between each pair of analysed case studies in the space of the keywords are computed in order to ascertain whether distinguishing methodologies, techniques, and challenges can be related to the three distinct categories of implicit, formal, and powerful. The outcomes of the analysis identified the methods and technologies that are more suited to modelling and processing specific forms of geosemantics categorised into implicit, formal, and explicit categories.

5. Conclusions

The contributions published in this Special Issue offer a panoply of techniques and approaches used to deal with GBDs by means of a variety of GeoAI methods. The approaches are varied with respect to the objectives of their studies, which include both social and environmental sensing, as well as with respect to the kind of GBD sources, genre, and formats. While remote-sensing data from satellites and sensors are used mainly for environmental applications, social media-georeferenced data, both textual and pictorial, are mainly used for social applications. In both domains, a current trend is to apply deep learning methods and to compare the results achieved with baselines or with more traditional machine learning algorithms.

Besides the mainstream deep learning methods, some bucking methods were also proposed by some of the papers, such as the use of transparent machine learning algorithms based on soft computing and fuzzy logic. This was motivated by the need for the analyst to have greater control over the automatic process in order to be able to understand the phenomenon and to explain it to stakeholders.

Some methodological proposals outlined the need to tackle new challenges with respect to GBD management, including the need for novel means for multisource GBD integration and transformation as well as uncertainty and imprecision management. Finally, from a meta-review of approaches, a synthesis is proposed in order to outline the most suitable GeoAI methods for managing GBD depending on their geosemantics.

We are also aware that the collected contributions and their topics do not exhaustively cover all of the challenges related to GeoAI. For example, other, unincluded challenging topics of GeoAI concern spatial-temporal and thematic solutions, which entails the ability to answer user questions regarding the retrieval of relevant information from heterogeneous, multisource GBD, thus satisfying user needs related to specific geographic areas and to a desired time range, such as "find a well-reputed pizza restaurant close to Milano railway station which is open on Monday evening". Another issue in the perspective concerning the reproducibility and replicability of experiments is the need for high-quality, labelled GBD benchmark collections that are freely available and allow the research community to compare the proposed methods. While this practice is well-established in the textual information retrieval field, it is still at its infancy in the geographic research community [2].

Finally, we believe we are still in the early stages of integrating and analysing multi-source and multimodal GBD using GeoAI methods, including geotagged voice and audio files, remote-sensing images and their derived products, and geotagged text annotations, which have been collected as natural and environmental observations in many citizen science projects. The application of GeoAI methods based on embedding representations may constitute a quantum leap in multimodal GBD integration.

Author Contributions: Both authors contributed equally to the conceptualization and writing of the editorial. All authors have read and agreed to the published version of the manuscript.

Funding: This research received no external funding.

Acknowledgments: We want to express our congratulations to the authors of the papers for their interesting works; to the anonymous referees whose key help made it possible to improve the contents of the papers; and finally to the editorial staff of IJGI for their excellent assistance in producing this Special Issue.

Conflicts of Interest: The authors declare no conflict of interest.

References

1. What is GeoAI? Available online: https://ecce.esri.ca/mac-blog/2018/04/23/what-is-geoai/ (accessed on 10 November 2022).
2. Janowicz, K.; Gao, S.; McKenzie, G.; Hu, Y.; Bhaduri, B. GeoAI: Spatially explicit artificial intelligence techniques for geographic knowledge discovery and beyond. *Int. J. Geogr. Inf. Sci.* **2019**, *34*, 625–636. [CrossRef]
3. Hernandez, L. ELISE Webianar: GeoAI—Presentation: Geospatial Data and Artificial Intelligence—A Deep Dive into GeoAI. 2020. Available online: https://joinup.ec.europa.eu/collection/elise-european-location-interoperability-solutions-e-government/document/presentation-geospatial-data-and-artificial-intelligence-deep-dive-geoai (accessed on 10 November 2022).
4. Elsaka, T.; Afyouni, I.; Hashem, I.; Al Aghbari, Z. Spatio-Temporal Sentiment Mining of COVID-19 Arabic Social Media. *ISPRS Int. J. Geo-Inf.* **2022**, *11*, 476. [CrossRef]
5. Kim, J.; Kang, Y. Automatic Classification of Photos by Tourist Attractions Using Deep Learning Model and Image Feature Vector Clustering. *ISPRS Int. J. Geo-Inf.* **2022**, *11*, 245. [CrossRef]
6. Yue, H.; Xie, H.; Liu, L.; Chen, J. Detecting People on the Street and the Streetscape Physical Environment from Baidu Street View Images and Their Effects on Community-Level Street Crime in a Chinese City. *ISPRS Int. J. Geo-Inf.* **2022**, *11*, 151. [CrossRef]
7. Cetin, Z.; Yastikli, N. The Use of Machine Learning Algorithms in Urban Tree Species Classification. *ISPRS Int. J. Geo-Inf.* **2022**, *11*, 226. [CrossRef]
8. Sun, Z.; Zhou, W.; Ding, C.; Xia, M. Multi-Resolution Transformer Network for Building and Road Segmentation of Remote Sensing Image. *ISPRS Int. J. Geo-Inf.* **2022**, *11*, 165. [CrossRef]
9. Zhou, L.; Zheng, C.; Yan, H.; Zuo, X.; Liu, Y.; Qiao, B.; Yang, Y. RepDarkNet: A Multi-Branched Detector for Small-Target Detection in Remote Sensing Images. *ISPRS Int. J. Geo-Inf.* **2022**, *11*, 158. [CrossRef]
10. Zheng, K.; Li, J.; Ding, L.; Yang, J.; Zhang, X.; Zhang, X. Cloud and Snow Segmentation in Satellite Images Using an Encoder–Decoder Deep Convolutional Neural Networks. *ISPRS Int. J. Geo-Inf.* **2021**, *10*, 462. [CrossRef]
11. Fosci, P.; Psaila, G. Soft Integration of Geo-Tagged Data Sets in J-CO-QL$^+$. *ISPRS Int. J. Geo-Inf.* **2022**, *11*, 484. [CrossRef]
12. Keskin, S.; Yazıcı, A. Modeling and Querying Fuzzy SOLAP-Based Framework. *ISPRS Int. J. Geo-Inf.* **2022**, *11*, 191. [CrossRef]
13. Bordogna, G.; Fugazza, C.; D'Aragona, P.T.A.; Carrara, P. Implicit, Formal, and Powerful Semantics in Geoinformation. *ISPRS Int. J. Geo-Inf.* **2021**, *10*, 330. [CrossRef]

International Journal of
Geo-Information

MDPI

Article

Spatio-Temporal Sentiment Mining of COVID-19 Arabic Social Media

Tarek Elsaka [1,2,*], Imad Afyouni [1], Ibrahim Hashem [1] and Zaher Al Aghbari [1]

[1] Computer Science, University of Sharjah, Sharjah 27272, United Arab Emirates
[2] Agricultural Research Center, Cairo 12619, Egypt
* Correspondence: teka@sharjah.ac.aelsa

Abstract: Since the recent outbreak of COVID-19, many scientists have started working on distinct challenges related to mining the available large datasets from social media as an effective asset to understand people's responses to the pandemic. This study presents a comprehensive social data mining approach to provide in-depth insights related to the COVID-19 pandemic and applied to the Arabic language. We first developed a technique to infer geospatial information from non-geotagged Arabic tweets. Secondly, a sentiment analysis mechanism at various levels of spatial granularities and separate topic scales is introduced. We applied sentiment-based classifications at various location resolutions (regions/countries) and separate topic abstraction levels (subtopics and main topics). In addition, a correlation-based analysis of Arabic tweets and the official health providers' data will be presented. Moreover, we implemented several mechanisms of topic-based analysis using occurrence-based and statistical correlation approaches. Finally, we conducted a set of experiments and visualized our results based on a combined geo-social dataset, official health records, and lockdown data worldwide. Our results show that the total percentage of location-enabled tweets has increased from 2% to 46% (about 2.5M tweets). A positive correlation between top topics (lockdown and vaccine) and the COVID-19 new cases has also been recorded, while negative feelings of Arab Twitter users were generally raised during this pandemic, on topics related to lockdown, closure, and law enforcement.

Keywords: Arabic tweets; COVID-19 pandemic; sentiment analysis; social data mining; spatio-temporal correlation

Citation: Elsaka, T.; Afyouni, I.; Hashem, I.; Al Aghbari, Z. Spatio-Temporal Sentiment Mining of COVID-19 Arabic Social Media. *ISPRS Int. J. Geo-Inf.* **2022**, *11*, 476. https://doi.org/10.3390/ijgi11090476

Academic Editors: Gloria Bordogna, Cristiano Fugazza and Wolfgang Kainz

Received: 1 June 2022
Accepted: 30 August 2022
Published: 2 September 2022

1. Introduction

Global digital statistics [1] reveal that there were more than 4.2 billion active social media users by January 2021, which is 90% of the total number of internet users. In addition, social networks have become a house for numerous real-life events that may occur in our everyday life. The global COVID-19 pandemic has been spreading worldwide, and related topics have been trending since then. Many scientists and companies have started working on challenges related to the processing and analysis of diverse types of health data, medical images, Bluetooth, and GPS data, as well as social data. From a data mining perspective, researchers have been trying to extract knowledge from people's opinions, thoughts, and feelings from social networks. Social data mining includes various associated fields such as Sentiment Analysis (SA) [2]. SA infers positive and negative mentions of people's thoughts, behaviors, and feelings based on their writings about trending topics [3]. The Twitter platform is well suited for analysing users' sentiments during the COVID-19 period, with over 353 million monthly active users [1]. Using data mining techniques, public opinion on COVID-19-related topics can be monitored and tracked in space and time.

From a different perspective, and according to the latest Internet world statistics, Arabic is ranked fourth among the ten most used languages over the Internet [4], with more than 250 million Internet users [5] originating from Arab countries. Arabic is identified by 22 Arabic-speaking countries as an official language [6]. Furthermore, millions of Arabic users use social media networks to communicate and contribute daily Arabic content over social media. Therefore, our focus in this paper is to analyse Arabic social content available on Twitter, and to investigate people's opinions and sentiments about the COVID-19 pandemic. Recent research works have primarily focused on analyzing social data by extracting trending topics, and inferring general sentiments from related topics, with a special focus on the English language. However, COVID-19 related sentiment analysis on Arabic social media has not been fully addressed. In addition, the few existing research works on Arabic social data do not consider the spatial-temporal aspect in sentiment analysis.

In this study, we focus on analyzing Arabic social content from Twitter related to the COVID-19 pandemic, to discover people's sentiments and correlations between COVID-19-related topics and subtopics, at different levels of spatio-temporal granularities. We aim to highlight correlations between insights extracted from social data and official health data records while investigating the impact of the global pandemic on multiple aspects with different spatial and temporal scales.

This study extends our previous work [7] by presenting a comprehensive social data mining approach for the Arabic language, which employs Arabic-specific word embedding techniques with a focus on the correlation between spatio-temporal social data and official health data. Our approach presents several unique contributions compared to existing works as follows:

1. We used the dataset gathered in previous work [7] that contains Arabic tweets related to COVID-19 from two publicly shared datasets within the time frame from January 2020 to November 2020 (about 5.5M tweets). Then, we enhanced our previous approach for a location inference technique from non-geotagged tweets based on user profiles and textual content, which increased the total percentage of location-enabled tweets. We developed our Geo-Database containing bilingual (English and Arabic) names of world countries, their capitals, and the famous towns in the Arab world.
2. We implemented several mechanisms for topic-based analysis using occurrence-based and statistical correlation approaches to examine the spatio-temporal distribution of trending topics related to COVID-19.
3. We conducted a correlation-based analysis between Arabic tweets and official health data collected from online platforms.
4. We extend our previous work [7] of sentiment analysis by developing a deep learning model with bidirectional representations from the unlabeled text by conditioning on both left and right contexts in all layers. We also enhanced our detection mechanism at many spatial granularity levels (regions, countries, and cities) and different topic scales. It leverages unique insights from users' feedback on top controversial topics (lockdown, vaccination, etc.).
5. We conducted a comprehensive set of experiments and visualized our results based on the generated geo-social dataset, sentiment analysis, official health records, and lockdown data worldwide.

This paper is organized as follows: Section 2 outlines a review of some related work. The description and implementation of the proposed methodology are presented in Section 3. The results and findings of the proposed methodology are discussed in Section 4. Section 5 presents concluding remarks and future research directions.

2. Related Work

Current literature on social data mining has witnessed considerable achievements from NLP and ML research fields [8]. Sentiment Analysis (SA) [2] expresses the users' opinions in various forms with diverse linguistic styles to extract subjectivity and polarity from text [9] to provide countless benefits such as supporting people to make their choices. From the early days of 2020, researchers began studying social media content related to COVID-19 that focused on English tweets about COVID-19 or other Latin languages, while few researchers investigated Arabic content. Some research works motivated topic analysis to illustrate the hot topics discussed on social media using a word embedding, word frequency, location frequency, language frequency, and character and word n-gram features weighted by TF-IDF. Meanwhile, other researchers applied feature extraction in feature-based sentiment analysis to determine sentiment polarity and forecast sentiment in social data [10]. Most of them used ML classifiers to verify results by semantic analysis. The following sections classify our review of most research studies on social streams, particularly in Arabic.

2.1. Data Collection and Classification

Recent research works have principally focused on analyzing social data by extracting trending topics and inferring general sentiments from related topics, with a special focus on the English language and less production on Arabic content. Many research works motivated collecting social data to be shared with the research community. In addition, they used their datasets in statistical analysis investigations such as Alanazi et al. [11] and Haouari et al. [12]. Some researchers such as Alharbi [13] identified a coronavirus dataset of Arabic tweets from three Saudi social streams and classified the dataset as conversations about precautionary steps taken by governments, conversations demonstrating social unity, and conversations endorsing government decisions. Additionally, some research works focused on analysis of tweets datasets for classification such as Hamdy et al. [14] who studied different types of tweets collected from Twitter from different perspectives of analysis and machine learning classification. They combined different machine learning models to classify tweets into related/not related to Coronavirus.

2.2. Geolocation Analysis

Some researchers worked on the location-enabled features of social data such as Qazi et al. [15] that introduced the GeoCoV19, a large-scale Twitter dataset related to the COVID-19 pandemic. They used the Nominatim (Open Street Maps) data at geolocation granularity levels to derive their geolocation information using a gazetteer-based method to extract toponyms from user location and tweet text. Likewise, Lamsal [16] introduced the COV19Tweets Dataset, a large-scale English language tweets dataset with sentiment ratings. They filtered the COV19Tweets Dataset's geotagged tweets to create the GeoCOV19Tweets Dataset contains only 141k tweets (0.045 percent).

2.3. Topic Analysis and Semantic Analysis

Alshalan et al. [17] used the ArCov-19 dataset [12], an ongoing dataset of Arabic tweets related to COVID-19, to find the hate speech in the Arab world, as well as the most common topics addressed in hate speech tweets. They used a pre-trained convolutional neural network (CNN) model to evaluate tweets for hate speech. Similarly, Alsafari et al. [18] built Arabic hate and offensive speech detection system because of an increasing proliferation of hate speech on social media. However, unfortunately, the collected data are not related to COVID-19. They applied four robust extraction algorithms based on four forms of hate: religion, race, nationality, and gender. They then labeled the corpus using a three-hierarchical annotation methodology, ensuring ground truth at each level by verifying inter-annotation agreement evaluated by applying ML classifiers. As Well, Hamoui et al. [19] examined the Arabic content on Twitter to see what the most popular topics were among Arabic users. They used Non-negative Matrix Factorization (NMF) to find the most common unigrams, bigrams, and trigrams in a dataset of Arabic tweets. They presented, discussed, and divided the final discovered topics into many categories.

Likewise, Al-Laith et al. [20] analyzed the emotional reactions of people during the COVID-19 pandemic using a rule-based technique to classify tweets. They examined six forms of emotion to discover citizens' worries. Furthermore, they created a framework for tracking people's emotions and correlating emotions with tweets mentioning some of the COVID-19 pandemic symptoms. Similarly, Bahja et al. [21] revealed the initial results of identifying the relevancy of the tweets and what Arab people tweeted about the COVID-19 feelings/emotions (Safety, Worry, and Irony). They used ML and NLP techniques to discover what Arab people talked about COVID-19 on Twitter. Meanwhile, Essam and Abdo [22] examined how Arabs are dealing with the COVID-19 pandemic on Twitter. They extracted specific keywords and n-grams to classify common themes in the compiled corpus. They conducted a lexicon-based thematic analysis to find that tweeters had high levels of affective conversation full of negative emotions.

Some research work focused on the sentiment analysis of social data such as Manguri et al. [23]. They offered a graphical representation of the data after the sentiment analysis. Further, Chakraborty et al. [24] demonstrated tweets comprising and how health organizations have failed to guide people around this pandemic epidemic using a model with Deep Learning (DL) classifiers. Furthermore, Kabir et al. [25] created a neural network model and trained using manually labeled data to detect distinct emotions in Covid-19 tweets at fine-grained labeling. They constructed a bespoke Q&A roBERTa model to extract terms from tweets predominantly responsible for the accompanying emotions. Moreover, Hussain et al. [26] developed and used an AI-based technique to analyze social-media public reaction concerning COVID-19 vaccines in the United Kingdom and the United States to understand public opinion and discover hot subjects. They employed NLP and DL algorithms to anticipate average feelings, sentiment trends, and conversation topics. In addition, low-resource languages have witnessed recent efforts for investigating sentiment analysis, trying to bridge the gap by manually collecting and annotating social media data. ALBANA is a deep learning-based sentiment analyzer that performs sentiment analysis of around 10K Facebook comments in the Albanian language [27]. Attention mechanism along with fastText word embedding model was used to discover the interdependence and meanings of words while employing a BiLSTM for sentiment classification. Furthermore, Imran et al. [28] examined how people from various cultural backgrounds responded to COVID-19 and how they felt about the ensuing steps that various countries took in response. They used deep long short-term memory (LSTM) models to estimate the sentiment polarity and emotions from extracted tweets have been trained to reach cutting-edge accuracy. They demonstrated an original and cutting-edge method for validating the supervised DL models using Twitter tweets that had been extracted.

2.4. Misleading Information Detection

Some researchers tried to handle the misleading information published on social media such as Alsudias and Rayson [29] who collected and examined Arabic tweets about COVID-19 to identify the topics using the k-means algorithm, to detect rumors, and to predict tweets' sources. They used ML algorithms to identify false, correct, and irrelevant information, with two sets of features word frequency and word embedding. In a similar manner, Elhadad et al. [30] presented the COVID-19 Twitter dataset (COVID-19-FAKES) in bilingual (Arabic/English). They gathered COVID-19 pre-checked facts from several fact-checking websites to create a ground-truth database to annotate their collected dataset. They used shared knowledge from the official websites and Twitter accounts as a source of accurate information. They used ML algorithms and feature extraction techniques to annotate Tweets in the COVID-19-FAKES dataset. Similarly, Hussein et al. [31]) created an effective strategy based on the AraBERT language paradigm for combating the Tweets COVID-19 Infodemic. They trained language models on plain texts rather than tweets since pre-trained language models are widely available in many languages and available plain text corpora are larger than tweet-only corpora, allowing for greater performance.

2.5. Discussion

Table 1 summarizes attempts to process COVID-19-related social data with important information such as the number of tweets contained in each dataset, the language of the dataset, the time frame the data was collected, techniques used in the research work, and the features used in that work.

Table 1. Summary of attempts to process COVID-19 social data.

Ref.	Purpose	Tweets	Lang.	Time Frame	Technique	Features
[11]	Describe Arabic tweets dataset on COVID-19	3,934,610	Arabic	1 January 2020–30 April 2020	Statistical Analysis	Tweets frequency
[12]	Present and analyze Arabic Twitter dataset	748 K	Arabic	27 January 2020–31 March 2020	Statistical Analysis	tweets frequency
[13]	Finetune a BERT model to classify multilabel tweets about crisis events	1.6 M	Arabic	2018–2020	LDA model and BERT model	Topic frequency
[14]	Study from different perspectives of analysis, and classification	3,934,610	Arabic	1 January 2020–15 April 2020	Classification and Clustering	Word2Vec
[15]	Present Twitter dataset, infer geo-information and analysis dataset	524 M	Multi-lingual	1 February 2020–1 May 2020	Statistical Analysis	tweet, location, and language frequency
[16]	Present and analyze COV19 Tweets Datasets and sentiment scores	310 M, 141 k Geo tweets	English	20 March 2020–17 July 2020	Statistical and Sentiment Analysis	unigrams and bigrams, and topic modeling
[17]	Identify hate speech related to the COVID-19 pandemic	547,554	Arabic	27 January 2020–30 August 2020	pre-trained convolutional neural network (CNN) model	TF-IDF vectors (unigrams and bigrams)
[18]	Build Arabic hate and offensive speech detection system	800,000	Arabic and English	April to September 2019	tweet annotation with ML classifiers	unigram, word, and char-ngrams, word embeddings and contextual word embedding
[19]	Examine the most popular topics raised among Arabic users	3,934,610	Arabic	1 January 2020–30 April 2020	Non-negative Matrix Factorization (NMF)	TF-IDF and Topic Coherence-Word2Vec
[20]	Analyze the emotional reactions of citizens	300,000	Arabic	1 January 2020–30 August 2020	rule-based technique	Emotion frequency
[21]	Determine the relevancy of the tweets and people feelings/emotions	782,391	Arabic	16 February 2020–10 July 2020	ML and NLP	Frequency of themes labeling
[22]	Analysis of Arab tweets about COVID-19	1,920,593	Arabic	1 February 2020–30 April 2020	topics frequency and lexicon-based analysis	Most frequent features
[23]	Measure sentiment analysis	500,000	Arabic	9 April 2020–15 April 2020	Sentiment analysis	Word and tweet frequency
[24]	Demonstrate how people have tweeted regarding COVID19	226,668	Arabic	1 December 2019–31 May 2020	Sentiment analysis	Word Vector

Table 1. *Cont.*

Ref.	Purpose	Tweets	Lang.	Time Frame	Technique	Features
[25]	Detect distinct emotions in COVID-19 tweets	500 M	Arabic	5 March 2020–31 December 2020	neural network model	word vectors
[28]	detect sentiment polarity and emotion recognition	460,286	English	12 February 2020–30 April 2020	Deep long short-term memory (LSTM)	Linguistic Inquiry and Word Count
[29]	Identify topics, detect rumors, and predict tweets' source	1,048,575	Arabic	1 December 2019–30 April 2020	Cluster Analysis and Rumor Detection	Word frequency, count vector and TF-IDF
[30]	Annotate misleading Information dataset about COVID-19 Twitter dataset	3,047,255 English and 216,209 Arabic	English and Arabic	4 February 2020–10 March 2020	Statistical Analysis and ML	TF, TF-IDF- (N-gram, character level)
[31]	Analyse social-media public sentiment in the UK and the US towards COVID-19 vaccinations	300M	English	1 March 2020–22 November 2020	DL BERT	VADER and TextBlob

To summarize, researchers used analysis of social media content to learn more about how people react to the COVID-19 pandemic. The exiting work has been conducted with English data, while Arabic data receives fewer contributions. Moreover, the existing research works on Arabic social data do not consider the spatial-temporal aspect of the COVID-19-related content. Furthermore, the correlation between official health data and social media material has not been fully studied.

3. Research Methods

In this paper, we propose a method to automatically detect and process social datasets containing Arabic tweets related to the COVID-19 pandemic using ML models and topic detection and tracking techniques. Figure 1 shows the workflow of our methodology, while the below subsections describe each phase in more detail. Because we focused on analyzing the spatio-temporal social data in Arabic tweets related to the COVID-19 pandemic, accordingly our methodology started with data collection of Arabic tweets. We found two publicly shared Arabic datasets for COVID-19 tweets [11] (3,314,859 tweets) and [12] (2,111,650 tweets). Unfortunately, both have a low number of geo-tweets, which is about 2% of the total tweets as discussed in Section 2. Thus, we decided to merge both datasets (about 5.5M tweets) to feed our experiments. Then, we processed the merged dataset to infer locations from non-geotagged tweets, thus generating a new dataset of location-enabled tweets which contains about 46% (about 2.5M tweets) of the original combined dataset.

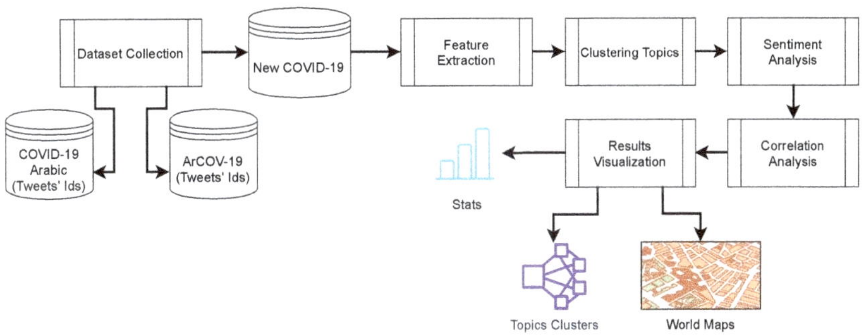

Figure 1. The Workflow of our methodology to analyze the social data.

Subsequently, useful features are extracted from COVID-19-related tweets. Then, classification techniques are applied to the extracted feature vectors to detect topics from the tweet's text and generate the hot topics. Then again, the sentiment analysis technique is applied with multiple perspectives to infer the sentiment and opinion polarity at many spatial (city, country, region) and temporal (day and month) levels. Next, correlation analysis between the collected tweets' data and the official health records is applied at different spatial granularities, such as country and region. Finally, we demonstrate the visual analytics based on a comprehensive set of experiments use the new Arabic COVID-19 dataset. The following subsections describe each step in our methodology.

3.1. Dataset Collection

The Twitter platform is well suited for studying the sentiment of users during COVID-19, with over 353 million monthly active users. Therefore we used Twitter because it focuses on becoming a broadcast platform similar to a real-time news station, with posts by default being global-readable to link people to the rest of the world. Most of the public Arabic datasets related to COVID-19 contain a low number of location-enabled tweets, such as Alanazi et al. [11] and Haouari et al. [12]. Both datasets are collected using Twitter Streaming API that matched Arabic tweets with the set of COVID-19-related keywords widely used by people, news media, and official organizations such as Coronavirus, Corona, and Pandemic.

We collected tweets from both datasets and from which we built a new geo-tagged dataset. In addition, we defined the period from 1 January 2020, to 30 November 2020, to filter tweets on the merged dataset. Unfortunately, based on the privacy policy of Twitter, both shared datasets contain only tweet IDs. Therefore, we hydrated (recollected) the dataset to get the full tweet objects from Twitter using the TWARC and Hydrator Python libraries developed for this purpose, which helps to generate JSON files for each day's tweets. The new dataset contains more than 5.5M tweets (5,054,141 tweets are unique and about 2.5M geo-tweets) with an average of Words per Tweet equal to 21. Table 2 presents the statistics of the monthly distribution of unique tweets, unique words, unique hashtags, and unique user IDs.

Table 2. Statistics of the monthly distribution of dataset contents.

Month	COVID-19 Arabic	ArCOV-19	New Dataset	Total Words	Unique Words	Unique Hashtag	User IDs
January	208,974	130,002	338,976	7,179,808	103,823	5046	103,823
February	383,474	178,095	561,569	11,311,723	188,158	9207	188,158
March	1,479,692	430,235	1,909,927	41,123,366	615,374	23,820	615,374
April	1,307,424	287,754	1,595,178	32,870,009	516,120	18,777	516,120
May	0	251,670	251,670	5,575,970	89,974	11,619	89,974
June	0	212,380	212,380	4,567,400	81,060	10,679	81,060
July	0	192,754	192,754	4,063,514	68,352	9340	68,352
August	0	166,825	166,825	3,487,262	64,475	8069	64,475
September	0	113,199	113,199	2,323,648	47,711	5553	47,711
October	0	105,634	105,634	2,144,079	43,297	5050	43,297
November	0	93,958	93,958	1,915,613	39,354	4477	39,354
Total	3,379,564	2,162,506	5,542,070	116,562,392	1,224,684	48,954	1,224,684

Figure A1 illustrates the monthly distribution of tweets and hashtags during the period from February to April 2020 in which most countries around the world were in COVID-19 Lockdown. One can notice the general trend of tweeting about COVID-19 was at its peak in March and then gradually reduced to normal about normal levels. It shows that the general trends of the number of tweets and hashtags were similar.

3.2. Features Extraction

We developed several processes for the Feature Extraction module, as shown in Figure 2. The first process "Prepare Dataset" contains several sub-processes: "Clean Dataset", "Filter Fields", and "Prepare Arabic Text" The "Clean Dataset" process removes null values from the tweet's object. On the other hand, the "Filter Fields" process removes the unnecessary fields from the tweet's metadata. Subsequently, to prepare the Arabic text in each tweet's object for further text analysis, the process "Prepare Arabic Text" has been applied to perform the following processes. Figure A2 provides a sample of Arabic tweets with their English translation after applying all of the following:

- Convert HTML to normal text, which removes the HTML tags.
- Remove links (remove all hyperlinks of advertisements, retweets, etc.)
- Remove diacritics (remove Arabic diacritics)
- Remove punctuation (remove Arabic punctuation characters)
- Normalize and Tokenize Arabic text (We used our own Python code for tokenization)
- Remove Stop Words (remove Arabic stop words based on a list contains 750 words that published by Alrefaie [32])
- Remove empty lines (remove extra empty lines and extra white spaces).

Figure 2. Reprocess COVID-19 tweets dataset.

Each tweet object contains several metadata fields that hold much information such as tweet text, hashtag, and user ID. The available features can be incorporated into our ML model with relative ease. The geo-location information, such as "Place" and "User Information", are important pieces of information that define the originating location of the tweet. Unfortunately, it needs further work because it is based on the user's choice to enable or disable the "Place" option in the Twitter settings. Hence, the number of geotagged tweets is so small compared to the total number of tweets. In the COVID-19 Arabic dataset [11] the number of geotagged tweets was 64,705 tweets (2%). Similarly, the ArCov-19 [12] dataset has only 50,856 tweets (2.4%). Therefore, we develop an algorithm to "generate Location-Enabled Tweets" from the non-geotagged tweets, as illustrated in Algorithm 1. The objective of the algorithm as shown in Algorithm 1 is to extract the location-enabled information. It required using the GeoDB, a manually developed Geo-location database that contains Bilingual names (English and Arabic) of world country names, capital cities, and famous towns in the Arab region. This approach analyzes all geo fields found with the tweet's object such as "Place name", "Country", and "User location" to extract the tweet source. Unfortunately, the user location information is an optional field that may is manually entered by the users. Therefore, it may be written in many languages or contained misinformation.

Our approach worked on two levels to extract chrononyms and astionyms (chrononyms: proper names of regions or countries; astionyms: proper names of towns and cities [33]) from user location. The first level tries to infer the country name from the information written in the user location metadata ([user][location]). It queries the GeoDB to extract the country name matching the ['user'] ['location']. Meanwhile, the second level works only if the first level failed, and it tries to infer the country name based on detecting the city name from the information in user location metadata and then retrieves the country name containing that city.

As a result of applying the Location Extraction algorithm, we increased the size of the experiments' dataset from about 115K (2%) up to 2.5M geotagged tweets (46%). Figure 3 illustrates the percentage of the geotagged tweets before and after applying our approach. Meanwhile, Figure A3 presents the monthly distribution of geo/non-geo tweets in our new Geo-Tweets dataset. Finally, using the occurrence-based approach, we run a process to numerically analyze the location-enabled Arabic tweets in the new geo-tweets' dataset (2.5M tweets). Most of the Arabic tweets are coming from Arab users living around the world, and most of them live in the Arab region. As a result, we compared the two regions (Arab and non-Arab) for originating the Arabic tweets that feed our experiments. This process extracted all information, which presents the proportion between Arabic tweets and top Hashtags tweeted by users in Arab and non-Arab regions. Figure 4 illustrates the comparison between Tweets and Hashtags (Top percentages only, more than 1%) in Arab and non-Arab countries during the time frame from January to November of 2020. Note that the top two countries for both tweet generation and hashtags in the Arab regions are Saudi Arabia and Kuwait. On the other hand, UK and France were the top two countries in the non-Arab region.

Algorithm 1: Location Extraction

```
Input:
TD: Tweets Dataset is the list of all tweets,
TP: Tweets place field, [place][name]
TC: Tweet Country field, [place][country]
TCC: Tweet Country Code field, [place][country code]
TCO: Tweet Coordinates field, [place][bounding box][coordinates]
TUL: Tweet User Location field, [user][location]
CCDB: Countries and Cities Database
GeoDB: Geo-Location Database
Output:
TW: Tweet Source
begin;
TD = loadTweetsCorpus();
for each tweet in TD do
    if TP is not None then
        country = TC;
        if country is not None then
            place name = TP;
            country code = TCC;
            coordinates = TCO;
        end
    else
        if TP is None OR country is None then
            Select country from CCDB where TUL = country;
            if country is None then
                Select country from CCDB where TUL = city;
            end
            country code, coordinates = retrieveData(GeoDB);
        end
    end
end
```

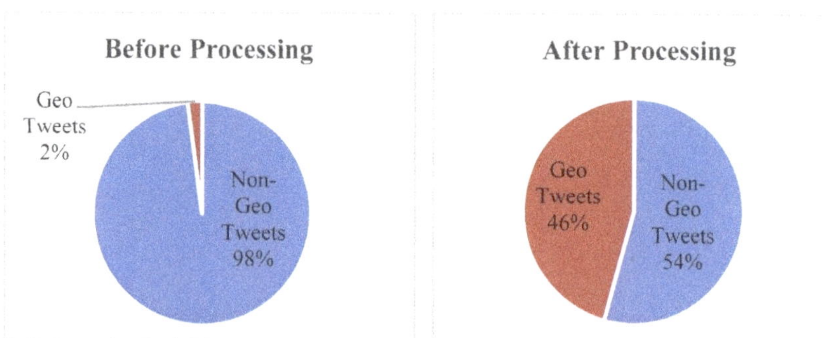

Figure 3. Percentage of the Geo-tagged tweets in the new COVID-19 Tweets Dataset after applying the Location Extraction algorithm.

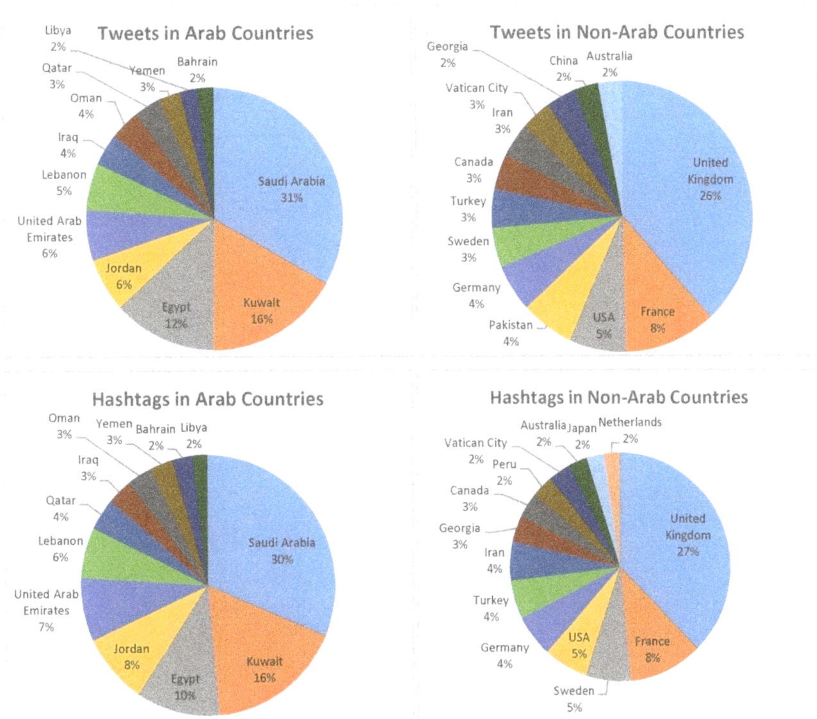

Figure 4. Tweets and Hashtags in Arab and non-Arab Countries.

3.3. Topic Clustering

In this section, we conducted some experiments to present an initial outcome of dataset distribution analysis. Our Tweets dataset was formed from two datasets originally collected using a list of keywords (about 45 unique keywords). We applied the AraVec tool to extend the list of keywords based on semantic matching. AraVec is a word embedding open-source project that aims to provide free and efficient word embedding models to the Arabic NLP research community [34]. The new list contains more than 1600 keywords. Then, we filtered the list (160 unique topics only) by removing topics non-related to COVID-19 hot topics such as names of countries and cities. Then, we applied an occurrence-based technique to each tweet text to find the best-matched keywords that represent the topics illustrated by tweets.

We used the FuzzyWuzzy Python library (an open-source string similarity matching library) to determine string similarity as a degree out of 100. It uses the Levenshtein Distance that calculates the differences between sentences. It also uses the Fuzzy C-Means (FCM) clustering method to generate a recommender system predicated on homogeneous attribute measures [35]. Figure 5 presents the monthly distribution of the top topics during the time frame from January to November 2020. Meanwhile, Figure 6 illustrates the comparison between the top 20 topics tweeted by users in Arab and non-Arab countries. Figure A4 displays the word cloud of the top topics. Figure A5 shows the comparison between top five topics in Arab and non-Arab countries. While, Figure A6 illustrates the monthly top topics distribution. The top five topics in the top 10 countries is illustrated in Figure A7. Figures A8 and A9 present the distribution of some main topics (Vaccine and Treatment) in Arab and non-Arab countries.

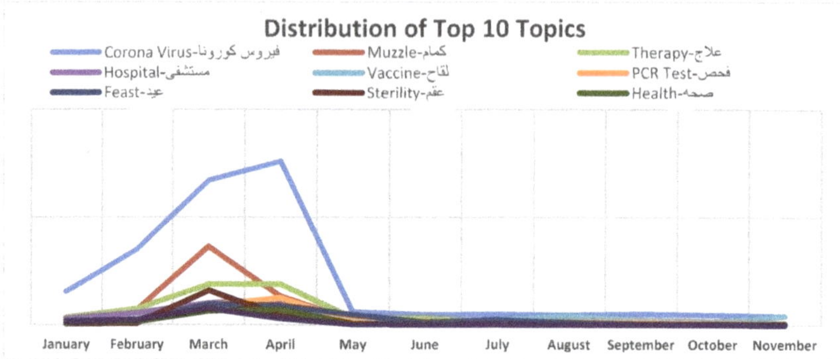

Figure 5. Monthly distribution of Top 10 Topics in the Tweets Dataset.

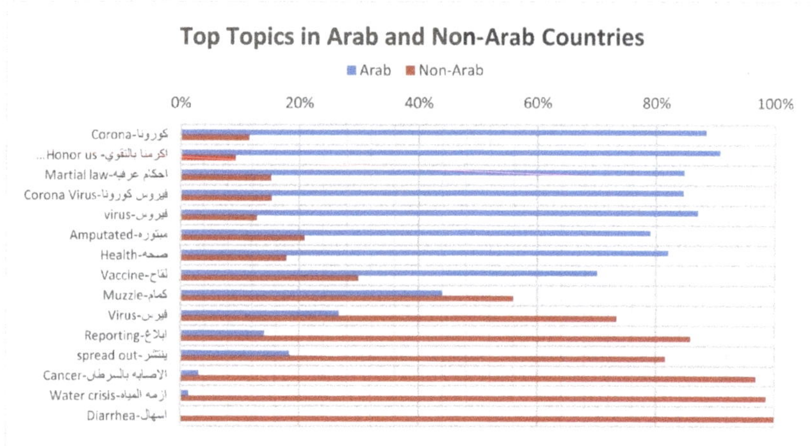

Figure 6. Top topics distributed globally.

Subsequently, to have a higher-level vision of topics found in Arabic tweets, we have used clustering techniques to cluster the 140 top topics extracted from our experiments into main topics. We applied the FCM clustering method that extracts 14 clusters to represent main topics. Table 3 represents the main high-level topics and their coverage. The coverage of a topic is calculated from the distribution of top topics in Arabic tweets in our Geo-dataset.

Figure 7 shows the percentage of main topics distribution in Arab and Non-countries. It illustrates that the top topics in Arab countries are Corona, Prayer, Lockdown, Hospital, and vaccines. Meanwhile, in non-Arab countries, the top topics are Sterilization, Organization, Feeling, Disease Symptoms, and Treatment. Figures A8 show Treatment topics tweeted in the Arab countries more than Vaccine topics during the first half of the year 2020 and vice versa. Meanwhile, Figure A9 displays the Vaccine topics appeared on social media in non-Arab countries from the second quarter of 2020.

Table 3. Main topics clusters.

Main Topic	Coverage	Description	Example of Top Topics
Corona	55.74%	Topics related to Coronavirus	New injuries, checkup, Virus, Corona, COVID-19
Prayer	25.68%	Topics related to prayers to God usually raised by Arab people	Save us, heal me by your ability, Protect from your torment, Your generosity, and mercy
Lockdown	13.90%	Topics related to actions done by governments against the COVID-19 pandemic	Curfew, School closure, Separate service, Closing
Hospital	3.41%	Topics related to tweets discuss entering hospitals and their procedures	Quarantine Hospital, Isolation Hospital, Oxygen, Anesthesia
Disease's Symptoms	0.44%	Topics related to symptoms of Coronavirus	fever, cough, tiredness, headache, sore throat, diarrhea
Diseases	0.23%	Topics related to other diseases rather than Coronavirus	Pneumonia, Kidney failure, Nervous breakdown, Heart failure
Feeling	0.16%	Topics related to feelings about Coronavirus	kindness, laugh, discontent
Treatment	0.12%	Topics related to treatment of Coronavirus	Enzymes, Tamiflu, Stem cells, Hydroxy chloroquine
Religious	0.11%	Topics related to religious activities	Fasting, Fasting, Pilgrimage
Vaccine	0.11%	Topics related to all talks about vaccine production or distribution	Vaccine, immunization, Serum
Sterilization	0.02%	Topics related to sterilization actions	Antiseptic, Chlorine, Wash off
Organization	0.01%	Topics related to organizations mentions	Health organization, Reuters
Miscellaneous	0.07%	Miscellaneous topics	Rationalization, make a complaint, Electricity shut down

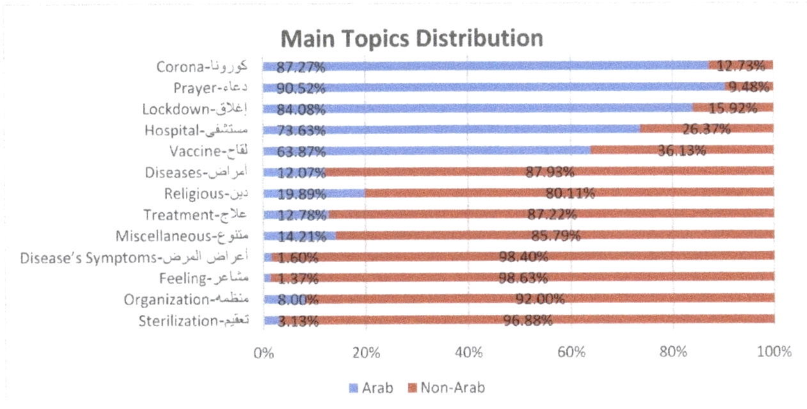

Figure 7. Main topics in Arab and non-Arab countries.

3.4. Sentiment Analysis

Social data mining has recently become an attractive field of research that relates to the method of automatically extracting valuable information from computerized textual data [36]. It includes various associated fields such as question answering systems, text summarization, and SA [2]. Generally, research on social data mining and SA can be classified into supervised and unsupervised approaches [37]. Supervised (or corpus-based) approaches require labeled datasets, usually expensive to collect and label. On the other hand, the unsupervised approach (recognized as the lexicon-based approach) depends on extracting features for classification and clustering purposes. For instance, unsupervised sentiment analysis employs sentiment lexicons built with the assumption that words have prior sentiments. Consequently, because of the variety and wide distribution of informal writing in languages, it is a time-consuming and challenging task to create such lexicons. Many ML classifiers can be applied to improve the performance of SA on both approaches such as Decision Tree (DT), Support Vector Machine (SVM), Multinomial Naïve Bayes (NB), or deep neural networks, among others.

We used the lexicon-based approach in the SA processes with the social network data (tweets from Twitter) as an opinion resource. We built our Arabic sentiment lexicon by merging nine Arabic lexicons (Bing Liu Lexicon; NRC Emotion Lexicon; MPQA Subjectivity Lexicon; SemEval-2016 Arabic Lexicon; AEWNA Lexicon; NileULex Lexicon [8,38–42] previously tested by the research community. The Arabic sentiment lexicon contains annotation for Arabic words such as "positive", "negative", or "neutral". We used this lexicon to extract the feature of the tweet's polarity as the numbers of positive, negative, and neutral words. Then, we used polarized bag-of-words features, representing a tweet as a set of words without order, each word being a feature. Then combined these features with the number of words by polarity. Principally, the Bag of Words (BoW) technique is a word-frequency approach that counts positive, negative, and neutral words in each tweet to assign the tweet polarity as "positive" (if positive count—negative count > zero), "negative" (if positive count—negative count < zero), otherwise it is "neutral". Subsequently, we applied four ML classifiers—Decision Tree (DT), Multinomial Naïve Bayes (MNB), Linear Support Vector Machine (SVM), and Random Forest (RF)—to check the classification performance of the sentiment analysis.

To apply the sentimental analysis processes, we developed a Sentiment Extraction algorithm for the SA processes to run in our experiments. Algorithm 2 starts with loading the Arabic corpus of the text collected from Arabic tweets, which are extracted from the geo-tweets' dataset. Subsequently, the approach prepares the Arabic corpus by applying many sub-processes: (1) filtering tweets to throw away the unnecessary fields and use the necessary fields (such as: "id", "text", "country code"). (2) Remove duplicated tweets. (3) Clean tweets' text by removing special characters such as (#, &, %), hyperlinks, punctuation, numbers, Unicode characters, non-ASCII characters, and unnecessary white spaces. (4) Normalize Arabic characters (Alef, Yaa, ...) in tweets' text. (5) Remove Arabic stop words such as the Arabic prepositions (Men, Ela, Fe, ...). (6) Remove duplicated words and single-character words as well. (7) Tokenize Tweets' text. (8) Stem tweets' words to rooted words.

Subsequently, the algorithm will run on each tweet's words to get the tweet's polarity by using our Arabic sentiment lexicon to get polarity for each word then run the BoW technique to set the polarity for each tweet. Finally, the algorithm runs four ML classifiers to improve the accuracy of the SA process. The algorithm extract features and labels from the Arabic polarity corpus. Then, it divides the corpus into train and test datasets by splitting with ratios 80% and 20% consequently. Next, the algorithm vectorizes the training dataset using the TF-IDF Vectorizer. Following this, the algorithm predicts the features using the classifier model applied to the test dataset. Finally, a confusion matrix with the classifier's scores is outputted such as the precision, recall, f1-score, and accuracy values.

```
Algorithm 2: Sentiment Extraction
    Input: TC: Tweets Corpus,
    AraLex: Arabic Lexicon,
    CL: ML Classifiers' List
    Output: Sentiment Results
begin
TC = loadTweetsCorpus();
RemoveDuplicatedTweets(TC);
// Prepare corpus;
for each tweet in TC do
    FilterTweetsFields(tweet);
    CleanTweets(tweet);
    NormalizeTweets(tweet);
    RemoveArabicStopWords(tweet);
    RemoveduplicatedWords(tweet);
    TokenizeTweetsText(tweet);
    StemTweetsWords(tweet);
    // Get tweet's polarity;
    GetTweetPolarity(tweet, AraLex);
    SetTweetPolarity(tweet);
end
for each classifier in CL do
    ExtractFeaturesLabels(TC);
    Train, Test = SplitDataset(TC);
    for n in uni-gram do
        Vectorising(Train, TF-IDF);
        Prediction = Predict(CL, Test);
        ConfusionMatrix = (precision, recall, f1-score);
    end
end
```

4. Results and Discussion

Figure 8 illustrates the monthly results from the sentiment analysis of COVID-19 Arabic tweets. Meanwhile, Figure A10 displays the top scores evaluating sentiment analysis in Arab countries. It shows Arab countries have the most positive/negative/neutral feelings based on users' opinions.

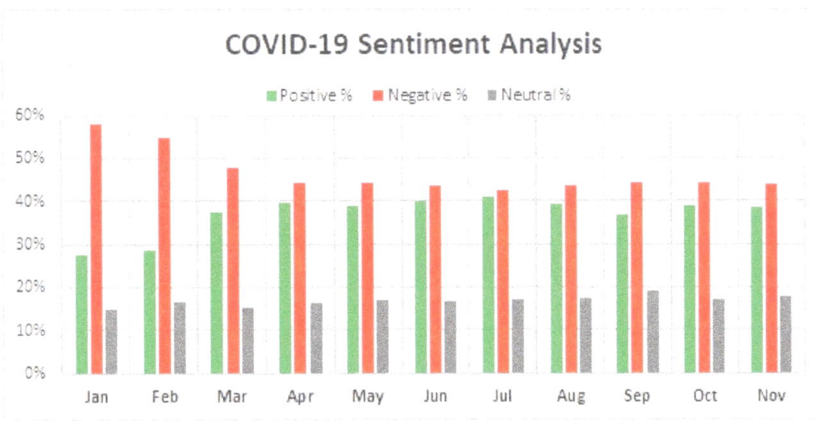

Figure 8. SA of COVID-19 Arabic Tweets.

Moreover, we trained four classic ML algorithms (DT, MNB, SVC, and RF) using our Geo COVID-19 dataset. Next, we evaluated the results with four sentiment-labeled Arabic datasets collected from Twitter, as shown in Table 4. Furthermore, we employed an NLP pre-training model dubbed Bidirectional Encoder Representations from Transformers (BERT), which was developed to pre-train deep bidirectional representations from the unlabeled text by conditioning on both left and right contexts in all layers [43]. We used the Arabic-BERT model (developed by Safaya et al. [44]) trained on our Geo COVID-19 dataset and evaluated with the same four datasets. We run certain model modification iterations, assess using the other datasets, and then examine how it affects the overall performance.

Table 4. Sentiment-labelled Arabic satasets.

Dataset Name	Total Tweets	Pos. Tweets	Neg. Tweets
Arabic Sentiment Twitter Corpus (ASTC) [45]	58,751	29,849	28,902
Arabic Sentiment Analysis Dataset (SS2030) [46]	4252	2436	1816
100 k Arabic Reviews [47]	66,666	33,333	33,333
Arabic Speech-Act and Sentiment Corpus of Tweets (ArSAS) [48]	11,784	4400	7384

Figure 9 presents the comparison between results collected after applying ML algorithms trained and evaluated with five Arabic datasets. We used both Unigrams and Bigrams features with TF-IDF Vectorization. We found that scores are close for both, so we present unigrams only. It shows the Accuracy measure results of the traditional ML classifiers for unigram and TF-IDF feature representation. It shows that applying the SVM and RF classifiers achieved a high score on our Geo-COVID-19 dataset than using other datasets. The results achieved by some ML classifiers are also higher in performance. We applied the Arabic-BERT model and compared our findings using the five datasets that accomplished a higher accuracy than the others, especially with the BERT-mini model.

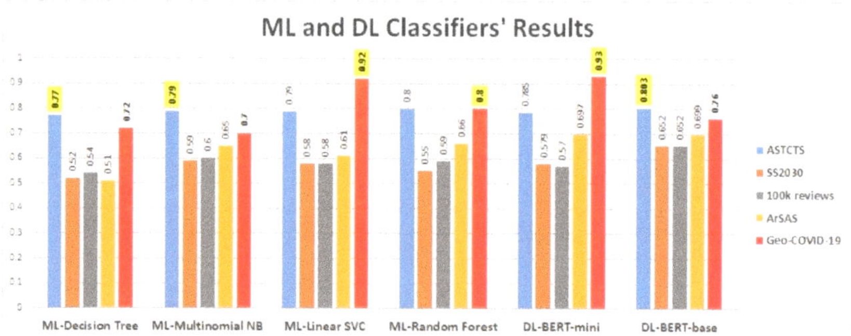

Figure 9. SA Classifiers' performance applied on Arabic Datasets.

4.1. Correlation Analysis

Daily, the official health records related to COVID-19 statistics are published by many official organizations such World Health Organization (WHO), European Centre for Disease Prevention and Control (ECDC), and Johns Hopkins University (JHU). We collected official COVID-19 records from ECDC and JHU for world countries during the time frame from January to November 2020. We implemented many mechanisms for the correlation-based analysis using occurrence-based to find the correlation between sentiment analysis, official health providers' data, lockdown information, and topic frequencies. Figure 10 illustrates the correlation between sentiment analysis of Arabic tweets related to COVID-19 and the COVID-19 new cases over the world. It shows that there were some high negative feelings at the beginning of the COVID-19 pandemic until the second quarter of the year 2020, subsequent it was in a steady state while cases increased. In addition, Figure 11 shows separately the correlation between feelings of the Arab tweets in the Arab countries and non-Arab countries and the COVID-19 new cases with some differences in both, especially in the last quarter of 2020.

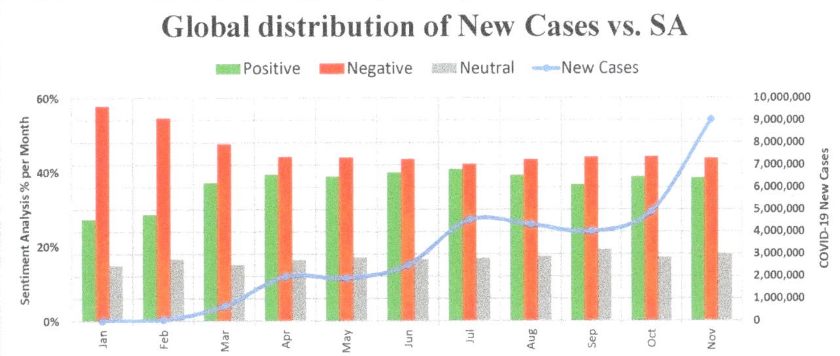

Figure 10. Correlation between SA and the Official Health Records.

Figure 11. Correlation between SA and the Official Health Records in Arab and Non-Arab Countries.

Similarly, Figure A11 illustrates a considerable difference in the correlation between feelings shown in Arabic tweets and the COVID-19 new cases announced in some countries, around the world. On average most feelings are equalized between positive and negative while the number of official new cases of COVID-19 are increased. In response to the COVID-19 outbreak, governments are imposing several measures against the COVID-19 pandemic such as {School closing}, {Workplace closing}, {Cancel public events}, and {travel restrictions}. On 23 January 2020, the first COVID-19 pandemic lockdown was initiated in Wuhan [49]. Since January 2020, most governments throughout the world enacted full or partial lockdowns to stop the virus from spreading, leaving millions stranded. Moreover, a third of the world's population is restricted in some way [50]. From the middle of March 2020, several Arab countries began implementing various types of lockdowns.

Some organizations, such as Oxford University, collect data on 20 indicators to inform a Risk of Openness Index, which aims to aid countries to understand whether it is safe to "open up" or "shut down" in their fight against the coronavirus [51]. We got data from the Oxford dataset related to the lockdown status worldwide. We used data of the "School closing", "Workplace closing" and "Cancel public events" indicators as the most important indicators to feed our mechanism applied in an experiment aimed to find the correlation between lockdown and the spatial-temporal data found in our Arabic tweets' dataset. Figures 12 and 13 displays the comparison between the correlation of the lockdown with COVID-19 new cases in Arab and non-Arab countries, which present the impact of lockdown on the spread of COVID-19 new cases. It shows a negative correlation between the number of COVID-19 new cases and the number of lockdown days in the Arab region— especially for the indicator "Cancel Public events". While, in the non-Arab region, there is a negative correlation between the number of COVID-19 new cases and the number of lockdown days—especially for the indicators "Cancel Public events" and "School Closing".

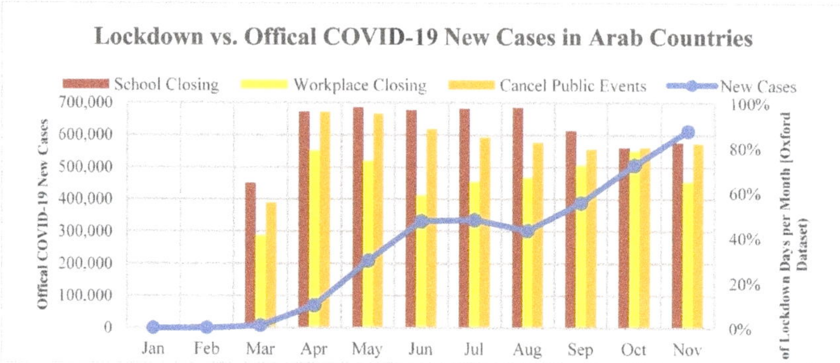

Figure 12. Correlation between lockdown and Official COVID-19 New Cases in Arab Countries.

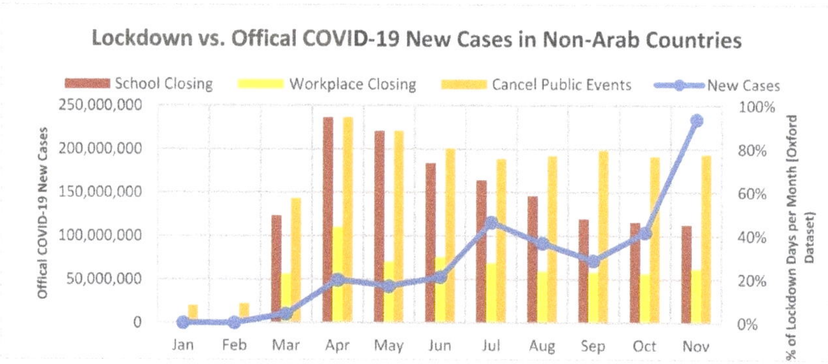

Figure 13. Correlation between lockdown and Official COVID-19 New Cases in Non-Arab Countries.

Figures 14 and 15 clarify the correlation between lockdown and the sentiment analysis of Arabic tweets tweeted by users during the time frame from January to November 2020, although the lockdown started late in Arab countries. The relationship shows different perspectives regarding the lockdown indicators and the sentiment of Arabic tweets related to COVID-19 in Arab and non-Arab countries. For example, they clarify a positive correlation between the positive feelings shown in Arabic tweets and the number of closing days of a school lockdown.

Back to the top-topics issue illustrated early in the above sections, we implemented another mechanism to define the correlation between sentiment analysis and topic-frequencies in a spatial-temporal manner. Subsequently, to have a wide vision of topics found in Arabic tweets, we clustered them into main top topics as discussed early in Table 3. Figures 16 and 17 illustrate the relationship between main topics and the sentiment analysis of Arabic tweets presented those topics in Arab and non-Arab countries. It shows the positive/negative feelings associated with each topic in that region.

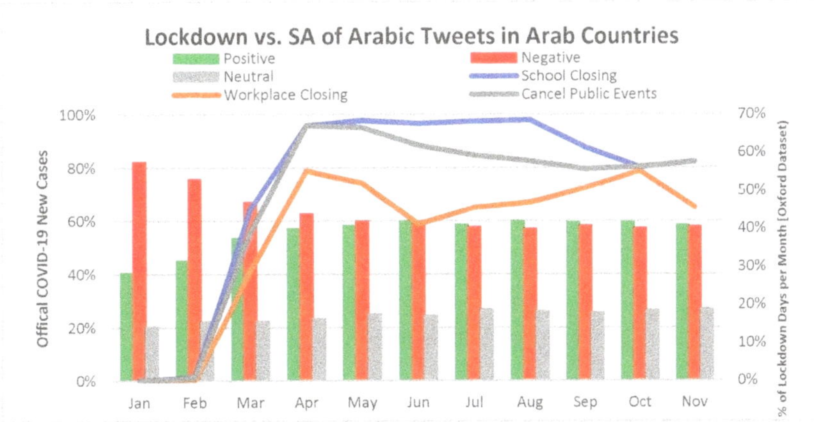

Figure 14. Correlation between Lockdown and SA in Arab Countries.

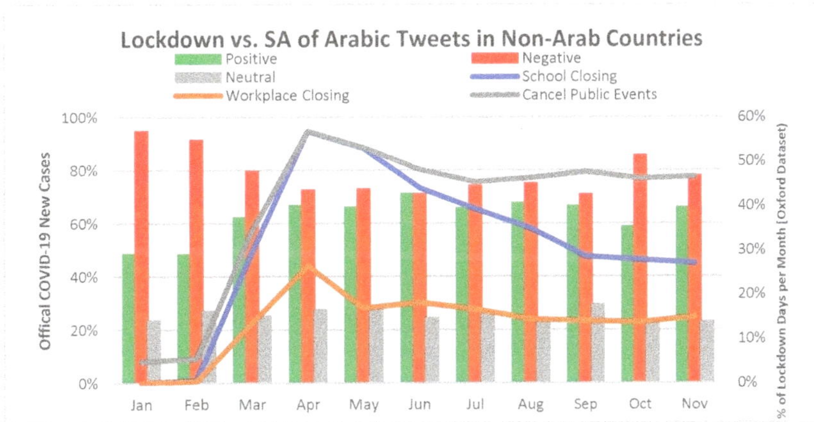

Figure 15. Correlation between Lockdown and SA in Non-Arab Countries.

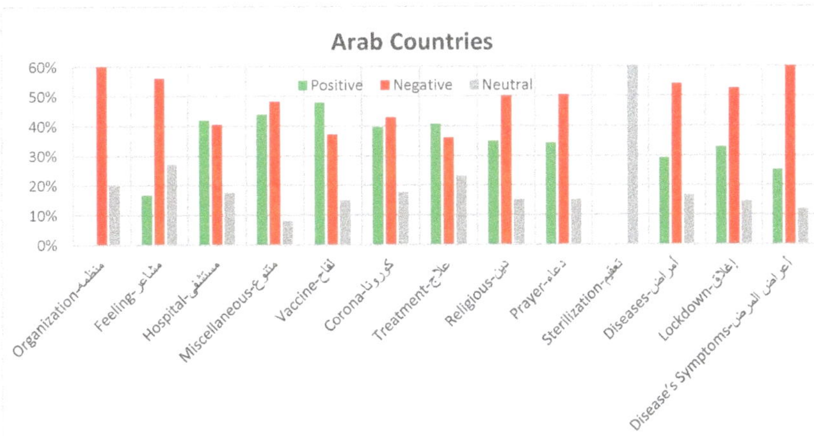

Figure 16. Correlation between main topics and SA in Arab countries.

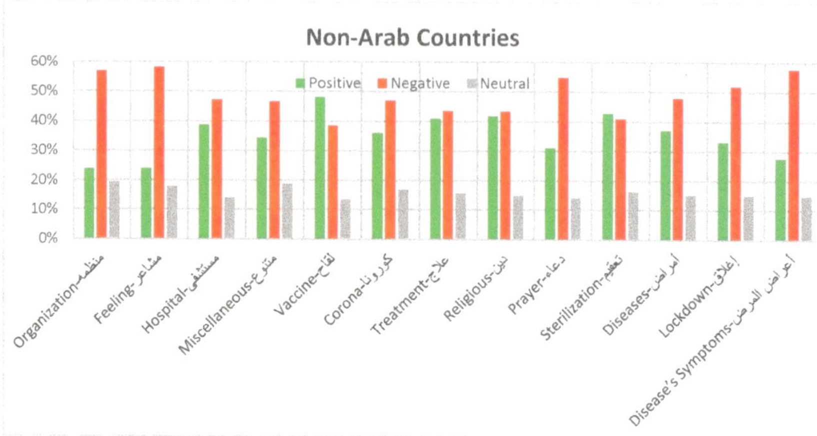

Figure 17. Correlation between main topics and SA in non-Arab countries.

We ran multiple experiments on our Arabic tweets COVID-19 dataset and the other data collected from previous experiments to explain the insight obtained from the analysis of increasingly large datasets. The results could play an essential role in representing our large-scale dataset with official data resulting early. We provided figures for visualizing data and findings in an approachable and stimulating way to enable us to extract information, superior to understand the data, and make more effective decisions. We tried to visualize the dataset components on world maps to learn COVID-19 pandemic distribution affects social media users globally. Figures A12–A14 illustrates the distribution of Arabic tweets, Arabic Hashtags, and Twitter users around the world. They clarify that most of them originated from Saudi Arabia and Egypt because they have the highest population as well the highest number of Internet users [1]. Meanwhile, Figure A15 displays the SA of Arabic tweets from our Geo-dataset over the world. It shows the average sentiment analysis of Arabic tweets is negative in most countries. Furthermore, Figure A16 illustrates the Word cloud of Sentiment Analysis over the world.

In addition, Figure A17 shows the distribution of COVID-19 confirmed cases globally, which emphasizes the known fact that the USA, India, and Brazil had the highest total number of COVID-19 confirmed cases. Finally, Figure A18 shows the distribution of the total number of lockdown days around the world based on the indicators used such as "Cancel Public Events" and "School closure". Specifically, it shows that most Arabic countries were in Lockdown for almost the year.

4.2. Discussion

In this research, we developed a novel location inference technique from non-geotagged tweets based on user profiles and textual content, which resulted in increasing the total percentage of geotagged tweets from 2% to 46% (about 2.5M tweets). Currently, we are conducting further research to improve the proposed location inference technique by using location inference from tweet textual content, such as the technique presented in the Arabic NER model using Flair embeddings [52]. This aims at increasing the percentage of produced geotagged tweets by utilizing an NLP process to manipulate the tweet's text. Moreover, evaluating our geo-dataset compared with four sentiment-labeled Arabic datasets (using four ML algorithms and one DL algorithm) shows that it accomplished a higher accuracy than the others, specifically with the DL model (BERT-mini).

In addition, some of the conducted correlation analysis shows that negative feelings of Arab Twitter users were raised during this pandemic. Moreover, we illustrated that there is a correlation between the discussed topics in the Arabic social media, such as lockdown and travel restrictions, which were enforced by governments, and the number of

COVID-19 new cases. Furthermore, the analysis showed that a positive correlation exists between the negative feeling of Arab users and the number of daily confirmed cases of COVID-19. Moreover, evaluating sentiment analysis in Arab countries shows that Saudi Arabia, Kuwait, and Egypt have the most positive/negative/neutral feelings based on users' opinions.

The correlation analysis in related work was mostly focused on time series analysis over health data, where we demonstrated in this work that spatio-temporal correlation with health and social data, can provide us with deeper insights on how people's sentiments differ over different topics and subtopics, and with different spatial scales, such as cities, countries, and regions. Finally, we visualized data and findings to understand the data components and distribute the effects of using social media related to COVID-19 globally.

5. Conclusions

This paper introduces a comprehensive social data mining approach for deriving COVID-19-related insights in the Arabic language, with a focus on the correlation between spatio-temporal social data and health data. In addition, it presented a sentiment analysis mechanism at multiple levels of spatial granularities and several topic scales. In addition, a technique to infer geo-information from non-geotagged tweets was developed, which increased the total percentage of location-enabled tweets from 2% to 46%, superior to most previous related works. To verify sentiment analysis performance, we applied sentiment-based classifications at many location resolutions (regions/countries) and some topic abstraction levels (subtopics and main topics) to derive people's opinions. Finally, we conducted many experiments and visualized our results based on the generated geo-social dataset, sentiment analysis, official health records, and lockdown data worldwide. Our findings show the great potential of integrating social data mining with other data sources, such as health data, to predict the evolution of such phenomena. In addition, such correlation can later be applied to other types of data such as contact tracing and GPS data, to provide an in-depth understanding of human behavior and the correlation between social and physical user interactions.

We intend to expand the dataset in the future to include more Arabic social contents to analyze the most recent periods when the social media users' focus has changed from COVID-19 in general to vaccinations. In addition, we plan to conduct further research to improve the proposed location inference technique by utilizing an NLP process to manipulate the tweet's text to increase the percentage of produced geotagged tweets.

Author Contributions: "Conceptualization", Imad Afyouni, Ibrahim Hashem and Zaher Al Aghbari; "Data curation", Tarek Elsaka; "Formal analysis", Tarek Elsaka and Imad Afyouni; "Investigation", Zaher Al Aghbari; "Methodology", Tarek Elsaka, Imad Afyouni and Ibrahim Hashem; "Software", Tarek Elsaka; "Supervision", Zaher Al Aghbari; "Validation", Ibrahim Hashem; "Visualization", Tarek Elsaka and Imad Afyouni; "Writing—original draft", Tarek Elsaka, Imad Afyouni, Ibrahim Hashem and Zaher Al Aghbari; "Writing—review & editing", Tarek Elsaka and Imad Afyouni All authors have read and agreed to the published version of the manuscript.

Funding: This research received no external funding.

Institutional Review Board Statement: Not applicable.

Informed Consent Statement: Not applicable.

Conflicts of Interest: The authors declare no conflict of interest.

Abbreviations

The following abbreviations are used in this manuscript:

SA	Sentiment Analysis
NLP	Natural Language Processing
ML	Machine Learning
CNN	Convolutional Neural Network

NMF	Matrix Factorization
FCM	Fuzzy C-Means
DT	Decision Tree
SVM	Support Vector Machine
SA	Sentiment Analysis
NLP	Natural Language Processing
ML	Machine Learning
CNN	Convolutional Neural Network
NMF	Matrix Factorization
FCM	Fuzzy C-Means
DT	Decision Tree
SVM	Support Vector Machine
MNB	Multinomial Naïve Bayes
BoW	Bag of Words
RF	Random Forest
WHO	World Health Organization
ECDC	European Centre for Disease Prevention and Control
JHU	Johns Hopkins University
ASTC	Arabic Sentiment Twitter Corpus
SS2030	Arabic Sentiment Analysis Dataset
ArSAS	Arabic Speech-Act and Sentiment Corpus of Tweets
BERT	Bidirectional Encoder Representations from Transformers

Appendix A. More Detailed Figures

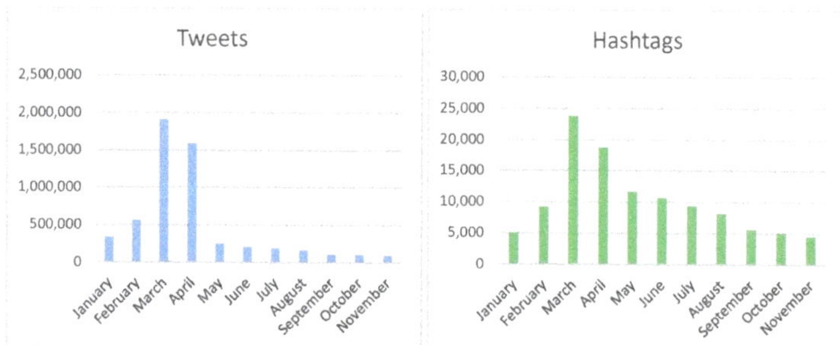

Figure A1. Monthly distribution of tweets and Hashtags.

Arabic Tweet	Translated Tweet
استشارية البريقة تعلن تماثل حالتين للشفاء من كورونا	The Brega consultant announces the recovery of two cases of corona
عظم الله اجرك واحسن الله عزاك اخوي طارق والله يرحمه ويغفر له ويرحم جميع المسلمين انا ابوي توفي الله يرحمه ويغفر له بفايروس كورونا احترزوا ارجوكم وطبقوا التباعد الاجتماعي	May God reward you, and may God bless you, my brother Tariq, and may God have mercy on him and forgive him and have mercy on all Muslims. My father passed away. May God have mercy on him and forgive him with the Coronavirus.
الاحد مؤتمر صحفي للمتحدث الامني ومتحدث وزاره الصحة حول مستجدات فيروس كورونا	Sunday a press conference for the security spokesman and the Ministry of Health spokesperson on the developments of the Coronavirus
الزكام الكحة السخونة من اعراض كورونا ترا الشم التذوق عادي بس هل هذا يصير من اعراض	Cold, cough, fever is some of the symptoms of the corona, you see the smell, the taste is normal, but is this a symptom?
اليه لتصنيف الائتمان السيادي يطلقها الاتحاد الافريقي وبنك التنمية الافريقي والامم المتحدة الاليه تأتي بعد ان خفضت شركات واصبحت دوله ذات تقييم مما رفع فائده اقتراضها في وقت انخفضت فيه مواردها بسبب كورونا	A mechanism for sovereign credit rating launched by the African Union, the African Development Bank and the United Nations The mechanism comes after companies reduced and became a rating country, which raised the interest of their borrowing at a time when their resources decreased due to Corona
الدول العظمى تفتح المدارس والدول العربية تغلق المدارس بذريعة علما بأن الاصابات في بلادهم كثيره هم دول متقدمة ونحن دول متخلفة بهذه الطريقة ستستمر الدول المتقدمة بتقدمها وتستمر الدول المتخلفة بتخلفها	Great countries open schools and Arab countries close schools on the pretext that the injuries in their countries are many. They have developed countries and we are backward countries. In this way, the developed countries will continue to advance and the backward countries will continue to lag.
عام دراسي موفق نتمناه للجميع اليوم يشرق عام دراسي مختلف بسبب ظروف تتجدد معها الطموحات والامنيات اللهم اجعلها بداية خير لجميع الطلاب والطالبات واجعل القادم اجمل مما مضى اللهم وفقهم واكتب لهم بداية جميله وسنه دراسية سعيدة واحفظهم من كل مكروه	A successful school year we wish everyone today a different school year shines due to circumstances with which aspirations and wishes are renewed.

Figure A2. Sample of Arabic tweets with English translation.

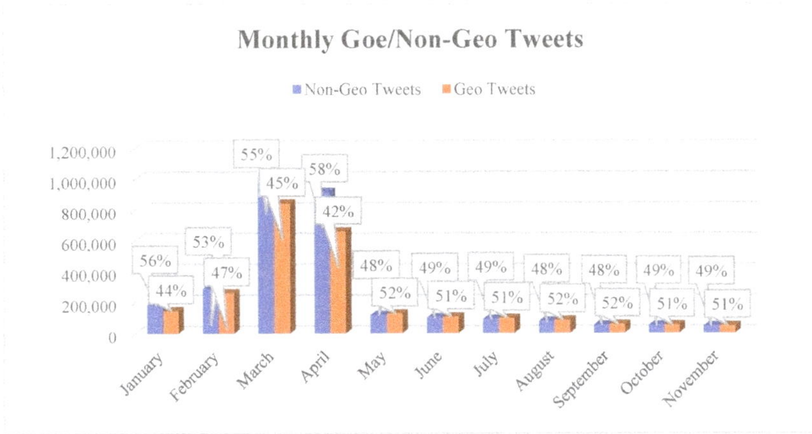

Figure A3. Monthly distribution of Geo and Non-geo tweets in Tweets Dataset.

Figure A4. Word cloud of the top topics in the COVID-19 Arabic Tweets Dataset.

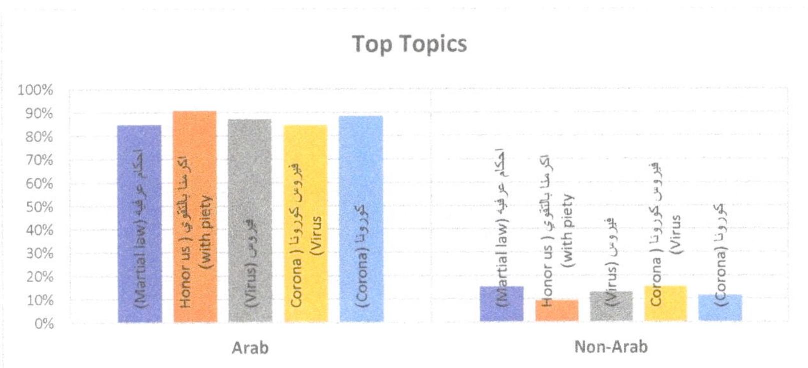

Figure A5. Top 5 Topics in Arab and Non-Arab Countries.

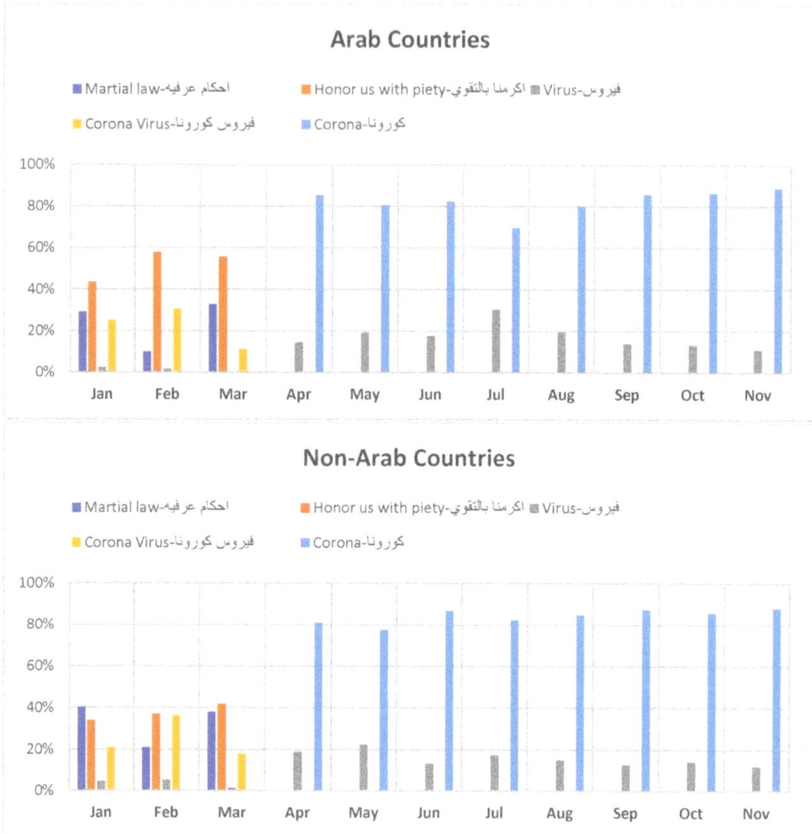

Figure A6. Monthly Top Topics in Arab and Non-Arab Countries.

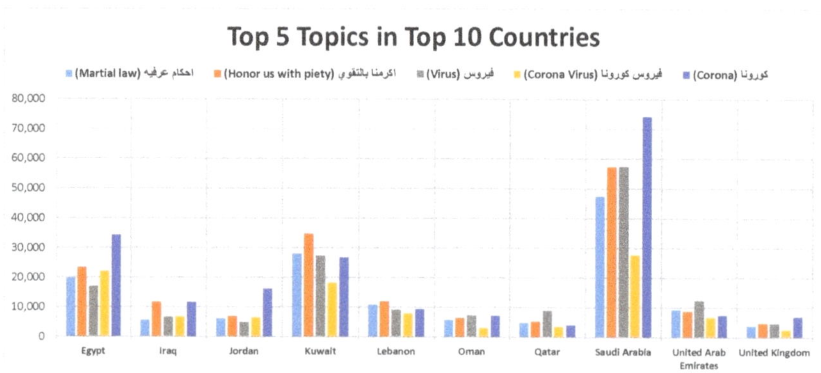

Figure A7. Top 5 Topics in Top 10 Countries.

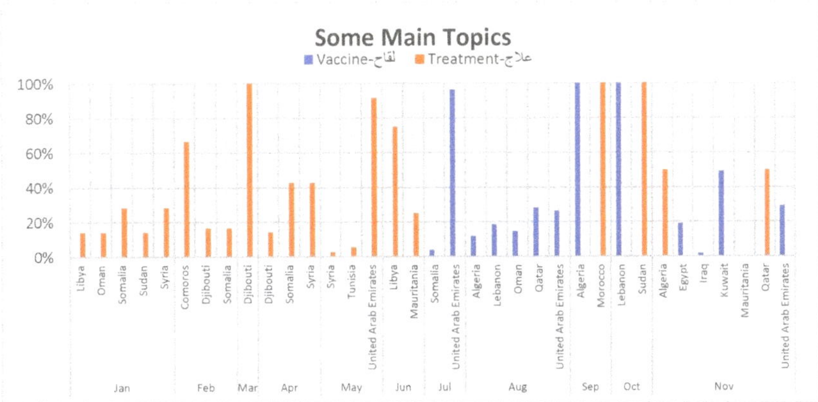

Figure A8. Frequency-Occurrence of Some Main Topics in Arab Countries.

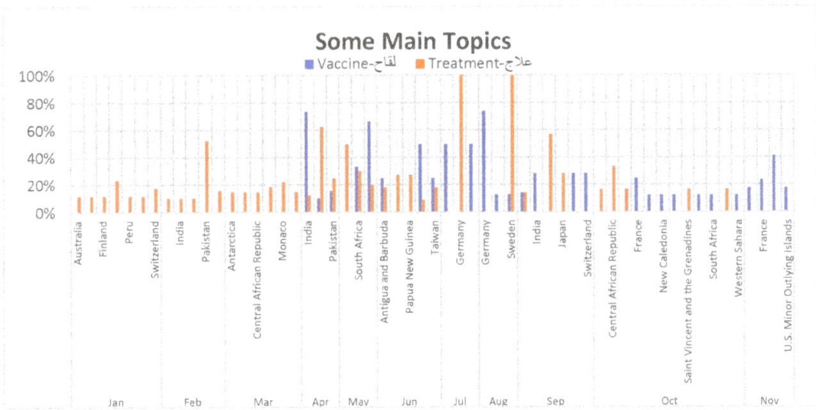

Figure A9. Frequency-Occurrence of Some Main Topics in Non-Arab Countries.

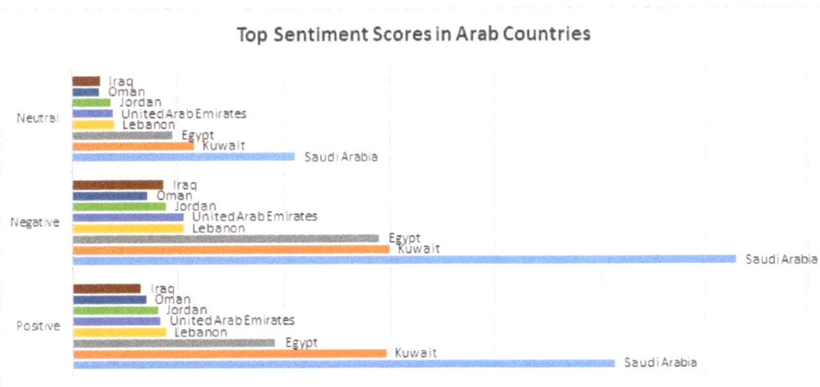

Figure A10. Top Sentiment (Positive/Negative/Neutral) in Arab Countries.

Figure A11. Correlation between SA and Official Health Records in some countries.

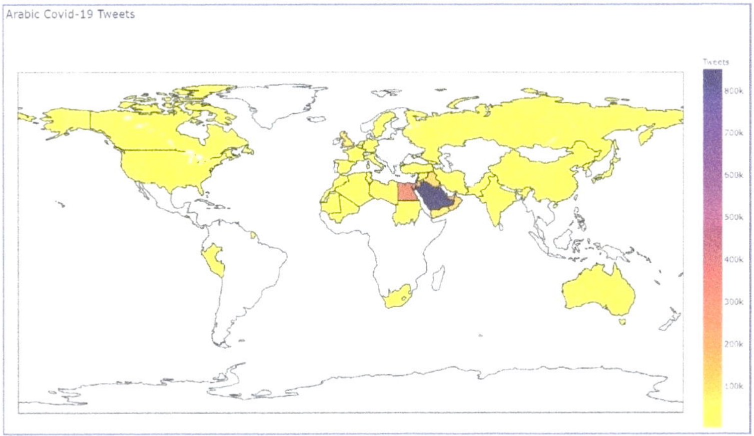

Figure A12. Distribution of Arabic Tweets.

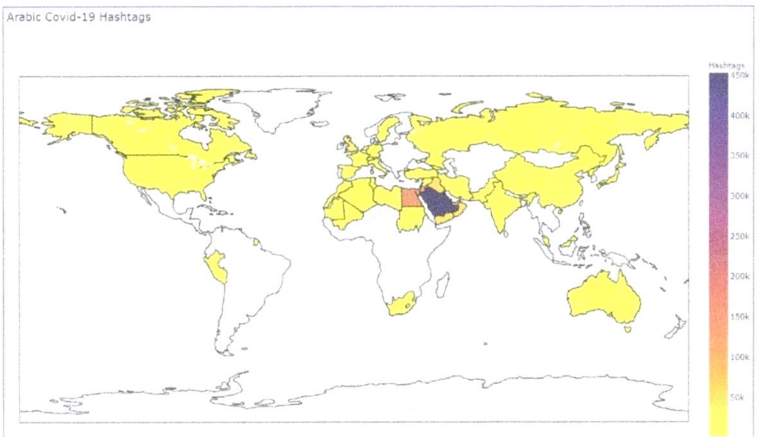

Figure A13. Distribution of Arabic Hashtags.

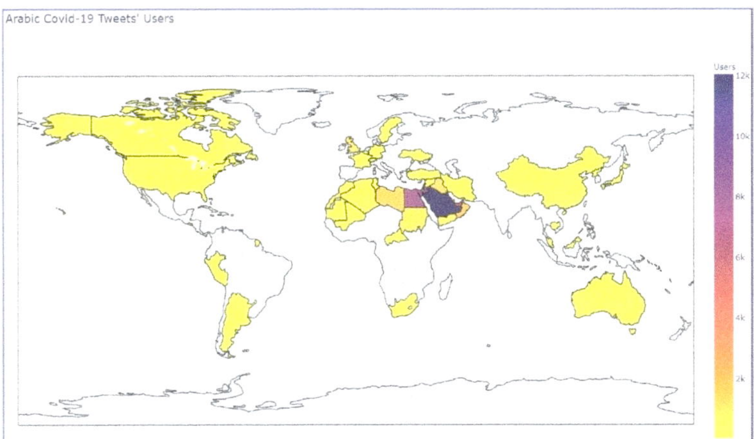

Figure A14. Distribution of Users.

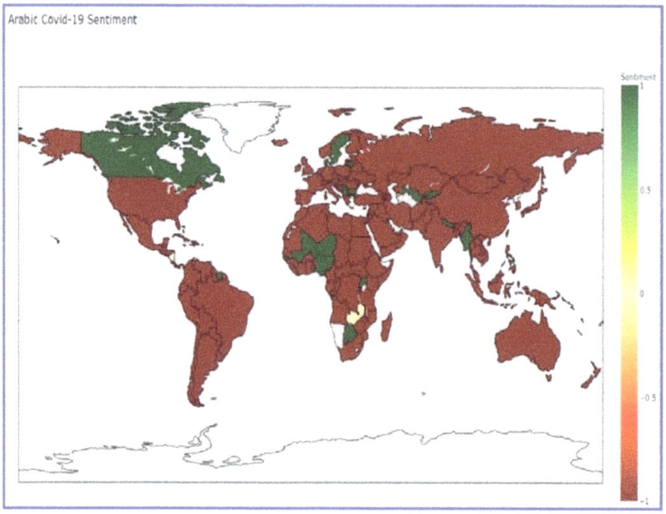

Figure A15. Sentiment Analysis of Arabic Tweets.

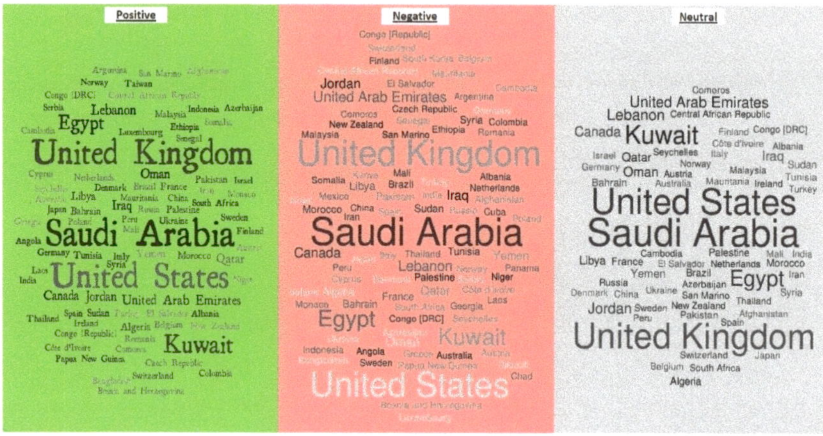

Figure A16. Word Cloud of Sentiment Analysis of Arabic Tweets over the world.

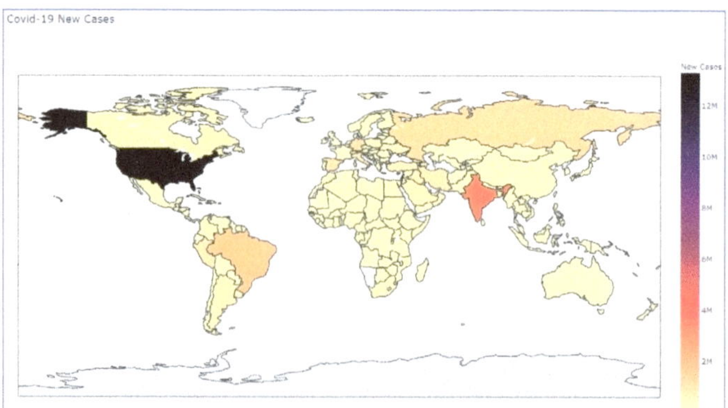

Figure A17. Distribution of COVID-19 Cases.

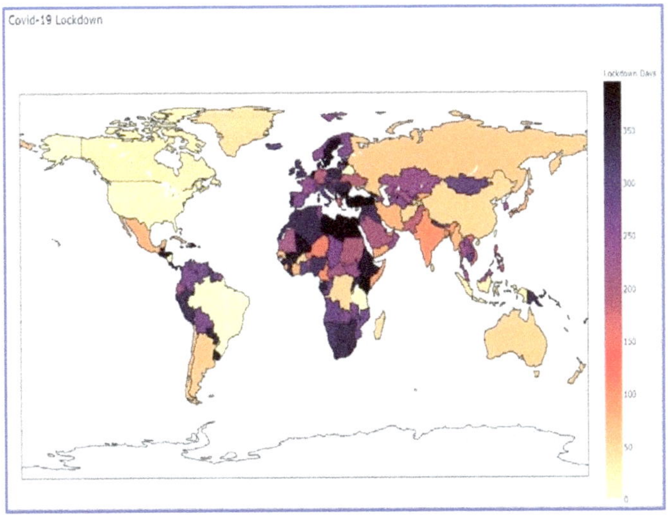

Figure A18. Covid-19 Lockdown Days.

References

1. Kemp, S. Digital 2021: Global Overview Report. 2021. Available online: https://datareportal.com/reports/digital-2021-global-overview-report (accessed on 5 October 2021).
2. Mohsen, A.M.; Hassan, H.A.; Idrees, A.M. A proposed approach for emotion lexicon enrichment. *Int. J. Comput. Electr. Autom. Control. Inf. Eng.* **2016**, *10*, 242–251.
3. Mostafa, A.M. Advanced Automatic Lexicon with Sentiment Analysis Algorithms for Arabic Reviews. *Am. J. Appl. Sci.* **2017**, *14*, 754–765. [CrossRef]
4. Miniwatts Marketing Group. Internet World Users by Language: Top 10 Languages. Available online: https://datareportal.com/reports/digital-2021-global-overview-report (accessed on 14 October 2021).
5. Miniwatts Marketing Group. Middle East Internet Statistics, Population, Facebook and Telecommunications Reports. 2021. Available online: https://eipss-eg.org/wp-content/uploads/2015/10/stats5.htm (accessed on 1 November 2021).
6. Salameh, Y. How Many Countries Speak Arabic around the World? 2020. Available online: https://www.tarjama.com/how-many-countries-that-speak-arabic-around-the-world/ (accessed on 22 June 2021).
7. Elsaka, T.; Afyouni, I.; Hashem, I.A.T.; AL-Aghbari, Z. Multi-scale Sentiment Analysis of Location-Enriched COVID-19 Arabic Social Data. In *Proceedings of the* International Conference on Discovery Science, Halifax, NS, Canada, 11–13 October 2021; Springer: Cham, Switzerland, 2021; pp. 194–203.
8. Ptáček, T.; Habernal, I.; Hong, J. Sarcasm detection on czech and english twitter. In Proceedings of the COLING 2014, the 25th International Conference on Computational Linguistics: Technical Papers, Dublin, Ireland, 25–29 August 2014; Dublin City University and Association for Computational Linguistics: Stroudsburg, PA, USA, 2014; pp. 213–223.
9. Badaro, G.; Baly, R.; Hajj, H.; Habash, N.; El-Hajj, W. A large scale Arabic sentiment lexicon for Arabic opinion mining. In Proceedings of the EMNLP 2014 Workshop on Arabic Natural Language Processing (ANLP), Doha, Qatar, 25 October 2014; Association for Computational Linguistics: Stroudsburg, PA, USA, 2014; pp. 165–173.
10. Asghar, M.Z.; Khan, A.; Ahmad, S.; Kundi, F.M. A review of feature extraction in sentiment analysis. *J. Basic Appl. Sci. Res.* **2014**, *4*, 181–186.
11. Alanazi, E.; Alashaikh, A.; Alqurashi, S.; Alanazi, A. Identifying and Ranking Common COVID-19 Symptoms From Tweets in Arabic: Content Analysis. *J. Med. Internet Res.* **2020**, *22*, e21329. [CrossRef]
12. Haouari, F.; Hasanain, M.; Suwaileh, R.; Elsayed, T. Arcov-19: The first arabic COVID-19 twitter dataset with propagation networks. In Proceedings of the Sixth Arabic Natural Language Processing Workshop, Kyiv, Ukraine, 9 April 2021; Association for Computational Linguistics: Stroudsburg, PA, USA, 2021; pp. 82–91.
13. Alharbi, A.; Lee, M. Kawarith: An Arabic Twitter Corpus for Crisis Events. In Proceedings of the Sixth Arabic Natural Language Processing Workshop, Kiev, Ukraine, 19–20 April 2021; Association for Computational Linguistics: Stroudsburg, PA, USA, 2021; pp. 42–52.
14. Hamdy, A.; Youssef, A.; Ryan, C. Arabic Hands-on Analysis, Clustering and Classification of Large Arabic Twitter Data set on COVID-19. *Int. J. Simul. Syst. Sci. Technol.* **2021**, *22*, 6.1–6.6.
15. Qazi, U.; Imran, M.; Ofli, F. Geocov19: A dataset of hundreds of millions of multilingual Covid-19 tweets with location information. *Sigspatial Spec.* **2020**, *12*, 6–15. [CrossRef]

16. Lamsal, R. Design and analysis of a large-scale COVID-19 tweets dataset. *Appl. Intell.* **2021**, *51*, 2790–2804. [CrossRef]
17. Alshalan, R.; Al-Khalifa, H.; Alsaeed, D.; Al-Baity, H.; Alshalan, S. Detection of Hate Speech in COVID-19–Related Tweets in the Arab Region: Deep Learning and Topic Modeling Approach. *J. Med. Internet Res.* **2020**, *22*, e22609. [CrossRef]
18. Alsafari, S.; Sadaoui, S.; Mouhoub, M. Hate and offensive speech detection on arabic social media. *Online Soc. Netw. Media* **2020**, *19*, 100096. [CrossRef]
19. Hamoui, B.; Alashaikh, A.; Alanazi, E. COVID-19: What Are Arabic Tweeters Talking about? In Proceedings of the International Conference on Computational Data and Social Networks, Dallas, TX, USA, 11–13 December 2020; Springer: Berlin/Heidelberg, Germany, 2020; pp. 425–436.
20. Al-Laith, A.; Alenezi, M. Monitoring People's Emotions and Symptoms from Arabic Tweets during the COVID-19 Pandemic. *Information* **2021**, *12*, 86. [CrossRef]
21. Bahja, M.; Hammad, R.; Kuhail, M.A. Capturing Public Concerns About Coronavirus Using Arabic Tweets: An NLP-Driven Approach. In Proceedings of the 2020 IEEE/ACM 13th International Conference on Utility and Cloud Computing (UCC), 7–10 December 2020; IEEE: Leicester, UK, 2020; pp. 310–315.
22. Essam, B.A.; Abdo, M.S. How Do Arab Tweeters Perceive the COVID-19 Pandemic? *J. Psycholinguist. Res.* **2021**, *50*, 507–521. [CrossRef]
23. Manguri, K.H.; Ramadhan, R.N.; Amin, P.R.M. Twitter sentiment analysis on worldwide COVID-19 outbreaks. *Kurd. J. Appl. Res.* **2020**, *5*, 54–65. [CrossRef]
24. Chakraborty, K.; Bhatia, S.; Bhattacharyya, S.; Platos, J.; Bag, R.; Hassanien, A.E. Sentiment Analysis of COVID-19 tweets by Deep Learning Classifiers—A study to show how popularity is affecting accuracy in social media. *Appl. Soft Comput.* **2020**, *97*, 106754. [CrossRef]
25. Kabir, M.Y.; Madria, S. EMOCOV: Machine learning for emotion detection, analysis and visualization using COVID-19 tweets. *Online Soc. Netw. Media* **2021**, *23*, 100135. [CrossRef]
26. Hussain, A.; Tahir, A.; Hussain, Z.; Sheikh, Z.; Gogate, M.; Dashtipour, K.; Ali, A.; Sheikh, A. Artificial intelligence–enabled analysis of public attitudes on facebook and twitter toward COVID-19 vaccines in the united kingdom and the united states: Observational study. *J. Med. Internet Res.* **2021**, *23*, e26627. [CrossRef]
27. Kastrati, Z.; Ahmedi, L.; Kurti, A.; Kadriu, F.; Murtezaj, D.; Gashi, F. A deep learning sentiment analyser for social media comments in low-resource languages. *Electronics* **2021**, *10*, 1133. [CrossRef]
28. Imran, A.S.; Daudpota, S.M.; Kastrati, Z.; Batra, R. Cross-cultural polarity and emotion detection using sentiment analysis and deep learning on COVID-19 related tweets. *IEEE Access* **2020**, *8*, 181074–181090. [CrossRef]
29. Alsudias, L.; Rayson, P. COVID-19 and Arabic Twitter: How can Arab World Governments and Public Health Organizations Learn from Social Media? In Proceedings of the 1st Workshop on NLP for COVID-19 at ACL 2020, Online, 5–10 July 2020; Association for Computational Linguistics: Stroudsburg, PA, USA, 2020.
30. Elhadad, M.K.; Li, K.F.; Gebali, F. COVID-19-FAKES: A Twitter (Arabic/English) dataset for detecting misleading information on COVID-19. In *International Conference on Intelligent Networking and Collaborative Systems*; Springer: Cham, Switzerland, 2020; pp. 256–268.
31. Hussein, A.; Ghneim, N.; Joukhadar, A. DamascusTeam at NLP4IF2021: Fighting the Arabic COVID-19 Infodemic on Twitter Using AraBERT. In Proceedings of the Fourth Workshop on NLP for Internet Freedom: Censorship, Disinformation, and Propaganda, Online, 6 June 2021; Association for Computational Linguistics: Stroudsburg, PA, USA, 2021; pp. 93–98.
32. Alrefaie, M. Arabic-Stop-Words [Github Repository]. 2016. Available online: https://github.com/mohataher/arabic-stop-words/blob/master/README.md (accessed on 15 December 2021).
33. Khomutnikova, E.; Gunbina, E.; Zhurkova, M.; Fetyukov, F. Semantics and etymology of english astionyms in the aspect of linguistic geography. In *European Proceedings of Social and Behavioural Sciences EpSBS*; European Publisher: London, UK, 2020; pp. 505–513.
34. Soliman, A.B.; Eissa, K.; El-Beltagy, S.R. Aravec: A set of arabic word embedding models for use in arabic nlp. *Procedia Comput. Sci.* **2017**, *117*, 256–265. [CrossRef]
35. Kapoor, A.; Singhal, A. A comparative study of K-Means, K-Means++ and Fuzzy C-Means clustering algorithms. In Proceedings of the 2017 3rd International Conference on Computational Intelligence and Communication Technology (CICT), Ghaziabad, India, 9–10 February 2017; IEEE: Ghaziabad, India, 2017; pp. 1–6.
36. Vijayarani, S.; Ilamathi, M.J.; Nithya, M. Preprocessing techniques for text mining—An overview. *Int. J. Comput. Sci. Commun. Netw.* **2015**, *5*, 7–16.
37. Duwairi, R.M.; Qarqaz, I. A framework for Arabic sentiment analysis using supervised classification. *Int. J. Data Min. Model. Manag.* **2016**, *8*, 369–381. [CrossRef]
38. Salameh, M.; Mohammad, S.; Kiritchenko, S. Sentiment after translation: A case-study on arabic social media posts. In Proceedings of the 2015 Conference of the North American Chapter of the Association for Computational Linguistics: Human Language Technologies, Denver, CO, USA, 31 May–5 June 2015; Association for Computational Linguistics: Stroudsburg, PA, USA, 2015; pp. 767–777.
39. Mohammad, S.M.; Salameh, M.; Kiritchenko, S. How translation alters sentiment. *J. Artif. Intell. Res.* **2016**, *55*, 95–130. [CrossRef]

40. Mohammad, S.; Salameh, M.; Kiritchenko, S. Sentiment lexicons for Arabic social media. In Proceedings of the Tenth International Conference on Language Resources and Evaluation (LREC'16), Portorož, Slovenia, 23–28 May 2016; European Language Resources Association (ELRA): Portoroz, Slovenia, 2016; pp. 33–37.
41. Kiritchenko, S.; Mohammad, S.; Salameh, M. SemEval-2016 Task 7: Determining Sentiment Intensity of English and Arabic Phrases. In Proceedings of the 10th International Workshop on Semantic Evaluation (SemEval-2016), San Diego, CA, USA, 16–17 June 2016; Association for Computational Linguistics: Stroudsburg, PA, USA, 2016; pp. 42–51. [CrossRef]
42. El-Beltagy, S.R.; Ali, A. Open issues in the sentiment analysis of Arabic social media: A case study. In Proceedings of the 2013 9th International Conference on Innovations in Information Technology (IIT), Al Ain, United Arab Emirates, 17–19 March 2013; IEEE: Al Ain, United Arab Emirates, 2013; pp. 215–220.
43. Devlin, J.; Chang, M.W.; Lee, K.; Toutanova, K. Bert: Pre-training of deep bidirectional transformers for language understanding. *arXiv* **2018**, arXiv:1810.04805.
44. Safaya, A.; Abdullatif, M.; Yuret, D. KUISAIL at SemEval-2020 Task 12: BERT-CNN for Offensive Speech Identification in Social Media. In Proceedings of the Fourteenth Workshop on Semantic Evaluation, Barcelona, Spain, 12–13 December 2020; International Committee for Computational Linguistics: Barcelona, Spain, 2020; pp. 2054–2059.
45. Saad, M. Arabic Sentiment Twitter Corpus. 2020. Available online: https://www.kaggle.com/mksaad/arabic-sentiment-twitter-corpus (accessed on 6 May 2021).
46. Alyami, S. Arabic Sentiment Analysis Dataset SS2030 Dataset. 2019. Available online: https://www.kaggle.com/datasets/snalyami3/arabic-sentiment-analysis-dataset-ss2030-dataset (accessed on 7 May 2021).
47. Khooli, A. Arabic 100k Reviews. 2020. Available online: https://www.kaggle.com/datasets/abedkhooli/arabic-100k-reviews (accessed on 8 May 2021).
48. Elmadany, A.; Mubarak, H.; Magdy, W. Arsas: An arabic speech-act and sentiment corpus of tweets. *OSACT* **2018**, *3*, 20.
49. Li, R.Y.M.; Yue, X.G. COVID-19 in Wuhan: Pressing realities and city management. *Front. Public Health* **2020**, *8*, 1079. [CrossRef]
50. Ullah, A.A.; Nawaz, F.; Chattoraj, D. Locked up under lockdown: The COVID-19 pandemic and the migrant population. *Soc. Sci. Humanit. Open* **2021**, *3*, 100126. [CrossRef]
51. Hale, T.; Petherick, A.; Phillips, T.; Webster, S. Variation in government responses to COVID-19. *Blavatnik Sch. Gov. Work. Pap.* **2020**, *31*, 2011–2020.
52. Megahed, M. Sequence Labeling Architectures in Diglossia—A Case Study of Arabic and Its Dialects. Master's Thesis, Humboldt University of Berlin, Berlin, Germany, 28 October 2021. [CrossRef]

International Journal of *Geo-Information*

Article

Automatic Classification of Photos by Tourist Attractions Using Deep Learning Model and Image Feature Vector Clustering

Jiyeon Kim and Youngok Kang *

Department of Social Studies, Ewha Womans University, Seoul 03760, Korea; jy.kim20@ewhain.net
* Correspondence: ykang@ewha.ac.kr

Abstract: With the rise of social media platforms, tourists tend to share their experiences in the form of texts, photos, and videos on social media. These user-generated contents (UGC) play an important role in shaping tourism destination images (TDI) and directly affect the decision-making process of tourists. Among UGCs, photos represent tourists' visual preferences for a specific area. Paying attention to the value of photos, several studies have attempted to analyze them using deep learning technology. However, the research methods that analyze tourism photos using recent deep learning technology have a limitation in that they cannot properly classify unique photos appearing in specific tourist attractions with predetermined photo categories such as Places365 or ImageNet dataset or it takes a lot of time and effort to build a separate training dataset to train the model and to generate a tourism photo classification category according to a specific tourist destination. The purpose of this study is to propose a method of automatically classifying tourist photos by tourist attractions by applying the methods of the image feature vector clustering and the deep learning model. To this end, first, we collected photos attached to reviews posted by foreign tourists on TripAdvisor. Second, we embedded individual images as 512-dimensional feature vectors using the VGG16 network pre-trained with Places365 and reduced them to two dimensions with t-SNE(t-Distributed Stochastic Neighbor Embedding). Then, clusters were extracted through HDBSCAN(Hierarchical Clustering and Density-Based Spatial Clustering of Applications with Noise) analysis and set as a regional image category. Finally, the Siamese Network was applied to remove noise photos within the cluster and classify photos according to the category. In addition, this study attempts to confirm the validity of the proposed method by applying it to two representative tourist attractions such as 'Gyeongbokgung Palace' and 'Insadong' in Seoul. As a result, it was possible to identify which visual elements of tourist attractions are attractive to tourists. This method has the advantages in that it is not necessary to create a classification category in advance, it is possible to flexibly extract categories for each tourist destination, and it is able to improve classification performance even with a rather small volume of a dataset.

Keywords: image feature vector; clustering; Siamese Network; automatic classification of tourist photos; deep learning model

Citation: Kim, J.; Kang, Y. Automatic Classification of Photos by Tourist Attractions Using Deep Learning Model and Image Feature Vector Clustering. *ISPRS Int. J. Geo-Inf.* **2022**, *11*, 245. https://doi.org/10.3390/ijgi11040245

Academic Editors: Gloria Bordogna, Cristiano Fugazza and Wolfgang Kainz

Received: 30 January 2022
Accepted: 7 April 2022
Published: 10 April 2022

1. Introduction

Recently, as anyone can access social media platforms anytime, anywhere using mobile devices, a large volume of texts and photos have been shared on the web to communicate with others. People are freely expressing their thoughts and feelings through text and photos on social media platforms. Along with this trend, the way in which tourists get information related to travel attractions and share their experiences is also changing. More and more tourists share their experiences in the form of texts, photos, and videos on social media, which serves as an information source for potential tourists [1]. Data posted on social network services (SNS) is steadily receiving social attention in that it is user-generated content (UGC). The tourism industry is also paying attention to UGC data to identify new tourism trends and analyze the image of tourist attractions perceived by tourists [2]. In particular, the image of tourist attractions plays an important role when people select their tourist destination and destination marketing organizations (DMO) perform tourism marketing [3–6].

In the past, DMOs have played a leading role in shaping the image of tourist destinations. However, due to the popularization of social media platforms in recent years, it has been recognized that the image of tourist attractions is formed by both UGC and the contents created by DMOs [7]. Among UGCs, a photo plays an important role in forming the image of tourist attractions in that it visually reproduces the places [8]. A photo reflects the mental image of the physical elements experienced by the photographers. In addition, a photo is a record of a moment to express a mental image of a place in a visual form [9]. Therefore, since these photos contain tourists' visual preferences for a specific area, they can reflect actual tourists' preferences more directly than a few experts [10]. In addition, potential tourists tend to visit tourist sites that have been exposed to them and take pictures of visual images that have been projected on them [11].

Paying attention to the value of photos, more and more studies have attempted to analyze photos on SNS taken by tourists and uncover attractive factors that contribute to the formation of the image of a tourist destination [1,4,6,12–14]. However, due to the limitations in technologies, studies on tourism destination images (TDI) using UGC photos encounter challenges in terms of both the volume of data and the interpretation of results. The most widely used method is a manual analysis where researchers observe their collected photos and manually classify them into specific categories. Since this methodology is a labor-intensive process, there is a limit to the number of photos that can be analyzed, which makes it difficult to comprehensively analyze tourist attractions.

As computer vision technologies have developed, several studies have identified TDI from a number of SNS photos using deep learning methods [15–20]. However, they have limitations in classifying photos that represent unique characteristics of tourist attractions. They use predetermined photo categories such as Places365 or ImageNet which are designed for general purposes, so they are not appropriate for identifying the uniqueness of individual attractions. To overcome these limitations, Kang et al. and Yoon and Kang analyzed the images by generating a tourism photo classification category according to a specific area and training the model with training datasets for each category [21,22].

Although these existing studies have presented valuable results with the combination of UGC photos and a deep learning model to extract tourism destination images, studies are still in their infancy. In particular, studies on extracting distinctive characteristics of individual tourism attractions are limited. They have focused on analyzing the TDIs of a nation or a city rather than individual tourist attractions. While they also partially explore individual tourist' attractions included in the region, their categories for photo classification which are based on a national scale or city scale are not appropriate for figuring out the unique properties of individual tourism attractions.

To solve this problem, we propose a method for automatically building categories for photo classification using clustering and a Siamese network. This reduces the burden on the process of creating categories corresponding to each tourism attraction. In addition, the clustering methodology provides the advantage of establishing categories based on a data-driven manner. Our framework consists of the following four parts. First, we collected TripAdvisor photos in reviews posted by foreign tourists in Seoul. Second, we embedded individual images as 512-dimensional vectors using a VGG16 network pre-trained with Places365 and reduced these vectors to two dimensions with t-SNE. Third, to create a category based on visual content that frequently appears in photos taken by tourists, clusters were extracted through HDBSCAN analysis and they were set an image category of an attraction. Finally, a Siamese Network was applied to remove noise photos within the cluster and classify photos according to the category.

2. Literature Review

2.1. Analysis of Tourist Attractions Using UGC Photos

With the popularization of mobile devices and the rise of social media platforms, the images of tourist attractions tend to be formed through photos and narratives shared online. The shared images of tourist attractions are continuously perceived and reproduced from person to person [12]. As content posted on social media platforms are exposed to many people, they tend to travel to destinations or attractions that frequently appear on the SNS. These visual images allow DMOs to get insights into tourist behaviors and perceptions for marketing. Compared to existing marketing tools, this type of marketing is recognized as an effective tool that quickly affects the decision-making process of a tourist while reducing costs [13]. Paying attention to the value of such photos, more and more studies are attempting to analyze photos taken by tourists and uncover attractive factors of tourist destinations. Before the rise of the deep learning method in a tourism context, the predominant method in the analysis of photos is to directly observe them one by one, which is a manual manner. Agustí et al. and Dinh identified the process of forming a tourism image in a specific area through the analysis of these photos [1,12]. Stepchenkova et al. analyzed the difference between the image generated by tourists and the image projected by DMOs [14]. In the case of direct visual observation, which requires a labor-intensive process, it is difficult to comprehensively analyze tourist destinations because there is a limit to the volume of photos that can be analyzed. In addition, there is another limitation that the research results may be dependent on researchers.

With the rapid development of computer vision and image processing technologies in recent years, several studies that analyze a large volume of photos using deep learning models are emerging in the tourism field. Most of the studies have applied a convolutional neural network (CNN) developed to solve the image classification problem. These studies have classified photos according to specific categories and uncovered tourists' perceptions of specific areas based on their classification ratios.

Most of the studies analyzed tourist photos using a pre-trained model with Places365 which is a dataset specialized in place classification problems [15–20]. However, according to Kim et al., when using a pre-trained model, there is a problem of misclassifying unique objects or scenes that appear in photos of local tourist attractions [17]. To overcome this limitation, other studies that have transferred models to training datasets specialized in research areas have emerged. Kang et al. and Yoon and Kang analyzed tourism images in a specific area through transfer learning of a deep learning model after constructing categories and datasets specialized in that area without using a pre-trained model [21,22]. These studies, which have used categories to classify tourism images, have limitations in that it is difficult to properly identify the features of local tourist attractions and it takes a lot of time and effort to build training datasets manually.

2.2. Application of Deep Learning-based Image Embedding and Clustering

Clustering, one of the representative unsupervised learning methodologies, enables us to discover hidden patterns and structures in data. In order to perform clustering analysis, a process of extracting features of image and converting them into vectors is required. Before the rise of the deep learning method, image embedding algorithms that extract fixed feature points from the image, such as scale-invariant feature transform (SIFT), speeded up robust feature (SURF), and binary robust independent elementary features (BRIEF), have been used in this process [23]. With the rapid development of computer vision technology, CNN-based auto-encoders and embedding model based on the CNN network have been widely used. The latter methods transform images into vectors through feature maps in the CNN network pre-trained on specific datasets such as Places365 or ImageNet [24]. Image clustering provides a way for discovering hidden structures or patterns of data in various fields.

Tapaswi et al. embedded faces based on a deep learning model to determine the number of characters appearing in the video, then identified the number of clusters through hierarchical cluster analysis and derived the number of characters [25]. Gu et al. identified New York's fashion trends by embedding street fashion images and applying agglomerative hierarchical clustering by year [26]. Castellano and Vessio converted the artwork image into a feature vector with the DenseNet121 network, applied K-means clustering and auto-encoder to find the cluster, and analyzed the painting style of the artwork through cluster results [27].

2.3. Application of Siamese Network with Image Embedding

Siamese Network is particularly used in the research of medical care, palm print, face recognition, object tracking, etc. where it is difficult to obtain a large volume of data. Siamese Network has been applied to solve these problems. When two or more input images are given, Siamese Network learns the similarity between them and expresses it as a numerical distance. That is, if the input images are similar to each other, the distance is close, and if the input images are different, the distance becomes far. The same principle can be applied to image classification in various fields.

Schroff et al. developed a FaceNet model learned with triplet loss for face recognition based on the Siamese Network structure [28]. The FaceNet model embeds an input face image in 128 dimensions, and then distinguishes between photos of a person's face and photos that do not, through the distance between the embedded vectors. Zhong et al. developed a palm print recognition model using Siamese Network based on the VGG16 network [29]. In this study, they used Siamese Network to convert a long text image given as input data into a 500-dimensional vector and compared the distances between the two long text image vectors to determine whether they were identical. Mehmood et al. developed a model for early detection of Alzheimer's by utilizing the VGG-16 network-based Siamese Network [30]. They trained a model using MRI datasets classified into four types according to the severity of Alzheimer's and classified the progression of Alzheimer's by comparing the distance between embeddings. Bertinetto et al. developed an object tracking model based on Siamese Network [31]. They embedded the image containing the object to be tracked and the image to find the object into Siamese Network and identified the location of a specific object in the image by comparing the similarity between embeddings. As such, these studies are widely used, especially in cases where it is difficult to secure sufficient datasets such as long palm print, face recognition, and disease diagnosis. Even in the tourism field, if the scale of the research target is narrowed down to a specific tourist area, it may be difficult to secure sufficient data. Therefore, this study also intends to use the Siamese Network model to improve classification performance with a rather small volume dataset.

3. Materials and Methods

3.1. Research Process

The analysis process of this study is shown in Figure 1. First, we extracted the photos posted on TripAdvisor and then selected the photos posted by foreign tourists, excluding Koreans using the publisher's country of origin and the language used in writing the review. Second, we embedded individual images as 512-dimensional feature vectors using a VGG16 network pre-trained on the Places365 dataset. Third, we reduced each vector to two-dimension with t-SNE, employed HDBSCAN to cluster these photos, and set this as an image category. Fourth, we used Siamese Network to remove noise images not included in categories and classify photos according to the category.

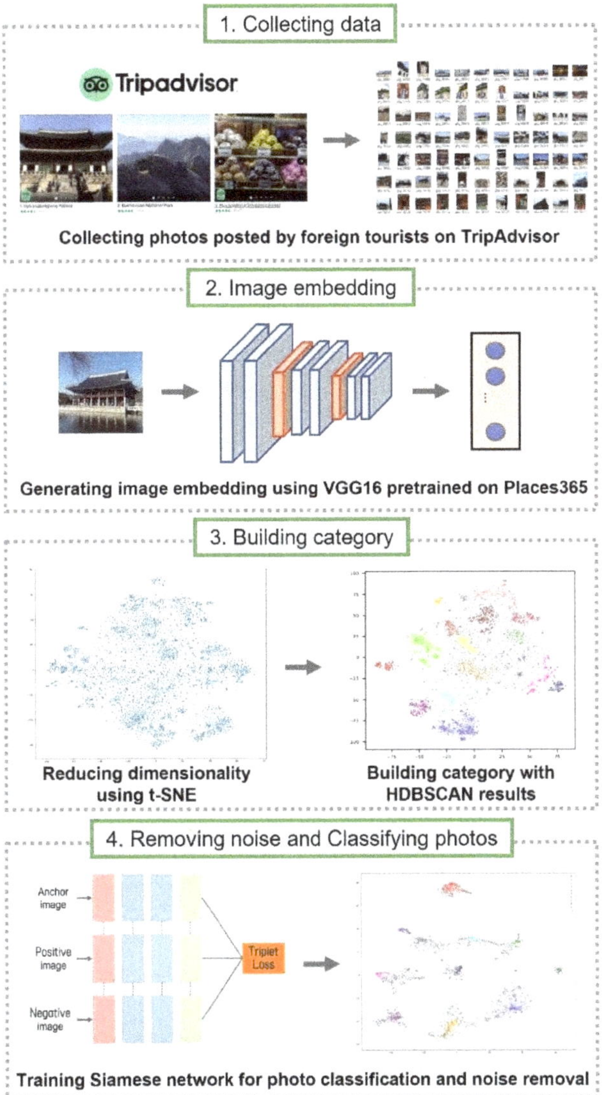

Figure 1. Research Process.

3.2. Collecting Data

TripAdvisor (www.tripadvisor.com, accessed on 19 September 2021) is the world's largest travel information platform with more than 260 million monthly users [32]. TripAdvisor provides reviews written by visitors to tourist attractions, hotels, and restaurants around the world. When you log on to TripAdvisor and search for a country or city, you can find information about popular tourist attractions, accommodations, restaurants, and activities in that area. If people search for Seoul on TripAdvisor, they can find a list of famous tourist attractions in Seoul under the heading 'Top Attractions in Seoul'. If people click on each tourist attraction, they can check the reviews posted by tourists who have visited the tourist attractions. In each review, they can identify the user nickname, region of origin, the number of posts, star rating, review title, date of visit, type of visit, textual body, attached photos, and date of creation. In this study, we collected reviews of 'Gyeongbokgung Palace' and 'Insadong' using Python, and collected photos attached to the review, date of visit, and nationality data of the publisher. 'Gyeongbokgung Palace' is a palace of the Joseon Dynasty located in the city center and is one of the most visited places by foreign tourists. 'Insadong', a distinctive area with a mixture of galleries, traditional restaurants, modern buildings, and shopping streets, is also a popular place for foreign tourists. In this study, we only selected the photos posted by foreign tourists using language filtering and the visitors' origin information. As the number of inbound tourists dropped sharply in 2020 due to COVID-19, all data before 2020 were used.

3.3. Image Embedding

To analyze image data using deep learning, it is necessary to map the image into an embedding space. In this case, when a vector is generated by arranging pixel values of an original image in a row, it is difficult to determine the similarity between images through distance measurement because the vector does not reflect the case where the same visual pattern exists at different locations in the picture. As an alternative to this, a method of extracting a visual pattern of an image and embedding it as a vector can be used.

In this study, we utilized a CNN-based embedding model where vectors were generated by reflecting the visual content of the photo. Therefore, vectors having similar visual contents are located close to each other in the embedding space and vice versa. The close distance between vectors means that the visual contents of the original images are similar. The CNN-based image classification model is largely divided into two parts: one is to learn the features of images and the other is to classify images based on the features. The former consists of a convolutional layer, an activation function, and a pooling layer, while the latter consists of a fully connected layer and a softmax layer. Since there is no need for the latter part in the embedding process, we replaced the fully connected layer at the top of the CNN model with a Global Max pooling layer. In this study, we employed a VGG16 network pre-trained on Places365 dataset for embedding. Places365 is a benchmark dataset created by extracting 365 categories from Place datasets consisting of a total of 10 million photos [33]. Since this study tries to analyze photos taken at tourist attractions, a pre-trained model with Places365 was used.

3.4. Dimension Reduction and Clustering

The major functions of this process are to extract visual contents that frequently appear in photos taken by tourists and to utilize them as a category. First, we reduced the 512-dimension to 2-dimension with t-SNE. Second, we employed the HDBSCAN clustering algorithm to cluster these embeddings. The clustering results showed tourists' visual preference for a tourism attraction that forms TDI.

The t-SNE is one of the nonlinear methods designed to reduce high-dimensional data to two or three dimensions based on probability distribution and visualize it [34]. This method was developed by supplementing the problems of Stochastic Neighbor Embedding [35] and is focused on maintaining a local structure when reducing dimensions. Here, maintaining the local structure means reducing the data so that the relationship can be kept even after the points that are close to each other in the high dimension are projected to the low dimension. In Equation (1), p_{ij} represents the probability that data points x_i and x_j existing in a high dimension are neighboring to each other. In Equation (2), q_{ij} represents the probability that x_i and x_j, which are low-dimensional points corresponding to y_i and y_j, are adjacent to each other. The cost function of t-SNE is calculated by Kullback-Leibler divergence, a function that calculates the difference between probability distributions of both a high and a low dimension in Equation (3) [36].

$$p_{ij} = \frac{exp(- \parallel x_i - x_j \parallel^2 / 2\sigma^2)}{\sum_{k \neq l} exp(- \parallel x_k - x_l \parallel^2 / 2\sigma^2)} \tag{1}$$

$$q_{ij} = \frac{exp(- \parallel y_i - y_j \parallel^2)}{\sum_{k \neq l} exp(- \parallel y_i - y_j \parallel^2)} \tag{2}$$

$$C = KL(P \parallel Q) = \sum_i \sum_j p_{ij} \log \frac{p_{ij}}{q_{ij}} \tag{3}$$

p_{ij}: joint probability that i and j are neighbors in a high dimension
q_{ij}: joint probability that i and j are neighbors in a low dimension
x_i, x_j: high-dimensional data points
y_i, y_j: low-dimensional data points counterparts of the x_i and x_j

We employed the HDBSCAN algorithm to identify clusters in embedding space and utilize them as a category. HDBSCAN has evolved from density-based spatial clustering with noise (DBSCAN), a density-based clustering algorithm [37]. DBSCAN finds a cluster of points over a certain density in the entire data point space [38]. Here, the certain density is defined as the value of Eps indicating the radius and m_{pts}, which is the minimum number of data points included in the Eps. DBSCAN has two drawbacks, the first being sensitive to parameters, and the second being that it cannot find clusters with different densities because it sets thresholds for density. HDBSCAN is an algorithm that compensates for the shortcomings of DBSCAN, and adds the concept of hierarchical clustering to DBSCAN. Since HDBSCAN finds clusters by defining only the minimum amount of data without using a fixed Eps value, it can extract various clusters with different density values.

3.5. Removing Noise Data and Classifying Photos

It may seem that clustering results could replace photo classification in that clusters are made up of similar visual contents. However, since HDBSCAN works based on density, the accuracy of clustering tends to be low at the edge of a cluster that has a relatively low density than the core. This is responsible for two problems. First, noise photos that are not related to the cluster may be included. Second, noise points located around the boundary of the cluster may not actually be noise.

To address these problems, we implemented a Siamese network to classify photos trained with our own dataset made up of photos taken in the research area. Siamese network consists of more than two identical subnetworks capable of learning patterns from input vectors [39]. The outputs generated through a Siamese network reflects the similarity between images. Although the model receives different photos as inputs, the weights in subnetworks are updated equally because they are combined by the loss function. This weight fixation means that visually similar images are located close to each other and vice versa.

In this study, triplet loss was used as a cost function for training the Siamese network, and semi-hard was used among triplet mining methods. The triplet loss function receives three types of input data: anchor, positive, and negative. There are three ways to construct input data: easy triplets, hard triplets, and semi-hard triplets. Schroff et al. revealed that the model trained using the semi-hard triplets' method is superior among them [28]. Figure 2 shows the architecture of the model used in the study.

Siamese network learns to position the images with similar visual content close to each other in vector space and vice versa. Based on this principle, photos can be classified, and noises can be detected. For this, a target set, a noise set, and a reference set are needed. The target set is a set of target photos to be classified and labelled according to category items. A noise set is a set of noise photos that do not belong to the category. The reference set consists of sample images of each category. Test sets and reference sets were made to consist of a similar number of photos for each category. The noise set was made with a similar number of photos to the test set. Photo classification and noise removal are performed through the following four steps: First, the distance between a target photo and the photo belonging to the reference set is calculated respectively. Second, the prediction label of the target image is assigned as a label of a reference photo having a minimum distance in Equation (4). These processes are repeated for all target images. Third, the distance between a noise photo and the photo belonging to the reference set is calculated, respectively. This step is repeated for all photos in the noise set. As a result, noise photos can be deleted by setting a threshold for the minimum distance. Fourth, the accuracy of prediction was evaluated as the minimum distance threshold was changed using the ROC curve. The threshold of the point showing the best accuracy was selected. The ROC curve is a graph showing how the performance of the classification system changes according to various thresholds. In this study X and Y axes of the ROC curve are True Positive Rate (TPR) and False Positive Rate (FPR). TPR is the percentage of cases where the true label matches the predicted label in the target set in Equation (5). FPR is the percentage of cases where noise is not classified as noise in Equation (6). The optimal threshold is the value of the point farthest from Y = X among points on the ROC curve in Figure 3.

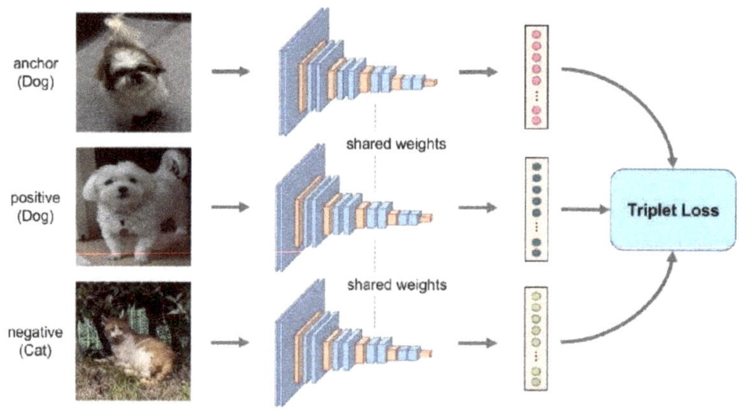

Figure 2. Model architecture based on semi-hard triplet Siamese Network.

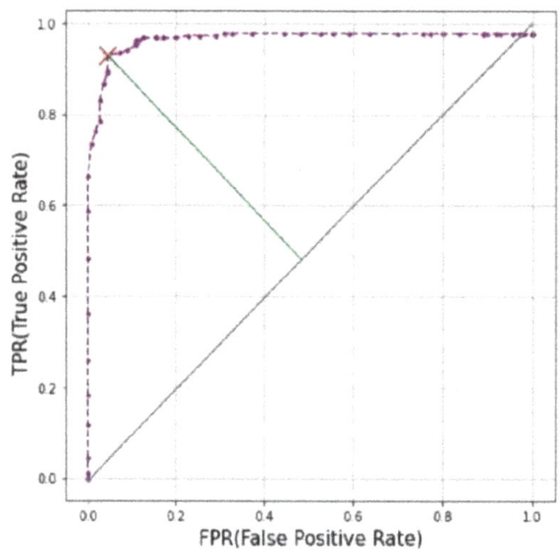

Figure 3. Example of ROC Curve.

$$l(t) = \underset{i}{\mathrm{argmin}}\left(d\left(t, S_{i,\,j}\right)\right) \tag{4}$$

$l(t)$: predicted label on a target photo
t: image embedding of a target image
$S_{i,j}$: Embedding of the jth sample image of category item i.

$d(x, y)$: Euclidean distance function between x and y

$$TPR = \frac{n(TL(i) = PL(i),\ i \in testset)}{n(testset)} \quad (5)$$

$$FPR = \frac{n(TL(i) \neq PL(i),\ i \in noiseset)}{n(noiseset)} \quad (6)$$

$TL(i)$: true label of a photo
$PL(i)$: predicted label of a photo

4. Results

4.1. Gyeongbokgung Palace

A total of 9940 photos were collected in 10,655 reviews registered on TripAdvisor on the 'Gyeongbokgung Palace' page. Out of a total of 9940 photos, we selected 8188 photos except for 715 reviews written in Korean. A VGG16 model pre-trained with Places365 embedded each photo in a 512-dimensional vector, and t-SNE reduced the vectors to two dimensions. We implemented HDBSCAN to cluster these vectors into several groups. The result is shown in Figure 4, and the number of photos for each cluster is shown in Table 1. Sixteen clusters were generated and 3824 points were classified as noise.

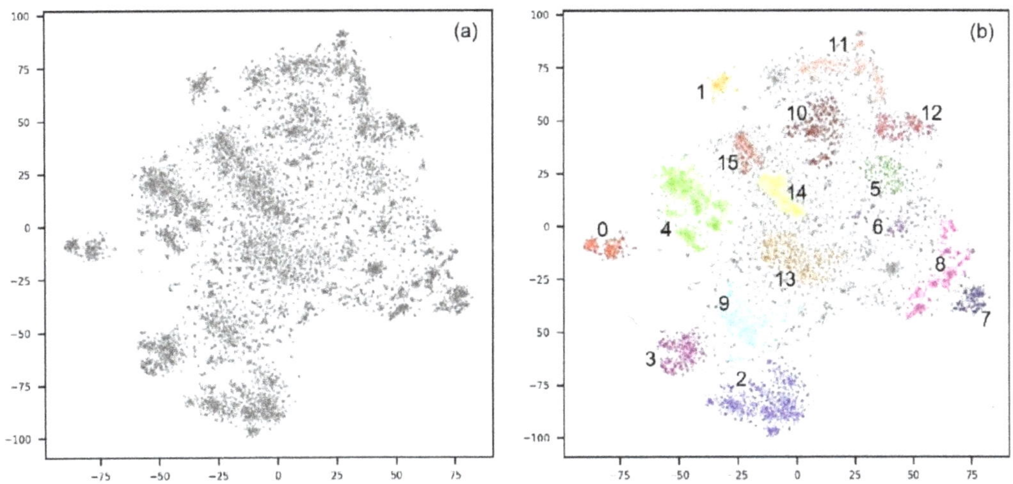

Figure 4. Gyeongbokgung Palace photos: (**a**) two-dimensional visualization; (**b**) result of HDBSCAN.

Table 1. Number of photos for each cluster generated as a result of HDBSCAN in Gyeongbokgung Palace.

Cluster	Number of Photos	Cluster	Number of Photos
0	189	8	412
1	124	9	381
2	715	10	380
3	358	11	391
4	737	12	239
5	199	13	456
6	146	14	235
7	173	15	183

After inspecting the photos in each cluster, we took two actions. First, if there were more than two clusters that contained the same visual contents, we integrated them into one. Second, if there was no similarity between the pictures that formed a cluster, that cluster was deleted because it is difficult to consider them as a meaningful cluster. Clusters 7 and 8 were integrated into one cluster because both were composed of photos of 'Throne'. Clusters 10 and 12 were also combined into one cluster because both consisted of photos of the 'Gate guard changing ceremony'. Clusters 14 and 15 were also integrated into one cluster because both were composed of photos of 'Heungnyemun gate' in the same way. On the other hand, clusters 11, 12, and 13 were not used to create a category because each cluster was made up of different photos. As a result, 10 categories were created as follows: 'Gyeonghoeru Pavilion', 'Geunjeongjeon Hall', 'Heungnyemun Gate', 'Gwanghwamun Gate', 'Hangwomen Pavilion', 'National Folk Museum', 'Throne', 'Hanbok (traditional dress of Korea)', 'Gate guard changing ceremony', and 'Tree'.

Siamese network enables us to classify photos according to the previously generated category and remove noise photos. For this purpose, we trained a Siamese network based on the VGG16 network using Gyeongbokgung Palace's photo dataset. In this process, the training dataset was composed of the photos in each cluster except for those that were incorrectly included in clusters. Table 2 shows the number of photos included in the training dataset for each category.

Table 2. Number of training photos by category in Gyeongbokgung Palace.

Category	Number of Training Photos
Gyeonghoeru Pavilion	365
Geunjeongjeon Hall	342
Heungnyemun Gate	270
Gwanghwamun Gate	196
Hangwonjeong Pavilion	174
National Folk Museum	101
Throne	156
Hanbok(traditional dress)	145
Gate guard changing ceremony	173
Tree	151

To improve the ability of the model for pattern extraction through nonlinearity, two convolutional layers were added to the basic structure of the VGG16 network. The model used in this process had weights pre-trained on Places365. These weights had been fine-tuned with our own dataset. Figure 5 shows the change in the loss value in the process of the model training. To prevent overfitting, we trained the model up to epoch 18.

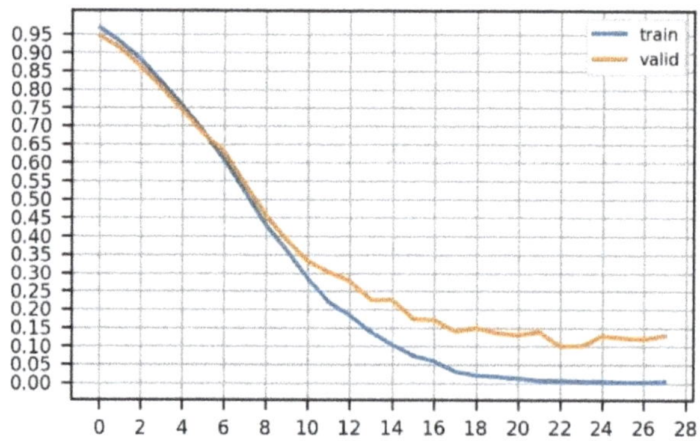

Figure 5. Loss graph of model training of Gyeongbokgung Palace.

To remove noise photos, it was necessary to identify the optimal threshold value from the ROC curve. Figure 6a shows the minimum distance between a target photo and the reference set as a histogram. Figure 6b shows the minimum distance between a target photo and the reference set as a histogram. Figure 7 represents the ROC curve, and 0.4 which is the threshold of the farthest point from Y = X, shown in red X on the graph, was selected as the optimal value. Figure 7 also shows that TPR was 0.928 and FPR was 0.045, which related to the accuracy of the model. Compared with Figure 4b, Figure 8 shows that data points belonging to the same cluster were close to each other and the distances between different clusters were farther apart. Figures 9 and 10 respectively show the number of photos and example photos finally classified by items in a category.

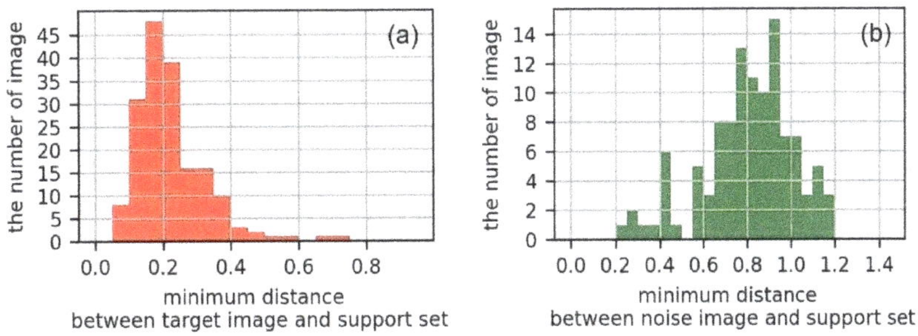

Figure 6. Histogram of Gyeongbokgung Palace: (**a**) minimum distance between a target photo set and reference set; (**b**) minimum distance between a noise photo and reference set.

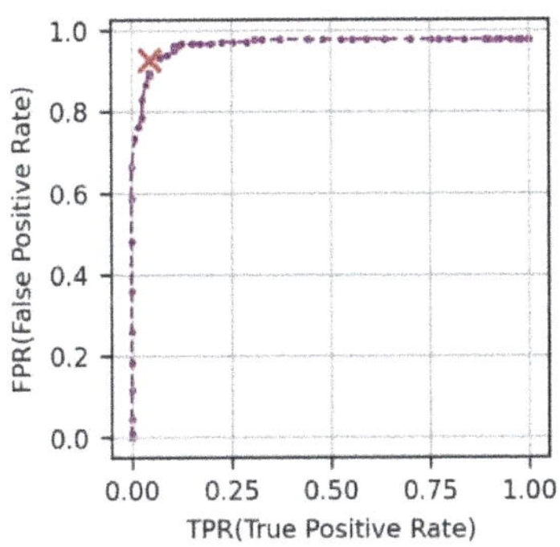

Figure 7. ROC curve of Gyeongbokgung Palace.

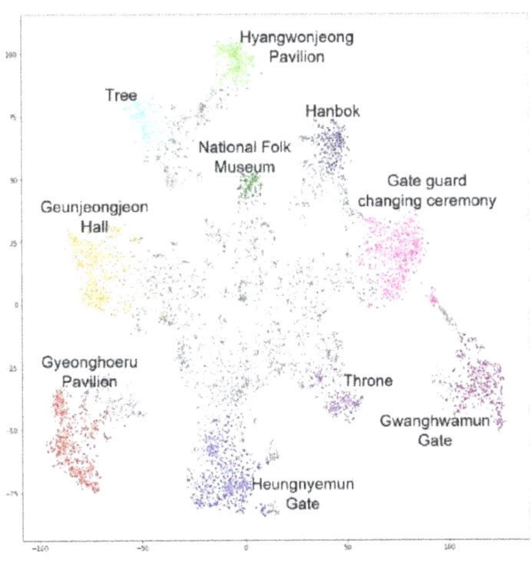

Figure 8. Datapoints classified with Siamese Network of Gyeongbokgung Palace.

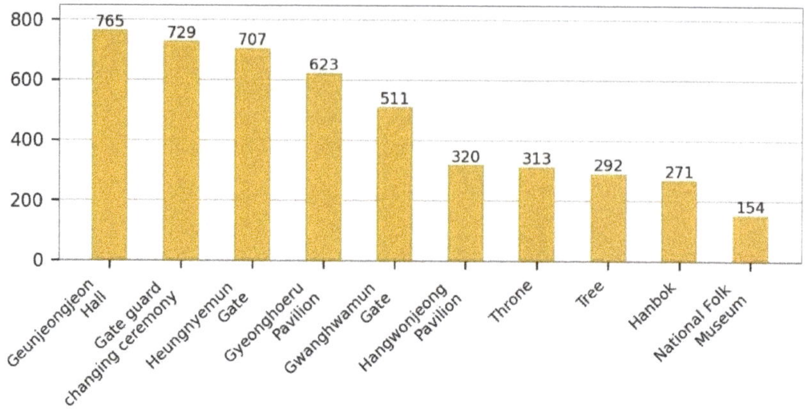

Figure 9. Number of photos by category in Gyeongbokgung Palace.

Figure 10. Example photos by category in Gyeongbokgung Palace.

4.2. Insadong

There are 6410 reviews registered on TripAdvisor's 'Insadong' page. Of these, 3695 photos were collected from 5915 reviews written in a foreign language. Each photo was embedded as a 512-dimensional vector, reduced to two dimensions using t-SNE, and clustered using HDBSCAN. Figure 11 shows the result and Table 3 shows the number of photos for each cluster. Of the total 3659 points, 2568 were classified as noise, and 5 clusters were generated in Figure 11b. Since the 134 photos belonging to cluster 4 represent different visual contents such as signs, souvenirs, murals, portraits, and food, the cluster was not considered in the building category. Four categories were finally created as follows: 'Ssamzigil', 'Insadong street', 'Food and Beverage', and 'Souvenir'.

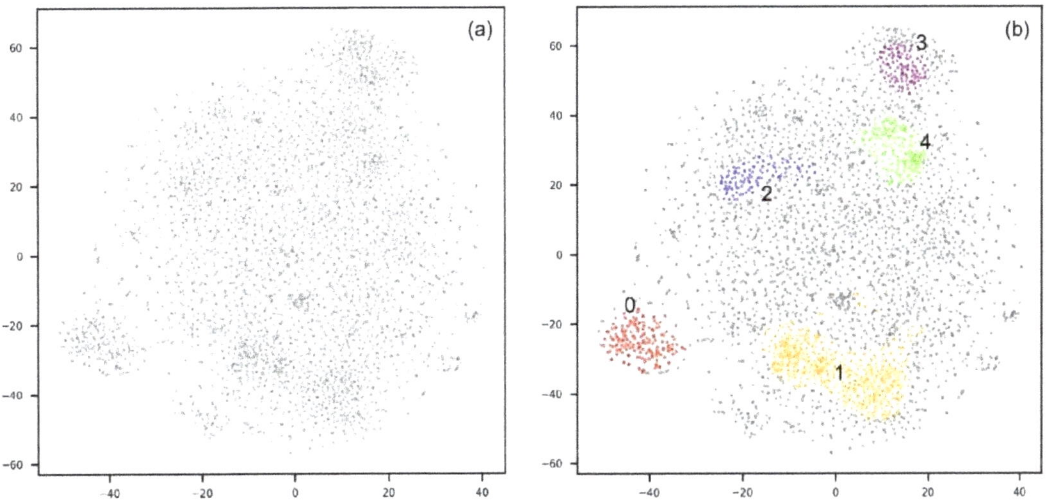

Figure 11. Insadong photos: (**a**) two-dimensional visualization; (**b**) result of HDBSCAN.

Table 3. Number of photos for each cluster generated as a result of HDBSCAN in Insadong.

Cluster	Number of Photos
0	158
1	478
2	133
3	168
4	134

Siamese network was used to classify photos according to the previously generated categories and remove noise photos. For this purpose, we trained the Siamese network on Insadong's photo dataset. At this time, the training dataset was organized using the photos composed of each cluster except for those which were incorrectly included in clusters. Table 4 shows the number of photos included in the training dataset for each category. The model used in 'Insadong' was also based on the VGG16 network. However, unlike the model used in 'Gyeongbokgung Palace', we fine-tuned the top four layers of the model without additional convolutional layers. Since 'Insadong' had a smaller training dataset than 'Gyeongbokgung Palace', overfitting may occur. Figure 12 shows the loss value change during model training, and the model trained up to epoch 10 was used to prevent overfitting.

Table 4. Number of training photos by category in Insadong.

Category	Number of Training Photos
Ssamzigil	150
Insadong Street	126
Food and Beverage	132
Souvenir	137

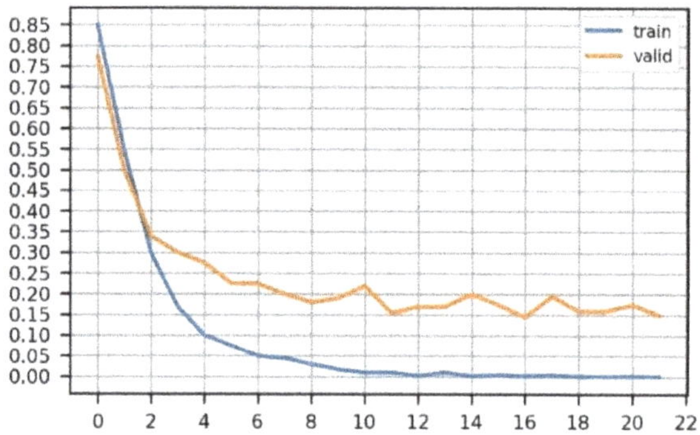

Figure 12. Loss graph of model training of Insadong.

To remove the noise photos included in the category, we examined the optimal threshold value through the ROC curve. Figure 13a shows the minimum distance between the test set and the reference set as a histogram. Figure 13b shows the minimum distance between the noise set and the reference set as a histogram. Figure 14 shows the ROC curve, and 0.32, the threshold value of the point farthest from Y = X, corresponding to the red X on the graph, was selected as the optimal value. Figure 14 also shows that TPR was 0.90 and FPR was 0.057, which related to the accuracy of the model. Figure 15 shows the trained model and data points classified by threshold. Compared with Figure 11b, Figure 15 shows that data points belonging to the same cluster were close to each other, and the distances between different clusters were farther apart. Figures 16 and 17 respectively show the number of photos and example photos finally classified by category.

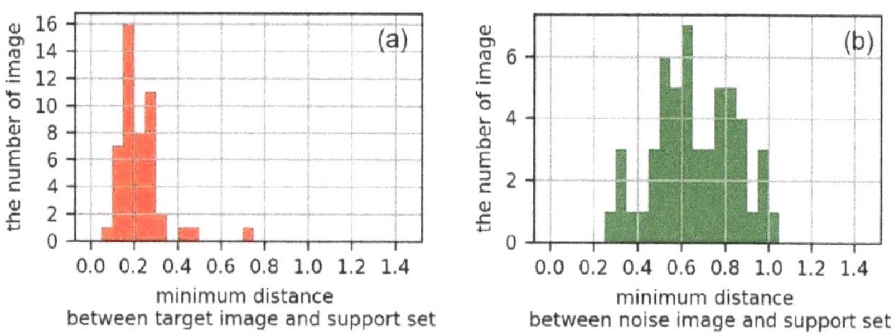

Figure 13. Histogram of Insadong: (**a**) minimum distance between a target photo set and reference set; (**b**) minimum distance between a noise photo and reference set.

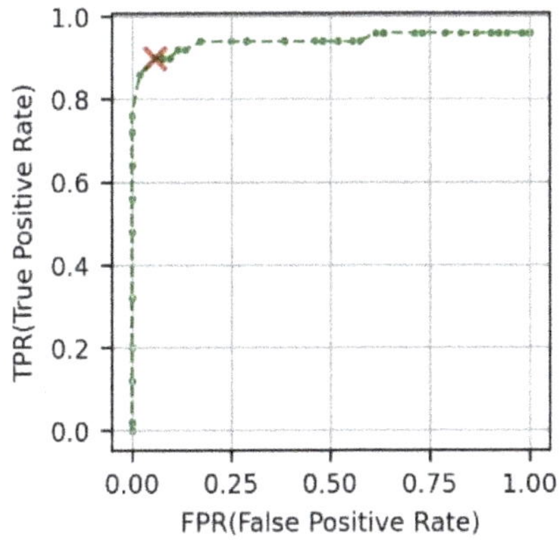

Figure 14. ROC curve of Insadong.

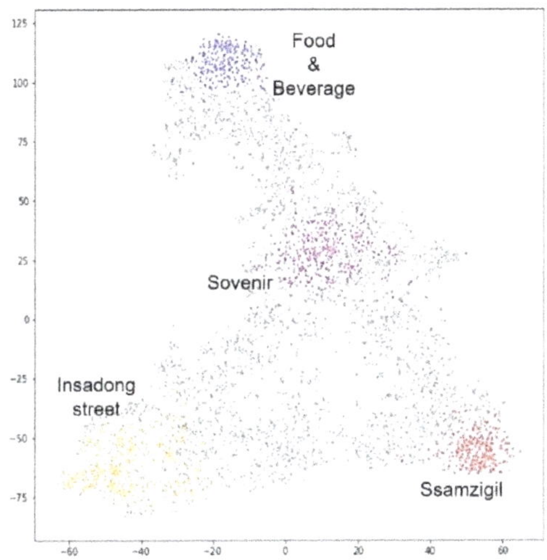

Figure 15. Datapoints classified with Siamese Network of Insadong.

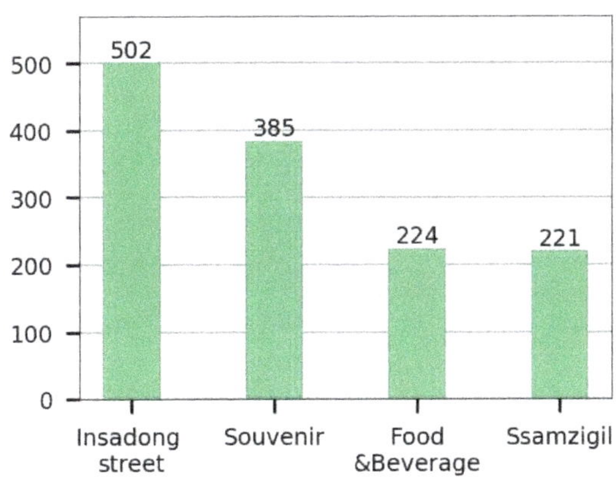

Figure 16. Number of photos by category in Insadong.

Figure 17. Example photos by category in Insadong.

5. Discussion and Conclusions

As the value of photos posted by tourists is recognized as more and more important in the tourism field, new approaches to analyzing tourist photos using deep learning technology are being attempted. The research methods that analyze tourism photos using recent deep learning technology are two-fold. The first method is that tourism images are analyzed after classifying tourist photos by predetermined photos classification categories such as Places365 or ImageNet. The second method is that tourism images are analyzed according to a tourism photo classification category generated on a city or national scale. In the former case, there is a shortcoming in that unique photos appearing in specific tourist attractions cannot be properly classified with a category designed for general purposes. In the latter case, there are limitations in that it puts a lot of time and effort into building a category and dataset and has difficulty detecting the locality of tourism attractions.

The purpose of this study is to propose a method for automatically building a category for each attraction by clustering photos and classifying them with a Siamese network, rather than classifying them into predetermined categories. In addition, this study attempts to confirm the validity of the proposed method by applying it to two representative tourist attractions in Seoul. This study has four steps to clarify the photo classification method for each tourist attraction and to confirm its validity. First, we collected tourist photos attached to reviews posted by foreign tourists on TripAdvisor. Second, we embedded photos as feature vectors in 512 dimensions using the VGG16 network pre-trained with Places365 and reduced them to 2 dimensions using t-SNE. Third, to create a category based on visual contents that frequently appear in photos taken by tourists, clusters were extracted through HDBSCAN analysis and they were set as an image category of an attraction. Fourth, we removed the noises in the cluster through the Siamese network and analyzed the image of tourist attractions by confirming the number of classified photos in each category.

Using the method proposed in this study, the Tripadvisor photos posted by foreign tourists in 'Gyeongbokgung Palace' and 'Insadong' in Seoul, Korea were analyzed. Gyeongbokgung Palace is a palace built during the Joseon Dynasty and is one of the representative tourist attractions located in the downtown area of Seoul. In Gyeongbokgung Palace, 10 categories were created as follows: 'Geunjeongjeon Hall', 'Gyeonghoeru Pavilion', 'Heungnyemun Gate', 'Hyangwonjeong', 'National Folk Museum', 'Throne', 'Hanbok(Korean traditional dress)', 'Gate guard changing ceremony' and 'Tree' 'Gwanghwamun Gate'. Through this, it was possible to check which destination images of 'Gyeongbokgung Palace' are preferred by foreign tourists. 'Insadong' is also one of the representative tourist attractions in the downtown area of Seoul. Insadong is an area known as an exhibition center for Korean traditional arts, antiques, and old ceramics that have been handed down generation after generation. In Insadong, four categories were created: 'Ssamzigil', 'Insadong street', 'Food and Beverage', and 'Souvenir'. Through this, it was possible to identify which images of Insadong are preferred by foreign tourists.

This study is differentiated from the existing studies in the following three aspects. First, since we make categories based on clustering results, features that make tourism destinations attractive can be identified more specifically and flexibly in a data-driven manner. Second, since we set the results of clustering analysis as categories, it is not necessary to manually build the training dataset. Third, to address the scarcity of data, we employ a Siamese network that can improve classification performance with a rather small volume dataset. In the case of the tourism field, if the research area is narrowed down to a specific tourist attraction, there may be a limit to the amount of data that can be used. However, since the data used in this study is the photos posted on TripAdvisor, there is a possibility that various photos may be less mixed because the categories are divided into tourist destinations, tourist attractions, and activities. Therefore, it is necessary to compare the photos posted on TripAdvisor with those from other SNS sites that are likely to post various photos for the same area. In addition, it is necessary to compare a category created by the proposed method with one by an existing method proposed by previous studies, which classifies tourist photos by predefined photo categories using Places365 or ImageNet.

Author Contributions: Conceptualization, Jiyeon Kim and Youngok Kang; Methodology, Software, Validation, Formal Analysis, Jiyeon Kim; Writing-Original Draft Preparation, Jiyeon Kim; Writing-Review & Editing, Youngok Kang; Supervision, Youngok Kang; Project Administration, Youngok Kang; Funding Acquisition, Jiyeon Kim. All authors have read and agreed to the published version of the manuscript.

Funding: This research was supported a grant from geospatial information workforce development program funded by the Ministry of Land, Infrastructure and Transport of Korean Government (1 March 2020).

Institutional Review Board Statement: Not applicable.

Informed Consent Statement: Not applicable.

Data Availability Statement: Not applicable.

Conflicts of Interest: The authors declare no conflict of interest.

References

1. Paül i Agustí, D. Characterizing the location of tourist images in cities. Differences in user-generated images (Instagram), official tourist brochures and travel guides. *Ann. Tour Res.* **2018**, *73*, 103–115. [CrossRef]
2. Arefieva, V.; Egger, R.; Yu, J. A machine learning approach to cluster destination image on Instagram. *Tour. Manag.* **2021**, *85*, 104318. [CrossRef]
3. Chiu, W.; Zeng, S.; Cheng, P.S.T. The influence of destination image and tourist satisfaction on tourist loyalty: A case study of Chinese tourists in Korea. *Int. J. Cult. Tour. Hosp. Res.* **2016**, *10*, 223–234. [CrossRef]
4. Mata, I.L.; Fossgard, K.; Haukeland, J.V. Do visitors gaze and reproduce what destination managers wish to commercialize? Perceived and projected image in the UNESCO World Heritage area "West Norwegian Fjords". *Int. J. Digit. Cult. Electron. Tour.* **2018**, *2*, 294–321. [CrossRef]
5. Song, C.-M.; Jeon, H.Y. A semiotic study of regional branding reflected in the slogans of Korean regions. *Soc. Semiot.* **2018**, *28*, 230–256. [CrossRef]
6. Zhao, Z.; Zhu, M.; Hao, X. Share the Gaze: Representation of destination image on the Chinese social platform WeChat Moments. *J. Travel Tour. Mark.* **2018**, *35*, 726–739. [CrossRef]
7. Sun, W.; Tang, S.; Liu, F. Examining perceived and projected destination image: A social media content analysis. *Sustainability* **2021**, *13*, 3354. [CrossRef]

8. Garrod, B. Understanding the relationship between tourism destination imagery and tourist photography. *J. Travel Res.* **2009**, *47*, 346–358. [CrossRef]
9. Pan, S.; Lee, J.; Tsai, H. Travel photos: Motivations, image dimensions, and affective qualities of places. *Tour Manag.* **2014**, *40*, 59–69. [CrossRef]
10. Han, S.; Ren, F.; Du, Q.; Gui, D. Extracting representative images of tourist attractions from Flickr by combining an improved cluster method and multiple deep learning models. *ISPRS Int. J. Geo-Inf.* **2020**, *9*, 81. [CrossRef]
11. Jenkins, O. Photography and travel brochures: The circle of representation. *Tour. Geogr.* **2003**, *5*, 305–328. [CrossRef]
12. Dinh, M.H. Photography in Tourism: Vietnamese Travelers' Photographs and Narrative-sharing on the Destination Image of Northwest Vietnam. Ph.D. Thesis, Auckland University of Technology, Auckland, New Zeeland, 2021.
13. Lim, Y.; Chung, Y.; Weaver, P.A. The impact of social media on destination branding: Consumer-generated videos versus destination marketer-generated videos. *J. Vacat. Mark.* **2012**, *18*, 197–206. [CrossRef]
14. Stepchenkova, S.; Zhan, F. Visual destination images of Peru: Comparative content analysis of DMO and user-generated photography. *Tour. Manag.* **2013**, *36*, 590–601. [CrossRef]
15. Zhang, K.; Chen, Y.; Li, C. Discovering the tourists' behaviors and perceptions in a tourism destination by analyzing photos' visual content with a computer deep learning model: The case of Beijing. *Tour. Manag.* **2019**, *75*, 595–608. [CrossRef]
16. Zhang, K.; Chen, D.; Li, C. How are tourists different? Reading geo-tagged photos through a deep learning model. *J. Qual. Assur. Hosp. Tour.* **2020**, *21*, 234–243. [CrossRef]
17. Kim, D.; Kang, Y.; Park, Y.; Kim, N.; Lee, J. Understanding tourists' urban images with geotagged photos using convolutional neural networks. *Spat. Inf. Res.* **2020**, *28*, 241–255. [CrossRef]
18. Payntar, N.D.; Hsiao, W.-L.; Covey, R.A.; Grauman, K. Learning patterns of tourist movement and photography from geotagged photos at archaeological heritage sites in Cuzco, Peru. *Tour. Manag.* **2021**, *82*, 104165. [CrossRef]
19. Law, S.; Shen, Y.; Seresinhe, C. An application of convolutional neural network in street image classification: The case study of London. In Proceedings of the 1st Workshop on Artificial Intelligence and Deep Learning for Geographic Knowledge Discovery, Redondo Beach, CA, USA, 7–10 November 2017; pp. 5–9.
20. Kang, Y.; Cho, N.; Yoon, J.; Park, S.; Kim, J. Transfer learning of a deep learning model for exploring tourists' urban image using geotagged photos. *ISPRS Int. J. Geo-Inf.* **2021**, *10*, 137. [CrossRef]
21. Kang, Y.; Yoon, J. Tourism scene analysis through CNN-based multi-label transfer learning. *KSIS* **2021**, *29*, 15–26.
22. Chen, M.; Arribas-Bel, D.; Singleton, A. Quantifying the characteristics of the local urban environment through geotagged Flickr photographs and image recognition. *ISPRS Int. J. Geo-Inf.* **2020**, *9*, 264. [CrossRef]
23. O'Mahony, N.; Campbell, S.; Carvalho, A.; Harapanahalli, S.; Hernandez, G.V.; Krpalkova, L.; Riordan, D.; Walsh, J. Deep learning vs. traditional computer vision. In Proceedings of the Science and Information Conference, Tokyo, Japan, 16–19 March 2019; pp. 128–144.
24. Nash, W.; Drummond, T.; Birbilis, N. A review of deep learning in the study of materials degradation. *npj Mater. Degrad.* **2018**, *2*, 1–12. [CrossRef]
25. Tapaswi, M.; Law, M.T.; Fidler, S. Video face clustering with unknown number of clusters. In Proceedings of the IEEE/CVF International Conference on Computer Vision, Seoul, Korea, 27–28 October 2019; pp. 5027–5036.
26. Gu, X.; Wong, Y.; Peng, P.; Shou, L.; Chen, G.; Kankanhalli, M.S. Understanding fashion trends from street photos via neighbor-constrained embedding learning. In Proceedings of the 25th ACM international conference on Multimedia, Mountain View, CA, USA, 23–27 October 2017; pp. 190–198.
27. Castellano, G.; Vessio, G. A deep learning approach to clustering visual arts. *arXiv* **2021**, arXiv:2106.06234.
28. Schroff, F.; Kalenichenko, D.; Philbin, J. Facenet: A unified embedding for face recognition and clustering. In Proceedings of the IEEE conference on computer vision and pattern recognition, Boston, MA, USA, 7–12 June 2015; pp. 815–823.
29. Zhong, D.; Yang, Y.; Du, X. Palmprint recognition using Siamese network. In Proceedings of the Chinese conference on biometric recognition, Urumqi, China, 11–12 August 2018; pp. 48–55.
30. Mehmood, A.; Maqsood, M.; Bashir, M.; Shuyuan, Y. A deep Siamese convolution neural network for multi-class classification of Alzheimer disease. *Brain Sci.* **2020**, *10*, 84. [CrossRef]
31. Bertinetto, L.; Valmadre, J.; Henriques, J.F.; Vedaldi, A.; Torr, P.H. Fully-convolutional Siamese networks for object tracking. In Proceedings of the European conference on computer vision, Amsterdam, The Netherlands, 8–16 October 2016; pp. 850–865.
32. Yoo, K.-H.; Sigala, M.; Gretzel, U. Exploring TripAdvisor. In *Open Tourism*; Springer: Berlin/Heidelberg, Germany, 2016; pp. 239–255.
33. Zhou, B.; Khosla, A.; Lapedriza, A.; Torralba, A.; Oliva, A. Places: An image database for deep scene understanding. *arXiv* **2016**, arXiv:1610.02055. [CrossRef]
34. Van der Maaten, L.; Hinton, G. Visualizing data using t-SNE. *J. Mach. Learn. Res.* **2008**, *9*, 2579–2605.
35. Hinton, G.; Roweis, S.T. Stochastic neighbor embedding. In Proceedings of the Neural Information Processing Systems, Vancouver, Canada, 9–14 December 2002; pp. 833–840.
36. Kullback, S.; Leibler, R.A. On information and sufficiency. *Ann. Math. Stat.* **1951**, *22*, 79–86. [CrossRef]
37. Campello, R.J.; Moulavi, D.; Sander, J. Density-based clustering based on hierarchical density estimates. In Proceedings of the Pacific-Asia conference on knowledge discovery and data mining, Gold Coast, QLD, Australia, 14–17 April 2013; pp. 160–172.

38. Ester, M.; Kriegel, H.-P.; Sander, J.; Xu, X. A density-based algorithm for discovering clusters in large spatial databases with noise. In Proceedings of the kdd, Portland, OR, USA, 2–4 August 1996; pp. 226–231.
39. Koch, G.; Zemel, R.; Salakhutdinov, R. Siamese neural networks for one-shot image recognition. In Proceedings of the ICML Deep Learning Workshop, Lille, France, 6–11 July 2015; pp. 160–172.

International Journal of *Geo-Information*

MDPI

Article

Detecting People on the Street and the Streetscape Physical Environment from Baidu Street View Images and Their Effects on Community-Level Street Crime in a Chinese City

Han Yue [1], Huafang Xie [1,*], Lin Liu [1,2] and Jianguo Chen [1]

[1] Center of GeoInformatics for Public Security, School of Geography and Remote Sensing, Guangzhou University, Guangzhou 510006, China; yuehan@gzhu.edu.cn (H.Y.); lin.liu@uc.edu (L.L.); chenjg@gzhu.edu.cn (J.C.)

[2] Department of Geography, University of Cincinnati, Cincinnati, OH 45221, USA

* Correspondence: 2112101040@e.gzhu.edu.cn

Abstract: The occurrence of street crime is affected by socioeconomic and demographic characteristics and is also influenced by streetscape conditions. Understanding how the spatial distribution of street crime is associated with different streetscape features is significant for establishing crime prevention and city management strategies. Conventional data sources that quantify people on the street and streetscape characteristics, such as questionnaires, field surveys, or manual audits, are labor-intensive, time-consuming, and unable to cover a large area with a sufficient spatial resolution. Emerging cell phone and social media data have been used to measure ambient population, but they cannot distinguish between the street and indoor populations. This study addresses these limitations by combining Baidu Street View (BSV) images, deep learning algorithms, and spatial statistical regression models to examine the influences of people on the street and in the streetscape physical environment on street crime in a large Chinese city. First, we collected fine-grained street view images from the Baidu Map website. Then, we constructed a Faster R-CNN network to detect discrete elements with distinct outlines (such as persons) in each image. From this, we counted the number of people on the street in every BSV image and finally obtained the community-level total amounts. Additionally, the PSPNet network was developed for pixel-wise semantic segmentation to determine the proportions of other streetscape features such as buildings in each BSV image, based on which we obtained their community-level averages. The quantitative measurement of people on the street and a set of streetscape features that had potential influences on crime were finally derived by combining the outputs of two deep learning networks. To account for the spatial autocorrelation effect and distributional characteristics of crime data, we constructed a set of spatial lag negative binomial regression models to investigate how three types of street crime (i.e., total crime, property crime, and violent crime) were affected by the number of people on the street and the streetscape-built conditions. The models also controlled the effect of socioeconomic and demographic factors, land use features, the formal surveillance level, and transportation facilities. The models with people on the street and streetscape environment features had noticeable performance improvements, demonstrating the necessity for accounting for the effect of these factors when understanding street crime. Specifically, the number of people on the street had significantly positive impacts on the total street crime and street property crime. However, no statistically significant impact was found on street violent crime. The average proportions of the paths, buildings, and trees were associated with significantly lower street crime among physical streetscape features. Additionally, the statistical significances of most control variables conformed to previous research findings. This study is the first to combine Street View images and deep learning algorithms to retrieve the number of people on the street and the features of the visual streetscape environment to understand street crime.

Keywords: street crime; people on the street; streetscape; Baidu Street View image; spatial lag negative binomial regression

Citation: Yue, H.; Xie, H.; Liu, L.; Chen, J. Detecting People on the Street and the Streetscape Physical Environment from Baidu Street View Images and Their Effects on Community-Level Street Crime in a Chinese City. *ISPRS Int. J. Geo-Inf.* **2022**, *11*, 151. https://doi.org/10.3390/ijgi11030151

Academic Editors: Gloria Bordogna, Cristiano Fugazza and Wolfgang Kainz

Received: 11 January 2022
Accepted: 21 February 2022
Published: 22 February 2022

1. Introduction

According to environmental criminology, the physical context creates necessary conditions for the confluence of motivated offenders, suitable targets, and the absence of qualified guardians, which leads to crime occurrence [1,2]. Environmental characteristics are of great significance for understanding the spatial aggregation of crime. Therefore, researchers emphasize the understanding of crime formation mechanisms from geography. They claim it is valuable to reveal crime patterns and provide references for constructing crime prevention and control strategies [3].

While the significant role of the urban environment in crime has been widely accepted, data sources applied by previous studies to quantify streetscape characteristics are defective in some respects. Traditional data-gathering methods including questionnaire surveys [4], field surveys, and human auditing [5,6] are time-consuming and labor-intensive. These limitations make them only suitable for conducting studies at several scattered places and not applicable to large-scale research. Satellite remote sensing images are popular data used to extract built environment characteristics [7–9]. This kind of data could be applied to studying a large geographical area. However, these images capture information from a bird's eye view and cannot obtain street-level information from the perspective of human eyes. The low accessibility of large-scale detailed data limits our ability to systematically measure the urban environment in a quantitative way, finally leaving the influence mechanism of the visual streetscape context on crime not understood so well.

As a kind of geo-referenced big data, the emerging street view images (SVIs) offer an excellent chance for diving into a more in-depth look at the associations between the urban street context and crime. The most significant advantage of SVIs over other data is that they are captured by cameras set on top of cars driving along streets. Therefore, SVIs can be adopted to extract street environment features from pedestrians' views, and they have the potential to help reveal the most direct connection between streetscape conditions and crime. In addition, this type of data covers most major cities and is usually open accessed. SVIs are increasingly mentioned and used by many authors [5,10–14]. However, most existing research just used SVIs to detect basic physical elements such as roads, buildings, and vegetation. Based on the extracted information, researchers investigated how the built environment can help explain crime aggregations [10], whether the street-level visual environment can be used to classify locations with high-crime and lower-crime activities [11], and the environmental mechanisms behind crime diversity [12].

Combining SVIs and deep learning algorithms, this study investigates the effect of people on the street and streetscape features on street crime in a large Chinese city. SVIs are utilized to extract both physical elements (through a semantic segmentation network) and the number of people on the street (through an object detection network). The primary purposes of this study are then (1) how to extract and measure the number of people on the street, which is an important variable affecting the occurrence of street crime, (2) how to extract other streetscape environment elements using SVIs, and (3) what the associations between street crime and people on the street and streetscape conditions is.

We selected street crimes such as snatching and robbery as our crimes of interest because they are significant threats to people's property and personal safety. Additionally, most of the time, they occur in public spaces. They are more likely to be affected by human activities and environmental features in immediate regions.

2. Literature Review

2.1. Street Crime and People on the Street

The spatial aggregation of crime is a common phenomenon, and it has a sufficient theoretical and empirical basis. The routine activity theory suggests that the confluence of motivated offenders, appropriate targets, and lack of competent guardians results in crimes [1]. Additionally, the convergence of these three elements is significantly influenced by the spatiotemporal pattern of people's routine activities, such as traveling for work, school, and leisure [1].

People's daily activities, such as when, where, and what to do in a day, usually have regular rhythms. An individual stays more often in some areas, such as residences, workplaces, and favorite shops, while he or she has less chance of staying in other places. The regularity of people's behaviors results in various crowd gathering levels in space and time [15]. Business districts, for example, are densely populated during the day because people work there. However, these places are less crowded in the evening as people return home for sleep. Typical residential areas, however, usually have opposite patterns. They are less crowded during the day but highly crowded in the evening. The routine activity theory acknowledges that such different human activities result in different crime opportunities in different places and at different times. This phenomenon could be explained by the core insight of routine activity theory; that is, there is more significant potential for people to be victimized or to victimize others when they spend more time away from the protective environment of their households and families, whether for work, leisure, or shopping [16]. Researchers have been particularly concerned about the facilities that attract people in regard to interpersonal crimes. The ambient population attracted by such facilities enhances the likelihood of the encounter of offenders and victims. Previous crime research has investigated facilities like bars, subway stations, and parks. For example, Roncek and Bell analyzed the relationship between bars and block-level crimes. Controlling the influence of other factors, they found that blocks with bars experienced more crimes than those without bars [17]. One piece of research by McCord et al. showed that street robberies tended to occur around subway stations [18]. Groff and McCord analyzed the spatial correlations between parks and crimes and found that parks attracted crimes [19]. Kubrin et al. pointed out a tight association between lending agencies and property and violent crimes [20]. These facilities are not necessarily criminogenic by nature; the cluster of people in these areas leads to high crime rates [21]. Thus, the disparity of crime patterns in space and time is due to human activity differences [22].

A series of research has proved the significant association between the presence of people and crime, but the effect is inconclusive. For example, Boivin adopted a transportation telephone survey to examine the influence of the ambient population on crime in the Toronto region [23]. Respondents were asked about their visited locations on a typical weekday. Based on this, researchers inferred respondents' trip purposes (such as home, school, shop, work, and others). They then estimated daily population flows between different purposes. Their results demonstrated that the population size was positively associated with crime in some areas; however, the opposite effects were found in other regions which received visits mainly for shopping, school, and work. Vomfell et al. combined different sources of population activity (such as social media and taxi flow data) to predict crime at the census tract level [24]. After accounting for demographic factors, they found that dynamic population variables had stronger influences on the prediction of property crime than violent crime.

As noted above, studies have not yet concluded whether an increased human presence in a given area is associated with an increase or decrease in crime. Boivin explained that the effect of human presence on crime is greatly determined by the nature of the crime [23]. The simple presence of people is just enough to restrain some types of crime by their guardianship effect [25]. However, other research demonstrated that the ambient population provides targets for offenders; thus, people's presence will increase criminal chances [26].

2.2. Street Crime and Streetscape Physical Environment

According to environmental criminology, human activities (including criminal activities) are affected by the physical environment. Environmental characteristics are of great significance for understanding the spatial aggregation of crime [1]. Crime is caused by characteristics in the location and surrounding areas [27]. These characteristics create opportunities for potential offenders. When an offender finds an opportunity, and adequate monitoring is absent, he or she will commit a crime. Crime pattern theory also explains the spatial aggregation of crime. Both theories emphasize the impact of crime opportunities in places [28]. From this point of view, different places have different crime opportunities in the city. Some places can provide affluent crime opportunities. In comparison, some places have few crime opportunities, leading to the spatial heterogeneity of crime.

A series of empirical studies has proven the significant role of built environments having on crime. For example, regions with detached houses were attractive to burglars because the offenders could easily invade and escape from these houses through doors and windows [29]. In addition, high-rise residential buildings are also prone to burglary because these buildings are usually equipped with convenient access channels such as elevators and corridors. At the same time, their architectural structures are often complex, providing hiding conditions for perpetrators. Moreover, many residents living here create rich crime opportunities [30]. Yue et al. analyzed the spatial colocations between different POI types and burglary, electric bicycle theft, and robbery in Wuhan, China [31]. Their results demonstrated that e-bike thefts were most likely to occur around stores. Hotels and primary and secondary schools were less attractive for e-bike thefts. There were many robberies near banks and stores. In addition, bus stops were also attractive for robberies.

Many studies examined the association between street configurations and crime risks based on space syntax theory. For example, a study conducted by Jones demonstrated that when controlling for the effect of demographic factors, isolated and less accessible streets were more likely to suffer from crimes. At the same time, regions with high permeability were safer [32]. Other researchers confirmed these results, such as Shu [33] and Yue [34]. Hillier claimed that permeable urban design elements of regular road network structures (such as liner and well-integrated streets) were safer than closed and impermeable street layouts (such as cul-de-sacs) [35].

Easy accessibility, inadequate place management, and the presence of people could create opportunities for crime in a place. Additionally, the presence of physical disorder elements such as abandoned cars, vacant or dilapidated buildings, litter, and graffiti can also boost offenders' motivation to commit crimes [36–38]. Similarly, gangs, begging, loiterers, prostitution, unruly and rowdy teenagers, public drunkenness, and public drug use or dealing are disruptive behaviors that indicate social disorder. Signs of social disorder in a location could also raise crime levels in the immediate areas [39].

2.3. Data Sources and Methods Used in Related Research

Various types of data have been applied in previous research to analyze how the distribution of street crime varies with the volume of people on the street and the streetscape's physical conditions. Basic demographic information extracted from census data is a typical measurement of potential population exposure to crime in a region. Other similar data sources include daily travel surveys, activity surveys, and workday census surveys. These data sources have an apparent drawback: they are time-consuming and labor-intensive to collect. Survey data usually covers a small region, so it is not applicable for large-area studies. Additionally, the quality of the survey data also suffers from sample bias. Some studies also adopted human auditing to collect data for street scenes. Specifically, some researchers gathered information by on-the-spot investigation, while some researchers conducted online audits with the help of electronic maps. Researchers can collect information in as much detail as possible. However, this has low efficiency, limiting its use. Additionally, human auditing has an unavoidable subjectivity issue.

The emergence of various big data compensates for the defect of traditional data sources. In recent years, mobile phone data have been a typical measurement of ambient population, which is a proxy of the baseline population or population at risk of crime [40]. Mobile phone data usually cover a large area such as an entire city and have high time resolution. Therefore, they have been adopted by many researchers to explore the spatiotemporal patterns of human activity and social behaviors such as crime [41]. However, mobile phone data usually have a low spatial resolution, making them unable to differentiate the local population diversity. They cannot measure the actual baseline population, leading to unreliable research findings [42]. Additionally, mobile phone data are of low availability because telecommunication operators usually own them. Geotagged social media data have also been utilized to evaluate the relationship between human activity and crime. For example, Hipp et al. used Twitter posts (one tweet per Twitter user per spatiotemporal unit) to determine the ambient population. They found that the number of Twitter users was associated with crime, controlling for the guardianship level [43]. Routine activities of social media users allow researchers to capture population movement directly. However, social media has drawbacks, as only a tiny proportion of the population is on Twitter [44]. Metro smart card data, taxi trajectory data, and bicycle trajectory data also provide excellent opportunities for measuring the mobility of people [45–47]. However, they are incapable of capturing the movement of pedestrians, which is the main component of the ambient population.

Previous research has used data such as satellite remote sensing images to measure the characteristics of the built environment. For example, Patino et al. used remote sensing images to examine whether a neighborhood's design elements (such as land cover, structure, and texture descriptors) were associated with the homicide rate in Medellin, Colombia. Their results revealed that urban layouts in areas with higher homicide rates tended to be more crowded and cluttered [9]. Algahtany and Kumar utilized satellite images to evaluate urban expansion over a decade in Saudi Arabia, based on which they explored the associations between such expansion and crime. The results demonstrated a significant relationship between urban expansion and crime. Additionally, the associations were more remarkable in places with more significant urban growth [48]. Although remote sensing images usually cover large regions, their most significant limitation is that they are captured by satellites observing cities from the top view. Therefore, they cannot quantify the vertical dimensions of the street environment (such as the vertical surface of high buildings and street canyons). People perceive their surroundings from a horizontal view, while remote sensing images cannot accurately and comprehensively measure the streetscape composition complying with people's real scene perception. Therefore, remote sensing images are insufficient for digging for the profound influence of streetscape elements on criminal behaviors.

SVIs have a unique advantage in that they are taken by cameras set upon cars driving along streets. Therefore, they have the potential to capture systematic and fine-grained urban landscapes from pedestrians' points of view. Compared with traditional data sources, SVIs contain more information, including artificial elements like buildings and roads and natural elements like trees and the sky [49]. In recent years, SVIs have been used to evaluate the streetscape environments' effect on offenders' decisions about whether, where, and when to commit crimes. For instance, He et al. measured the associations between the physical features of the urban residential environment and violent crimes based on Google Street View (GSV) images [5]. Using an environmental audit tool developed based on GSV images, they collected environmental factors like physical incivility (e.g., property damage and abandoned buildings), territorial functioning features (e.g., yard decorations), and defensible space features. The results demonstrated that the relationship between the residential built environment and violent crime was significant and GSV images were reliable for capturing many aspects of the built environment. Hipp et al. used machine learning methods to extract environment features from GSV images [10]. The results demonstrated that measuring the built environment through GSV images was effective. Specifically, auto-oriented elements like vehicles and pavements were positively related to crime, defensible space elements like the presence of walls had negative associations with crime, and green space elements like vegetation had positive effects on crime. Khorshidi used a deep learning service to extract objects from GSV images. Based on this, they computed census block-level object diversities and modeled crime diversity as a function of environmental diversity, population diversity, and population size [12]. The results revealed that environmental diversity extracted from GSV images was more predictive of crime diversity than commonly used census measures.

The applicability of SVIs to crime research owes a great deal to the development of artificial intelligence technology. Modern image processing techniques such as deep learning networks can fetch precise and detailed elements from the urban space [50]. Fieldwork cannot obtain many elements. For example, it is hard for people to calculate the proportion of roads in a place [13]. SVIs are not only usable for extracting physical elements but are also applicable for measuring collective pedestrian volumes [51]. For example, Chen et al. conducted a large-scale empirical validation study. They found that pedestrian volumes estimated using SVIs can provide acceptable (Cronbach's alpha $\geq$ 0.70) or good (Cronbach's alpha $\geq$ 0.80) levels of accuracy compared with field observation [52]. Other studies also validated SVIs as an efficient and reliable data source for estimating street-level pedestrian volumes [52,53].

3. Study Area, Data, and Method

3.1. Study Area

This study took place in ZG city. (Under the terms of the confidentiality agreement, we cannot reveal the city's true name.) ZG city is located on the southern coast of China, and it is one of the most developed cities in China. There are 2643 communities in this city, and 737 of them are within the Outer Ring Expressway Area. These were selected as the research communities in this study.

3.2. Data

3.2.1. Crime Data

Three years (2017–2019) of official crime data were sourced from the public security bureau of ZG. We aggregated three crime types, including snatching, pickpocketing, and theft from the person, to form a general street property crime type. We aggregated robbery, intentional injury, and assault to form a general street violent crime type. Additionally, we aggregated street property crime and street violent crime to form a total street crime type. Figure 1 presents the spatial distributions of the number of total street crimes (Figure 1a), street property crimes (Figure 1b), and street violent crimes (Figure 1c).

3.2.2. Collect BSV Images and Extract Streetscape Features

Compared with the human audit approach of obtaining streetscape characteristics from SVIs [5,54], emerging computer vision technologies are time-efficient and objective. We first collected fine-grained BSV images from the Baidu Map website. Then, we combined two deep learning networks to extract both people on the street and other built environment elements from BSVs and included these measures into statistical models.

- Fetch BSVs from the Baidu Map Website

Baidu Street View (BSV) is a map service website providing visual information on streets in more than 600 cities in China. BSV images were captured by street view cars. The key components of a street view car are a GPS and fisheye lens. The GPS is used to record geographic locations when the car is driving on the street, and the fisheye lenses are used to collect 360° street view images. The most significant advantage of street view images over other data is that they are captured by cameras set on top of cars driving along streets. Therefore, street view images can be adopted to extract street environment features from pedestrians' views, and they have the potential to help reveal the most direct connection between streetscape conditions and crime.

We took BSV images as a proxy of the streetscape environment. Some basic information is required to collect the BSVs at a position, including the coordinates (longitude and latitude), azimuth angle (commonly called the heading angle), and pitch angle. We first generated sampling points along the street at a uniform interval of 20 m. Based on their coordinates, we collected fine-grained BSV images. There were 215,760 sample sites in the study region.

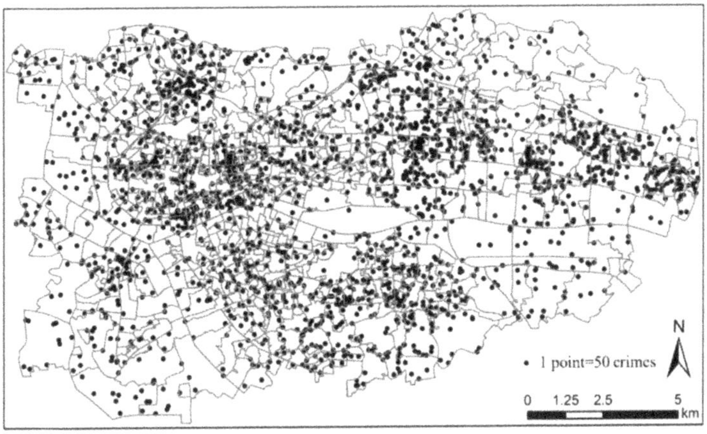

(a) Total street crime

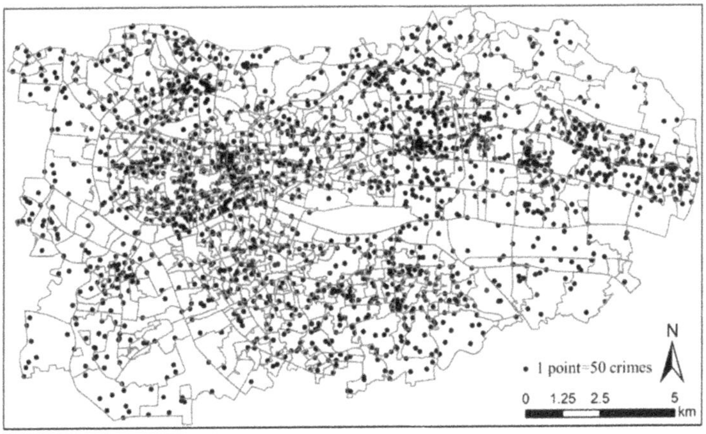

(b) Street property crime

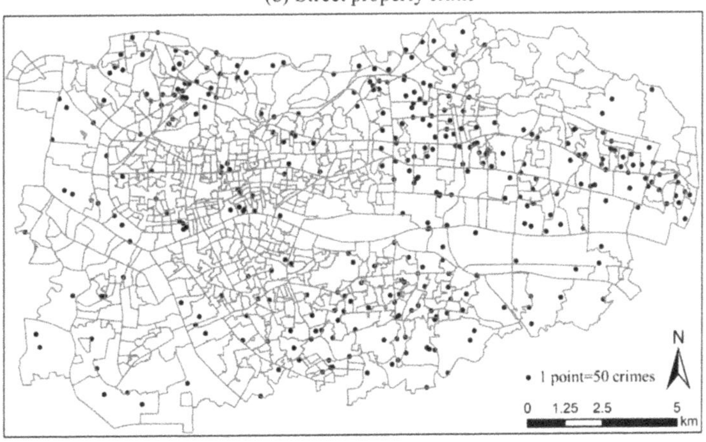

(c) Street violent crime

Figure 1. Dot density maps showing spatial distribution of the number of (**a**–**c**) in the study area as of 2017–2019. Dots were randomly placed in a polygon.

To be consistent with pedestrians' directions of eyesight, we collected BSV images in four horizontal directions at each sample site. Two directions were parallel to the street, and two directions were vertical to the street, as demonstrated in Figure 2. The pitch angle of each image was set to 0° to meet the way people experience the street environment. We then downloaded the BSV images through the Baidu Street View API (see Baidu Developer Platform). Finally, we collected a total of 863,040 images in the study region. The metadata show that they were all captured between 2017 and 2019, consistent with the crime time. As an urbanized region, the built environment in the study area did not changed dramatically during this short period, so the time differences of the BSV images were negligible. Each BSV image had a field of view of 90°, so four images together could capture the panorama of a site.

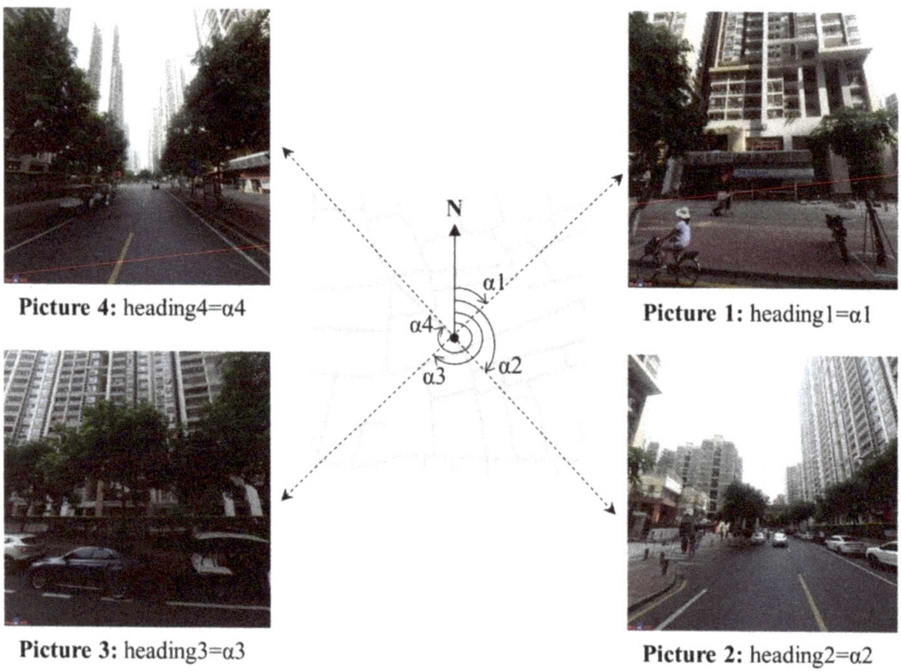

Figure 2. An example of calculating heading angles of four BSV images at a sample site. Heading angles of Pictures 2 and 4 are parallel to the street, capturing the front and rear views, while heading angles of Pictures 1 and 3 are vertical to the street, capturing the left-hand and right-hand views.

This study translated the BSV images into meaningful factors and then incorporated them into regression models. The urban streetscape is a complex system containing components of diverse shapes and sizes. Therefore, we combined two deep learning networks to extract different and complementary information from each BSV image.

- Object Detection Using the Faster R-CNN Network

Some objects like persons and cars are discrete elements with relatively fixed shapes and distinct outlines in an image. Therefore, it is practical to measure the count of identifiable objects. Faces and license plates are blurred in the Baidu Street View images; therefore, this study had no privacy or ethical issues. This study applied a pretrained Faster R-CNN network [55] to perform object detection for BSV images. This network was chosen because it reached a good balance between prediction accuracy and operational efficiency as a state-of-the-art deep learning network. Additionally, it was perfectly compatible with the high resolution of BSV images collected in this study (1024 × 1024 pixels).

The outputs of a Faster R-CNN network were a set of predicted bounding boxes. Each box had an associated score indicating the credibility of whether the box contained an object or not inside and a label determining which category the object belonged to. Based on the outputs of the Faster R-CNN network, we could count the number of objects in each category in an image and finally calculate their total amounts in each community.

This research retrieved the number of people on the street in a community according to the following formula:

$$Number\ of\ people\ on\ the\ street = \sum_{i=1}^{n}\sum_{j=1}^{4} Image_{p\_j} \tag{1}$$

where $Image_{p\_j}$ is the number of people in the image taken in the jth direction among the four directions at a position and n represents the total number of collecting points within a community.

The on-street population must be considered when studying street crime. However, on-street population sizes in different places are difficult to obtain. Pedestrian volume data have traditionally been collected through field observations, which has many methodological limitations, such as being time-consuming, labor-intensive, and inefficient. Various big data, such as mobile phone data, geotagged social media data, metro smart card data, and taxi and bicycle trajectory data, are incapable of capturing the movement of pedestrians. Assessing pedestrian volumes automatically from street view images with machine learning techniques can overcome such limitations, because this approach offers a broad geographic reach and consistent image acquisition. While SVIs have been recently used to estimate street-level pedestrian volumes [52,53], this approach has not been applied to crime research.

- Semantic Segmentation Using the PSPNet Network

Unlike objects with fixed shapes and distinct outlines, sky, grass, and roads may not have a definitive shape in an image. Therefore, object detection networks are not applied to these features. This study utilized a semantic segmentation network instead. After comparing several deep learning models, we chose the widely applied Pyramid Scene Parsing Network (PSPNet) [56]. Semantic segmentation models generate pixel-wise predictions and assign each pixel a category label. We measured the proportions of these features in the image.

By borrowing ideas from the green view index calculation formula developed by Li et al. [57], which measured the proportion of vegetarians in a location, we calculated the proportion of a class of objects in a community as follows:

$$Proportion\ of\ object = \frac{\sum_{i=1}^{n}\sum_{j=1}^{4} Image_{o\_j}}{\sum_{i=1}^{n}\sum_{j=1}^{4} Image_{t\_j}} * 100\% \tag{2}$$

where $Image_{o\_j}$ is the number of pixels belonging to one type of object in the image taken in the jth direction and $Image_{t\_j}$ is the total number of pixels in that image, while n represents the total number of collecting points within a community.

By combining the results of object detection and the semantic segmentation networks (see Figure 3), we finally derived eight quantitative measurements of streetscape features. They were the number of people on the street (per 1000), the average proportion of paths (%), the average proportion of roads (%), the average proportion of walls (%), the average proportion of buildings (%), the number of streetlamps (per 1000), the number of traffic lights (per 1000), and the average proportion of trees (%). The rest of the object categories were not included in the analysis as they were considered irrelevant to crime in an urban context. Figure 4 presents the spatial distributions of people on the street and the streetscape physical features retrieved by BSV images and deep learning methods.

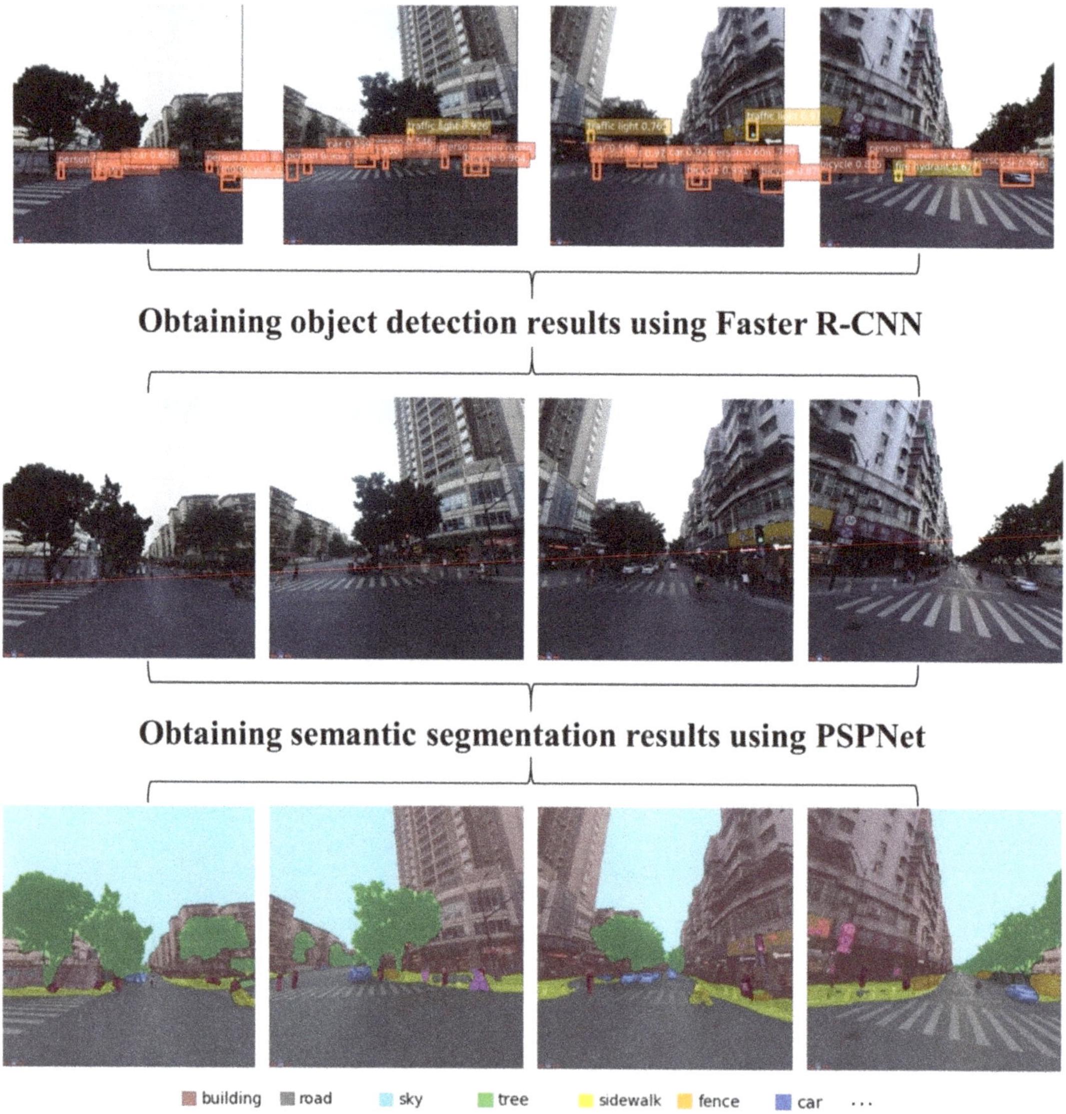

Figure 3. Obtaining object detection and semantic segmentation results of BSV images via Faster RCNN and PSPNet networks.

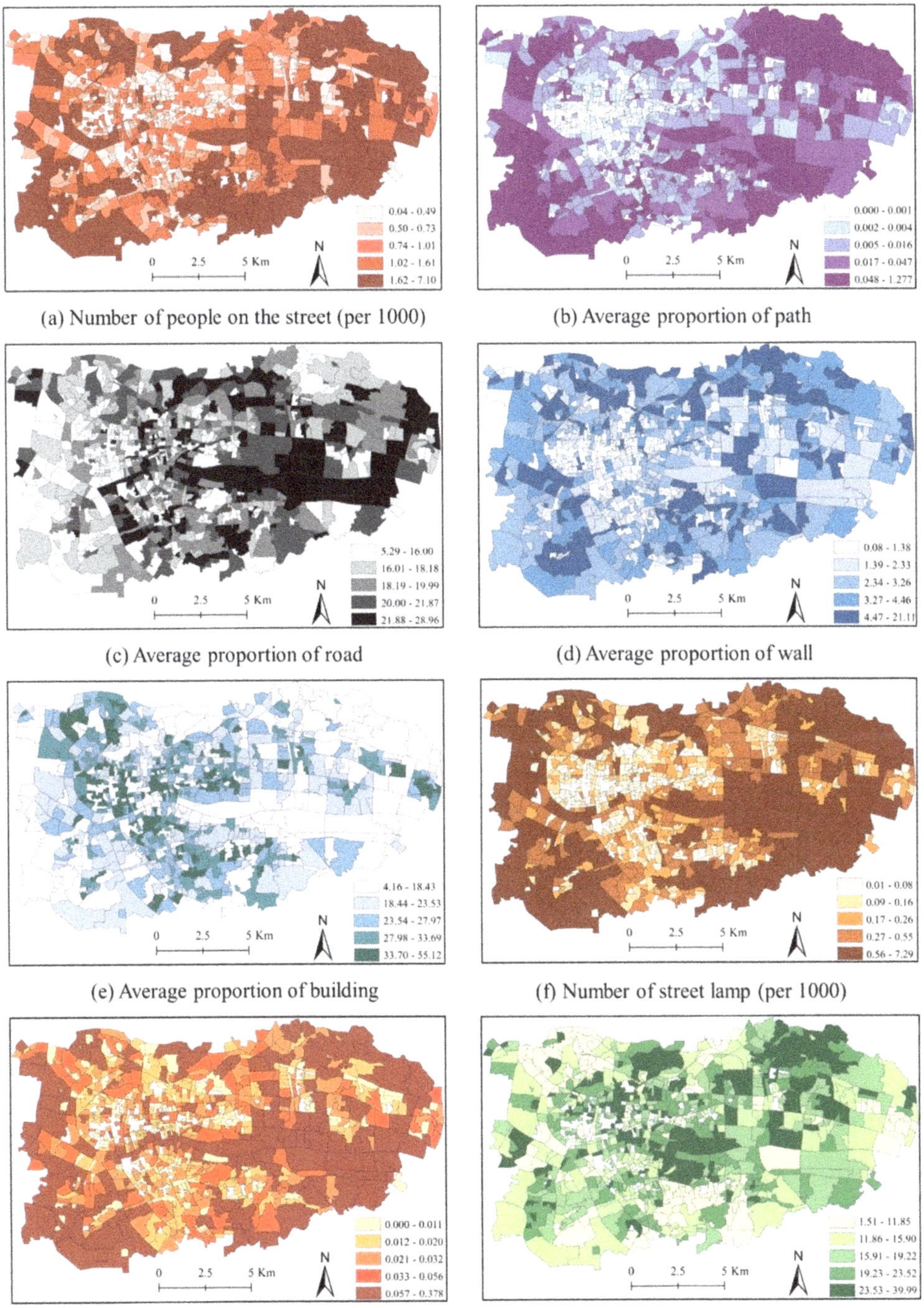

Figure 4. Spatial distributions of people on the street and streetscape physical features retrieved by BSV images and deep learning methods.

3.2.3. Control Variables

Data obtained from the Sixth Nationwide Census were used to retrieve socioeconomic and demographic factors. Land use features were extracted from Gaode Map. Based on the data provided by Daodaotong Map, we further acquired the features of surveillance and transportation facilities.

- Socioeconomic and Demographic Factors

We collected the socioeconomic and demographic factors, including the rate of young people, the rate of highly educated people, the rate of migrant people, and the rate of renters. Young people are the main perpetrators of crimes [15], so we used young people (aged 30–45 years) to indicate possible offenders. We used the rate of highly educated people to approximate the income level, which has a specific association with crime [15,58]. Studies proved that migrants were positively related to crime by increasing instability and disrupting social order [59]. Therefore, we obtained the rate of migrant people by calculating the proportion of people whose Hukou was not in ZG city. Similarly, we considered the rate of renters, as this is also a factor adverse to residential stability and social organization [59].

- Land Use Features

This study used the number of POIs in each community to proxy the number of point-level land uses. In addition, we calculated the mixture of POIs to measure the land use heterogeneity by the adjusted Herfindahl–Hirschman Index [60]:

$$Mix = 1 - \sum_{j=1}^{J} P_j^2 \quad (3)$$

where P_j is the proportion of the number of jth type POIs. A *Mix* close to 1 indicates a strong land use mixture, while a *Mix* close to 0 indicates a weak land use mixture. Some research claimed that a mixed land use pattern could weaken the informal control of residents and increase crime [61], while some research revealed that a complex land use composition attracts people and promotes activities, thereby curbing crime by increasing social control [15,62].

- Formal Surveillance

Police stations are the most basic level of governmental management institutions in China. They can act as a deterrent to crime [63]. This study used the number of police stations to proxy the formal surveillance levels.

- Transportation Facilities

This study adopted two transportation facility variables to measure traffic accessibility. They were the number of bus stops and the number of subway stations. The relationship between traffic accessibility and crime is complex. Some studies demonstrated that convenient transportation could promote pedestrian activities, enhance natural surveillance, and deter crimes [15], while some studies proved that transportation facilities attract targets and act as escape routes, thus providing opportunities for offenders [34,63].

Table 1 lists the summary statistics of the dependent and independent variables used in this study.

Table 1. Summary statistics of dependent and independent variables.

Variable	Mean	SD	Min	Max
Dependent Variables				
Number of total street crimes	155.28	183.79	5	1952
Number of street property crimes	126.53	155.67	3	1808
Number of street violent crimes	28.75	34.40	0	306

Table 1. *Cont.*

Variable	Mean	SD	Min	Max
Control Variables				
Rate of young people (%)	26.534	5.908	4.654	47.431
Rate of highly educated people (%)	14.316	12.565	0.000	86.245
Rate of migrant people (%)	42.235	21.494	0.203	98.518
Rate of renters (%)	30.636	23.441	0.000	100.000
Number of POI (per 1000)	0.323	0.283	0.004	2.451
Mixture of POI	0.826	0.074	0.341	0.904
Number of bus stops	1.769	2.121	0	15
Number of subway stations	0.111	0.351	0	3
Number of police stations	1.384	1.524	0	10
Streetscape Variables				
Number of people on the street (per 1000)	1.176	1.048	0.043	7.097
Average proportion of paths (%)	0.037	0.086	0.000	1.276
Average proportion of roads (%)	18.939	3.478	5.299	28.953
Average proportion of walls (%)	3.131	2.296	0.085	21.107
Average proportion of buildings (%)	26.218	8.753	4.166	55.116
Number of streetlamps (per 1000)	0.430	0.711	0.014	7.280
Number of traffic lights (per 1000)	0.038	0.042	0	0.377
Average proportion of trees (%)	17.611	6.705	1.519	39.980

3.3. Method

The dependent variables were community-level crime counts, which were overdispersed nonnegative integers. Therefore, negative binomial regression models were adopted in this study to model the associations between the street view variables and the number of crimes:

$$ln(Y_i) = \beta_0 + \sum_{k=0}^{k} \beta_k X_{ik} + \sum_{l=0}^{l} \beta_l X_{il} + \sum_{m=0}^{m} \beta_m X_{im} + \sum_{n=0}^{n} \beta_n X_{in} + \sum_{p=0}^{p} \beta_p X_{ip} \quad (4)$$

where Y_i represents the crime count in community i, the βs are regression coefficients estimated by the model, indicating the influences of independent variables on the dependent variable, and X_{ik}, X_{il}, X_{im}, X_{in}, and X_{ip} are independent variables of five categories (socioeconomic and demographic factors, land use features, formal surveillance, transportation facilities, and streetscape features, respectively).

We calculated the Moran's I indexes of the dependent variables to examine whether spatial autocorrelation effects existed. The results indicate that all three types of crime were autocorrelated in space. Therefore, we added a spatial lag into the modal as an independent variable to address the spatial autocorrelation issue. The spatial lag was calculated as follows:

$$Lag_i = \sum_{j=1, i\neq j}^{N} \frac{C_j}{N} \quad (5)$$

where Lag_i is the spatial lag of the dependent variable in community i, j is a neighbor of community i, N is the total number of neighbors of community i, and C_j is the number of crimes in community j. In short, the spatial lag of community i measured the average crime count of its neighbors. In this study, we used the Queen adjacency criterion to determine the neighbors of community i.

Figure 5 summarizes the workflow of the study, which included three steps: (1) generating sampling points along the street, based on which fine-grained BSV images were collected using the Baidu API, (2) extracting streetscape features using an object detection method (Faster R-CNN) and a semantic segmentation method (PSPNet), and (3) building regression models to determine the influences of the on-street population and streetscape physical environment on street crime, controlling for the effects of socioeconomic and demographic factors, land use features, and surveillance and transportation facilities.

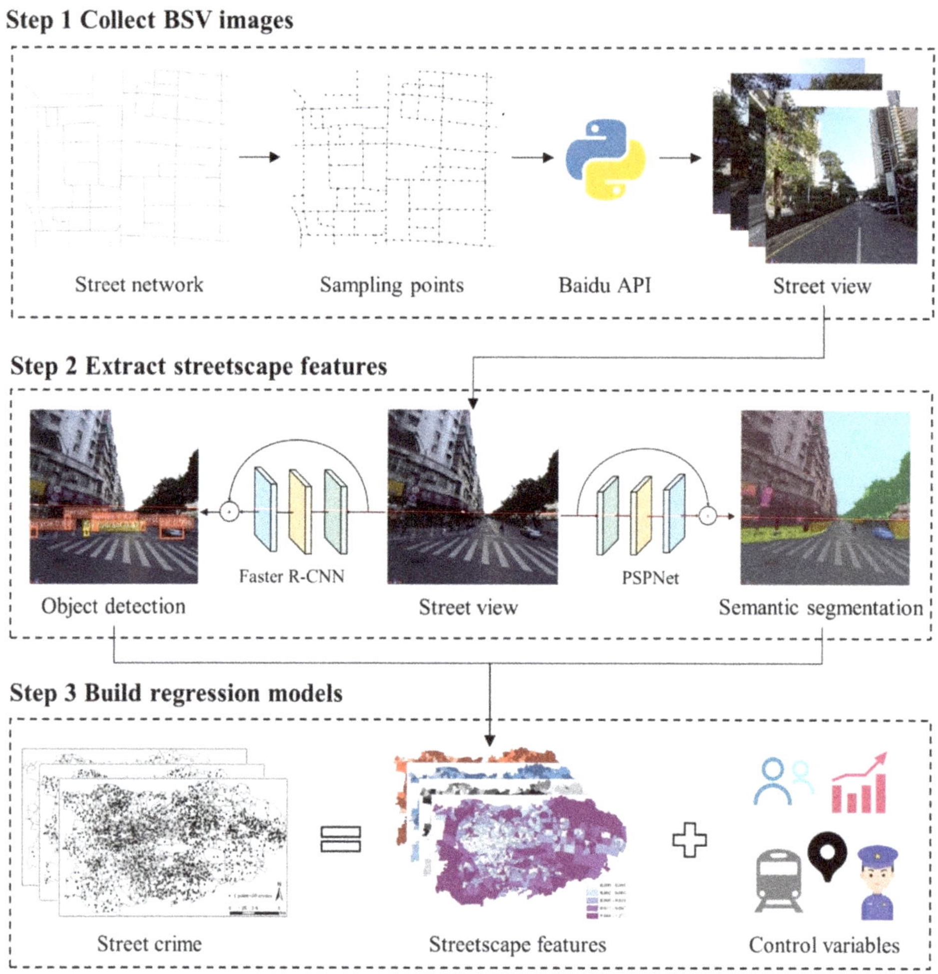

Figure 5. Workflow of this study.

4. Results

Before running the regression models, we checked all explanatory variables' VIF (variable inflation factor) values to check for multicollinearity. The results showed that the VIF values of all explanatory variables were much smaller than the commonly accepted threshold of 10 in crime research [64]. Therefore, the results in this study had no serious multicollinearity issues. Additionally, we standardized all explanatory variables before incorporating them into the regression models because the covariates had different units and significant disparities in magnitude. Standardization also makes it easy to compare the magnitudes of the impacts of different variables [65]. In order to assess the improvements of model performances after incorporating streetscape variables, we ran baseline models which did not contain the streetscape variables. We utilized log-likelihood and AIC to compare the model performances comprehensively. A larger log-likelihood value indicated a better model fit, while a smaller AIC value indicated a better one.

Table 2 presents the results of the spatial lag negative binomial regression estimations. Models (1, 3, and 5) are baseline models which only included the control variables, while Models (2, 4, and 6) are the full models which contained additional streetscape variables. Both the log-likelihood and AIC values demonstrated that adding street view variables improved the model fits. For total street crime, the log-likelihood of Model (2) (−3821.261) was larger than that of Model (1) (−3837.600), The AIC of Model (2) (7682.521) was smaller than that of Model (1) (7699.199). For street property crime, the log-likelihood of Model (4) (−3689.161) was larger than that of Model (3) (−3704.309), and the AIC of Model (4) (7418.322) was smaller than that of Model (3) (7432.617). For street violent crime, the log-likelihood of Model (6) (−2592.888) was larger than that of Model (5) (−2608.377), and the AIC of Model (6) (5225.775) was smaller than that of Model (5) (5240.753). Therefore, we discuss only the results of the full models in the following section.

Table 2. Results of spatial lag negative binomial regression models with all independent variables standardized.

Dep. Var.	Total Street Crime		Street Property Crime		Street Violent Crime	
Model	(1) IRR (Std. Err.)	(2) IRR (Std. Err.)	(3) IRR (Std. Err.)	(4) IRR (Std. Err.)	(5) IRR (Std. Err.)	(6) IRR (Std. Err.)
Control Variables						
Rate of young people	1.046 ** (0.019)	1.038 ** (0.019)	1.04 ** (0.02)	1.032 * (0.019)	1.08 *** (0.02)	1.073 *** (0.020)
Rate of highly educated people	1.040 ** (0.019)	1.062 *** (0.021)	1.04 * (0.02)	1.059 *** (0.021)	1.06 *** (0.02)	1.069 *** (0.021)
Rate of migrant people	1.032 (0.021)	1.024 (0.021)	1.03 (0.02)	1.019 (0.022)	1.04 * (0.02)	1.028 (0.022)
Rate of renters	1.087 *** (0.023)	1.086 *** (0.024)	1.09 *** (0.02)	1.091 *** (0.024)	1.05 ** (0.02)	1.055 ** (0.023)
Number of POIs	1.188 *** (0.033)	1.162 *** (0.033)	1.20 *** (0.03)	1.166 *** (0.034)	1.10 *** (0.03)	1.088 *** (0.029)
Mixture of POIs	1.015 (0.018)	1.027 (0.018)	1.01 (0.02)	1.025 (0.019)	1.00 (0.02)	1.013 (0.019)
Number of subway stations	1.005 (0.018)	1.000 (0.018)	1.01 (0.02)	1.005 (0.018)	0.97 (0.02)	0.963 ** (0.017)
Number of bus stops	1.034 * (0.020)	1.004 (0.023)	1.03 * (0.02)	1.005 (0.024)	1.04 ** (0.02)	1.014 (0.024)
Number of police stations	0.984 (0.016)	0.987 (0.016)	0.99 (0.02)	0.989 (0.017)	0.97 (0.02)	0.979 (0.017)
Streetscape Variables						
Number of people on the street		1.078 *** (0.029)		1.079 *** (0.030)		1.042 (0.028)
The average proportion of paths		0.955 ** (0.019)		0.957 ** (0.020)		0.960 * (0.021)
The average proportion of roads		1.019 (0.024)		1.015 (0.025)		1.037 (0.026)
The average proportion of walls		0.979 (0.021)		0.975 (0.021)		0.997 (0.022)
The average proportion of buildings		0.950 * (0.028)		0.951 * (0.029)		0.939 ** (0.029)
Number of streetlamps		0.962 (0.028)		0.957 (0.029)		0.982 (0.029)
Number of traffic lights		1.042 (0.029)		1.047 (0.030)		1.019 (0.029)
The average proportion of trees		0.935 *** (0.024)		0.935 ** (0.024)		0.928 *** (0.025)

Table 2. *Cont.*

Dep. Var.	Total Street Crime		Street Property Crime		Street Violent Crime	
Model	(1) IRR (Std. Err.)	(2) IRR (Std. Err.)	(3) IRR (Std. Err.)	(4) IRR (Std. Err.)	(5) IRR (Std. Err.)	(6) IRR (Std. Err.)
Spatial lag of dependent variable	1.918 *** (0.064)	1.851 *** (0.061)	1.946 *** (0.068)	1.882 *** (0.065)	1.997 *** (0.056)	1.939 *** (0.054)
Log-likelihood	−3837.600	−3821.261	−3704.309	−3689.161	−2608.377	−2592.888
AIC	7699.199	7682.521	7432.617	7418.322	5240.753	5225.775

Note: The dependent variables are the number of total street crimes, number of street property crimes, and number of street violent crimes. IRR = incidence rate ratio. ***, **, and * indicate significance at the 1%, 5%, and 10% levels, respectively. In parentheses, the standard errors are given. The intercept terms are not listed. The likelihood ratio test of $\alpha = 0$ demonstrates that negative binomial models are more suitable than standard Poisson models ($p < 0.001$).

The number of people on the street had the most considerable impact on the total street crimes and street property crimes among all the streetscape variables. Specifically, a one standard deviation increase in the number of people on the street was associated with a 7.8% (IRR = 1.078) increase in the number of total street crimes and a 7.9% (IRR = 1.079) increase in the number of street property crimes. The number of people on the street also positively influenced street violent crime, but the effect was not statistically significant. The average proportion of paths had a significant negative impact on three types of crime. A one standard deviation increase of this factor was associated with a 4.5% (IRR = 0.955) decrease in the number of total street crimes, a 4.3% (IRR = 0.957) decrease in the number of street property crimes, and a 4% (IRR = 0.960) decrease in the number of street violent crimes. Similarly, the average proportions of buildings and trees were also significantly and negatively associated with three types of crime. Specifically, a one standard deviation increase in the average proportion of buildings would result in a 5% (IRR = 0.950) decrease in the number of all street crimes, a 4.9% (IRR = 0.951) decrease in the number of street property crimes, and a 6.1% (IRR = 0.939) decrease in the number of street violent crimes. A one standard deviation increase in the average proportion of trees was associated with a 6.5% (IRR = 0.935) decrease in the number of total street crimes, a 6.5% (IRR = 0.935) decrease in the number of street property crimes, and a 7.2% (IRR = 0.928) decrease in the number of street violent crimes. The average proportion of roads and the number of traffic lights had positive relationships with the three types of crime, but these relationships did not reach statistical significance. The average proportion of walls and number of streetlamps had nonsignificant and negative correlations with the three types of crime.

As for the control variables, the rate of young people, the rate of highly educated people, the rate of renters, and the number of POIs had significantly positive associations with the three types of crime. The rate of migrant people, the mixture of POIs, and the number of bus stops had positive relationships with the three types of crime, but the effects were not significant. The number of subway stations had a significant negative association with street violent crime, and its correlations with the other two types of crime were insignificant. The number of police stations had negative associations with the three types of crime, but the effects were insignificant.

The spatial lags of the dependent variables had significant and solid positive associations with the numbers of all types of crime, revealing the spatial autocorrelation effect of crime events. Therefore, the spatial lag models used in this study were valid.

We used the k-fold cross-validation technique to validate the regression models used in this study. This technique first divides the total dataset into k parts of equal size, iteratively excludes one part (called the validation set) at a time, and predicts it with the parts not excluded (called the training set). At each step, an R^2 score can be calculated, measuring the prediction accuracy of the trained model. Figure 6 presents the results of the cross-validation R^2 score when k was set at different values. When k reached about 30 and above, the performance of the three models became stable, with the R^2 score reaching above 0.8, indicating that the models in this study were sufficiently accurate.

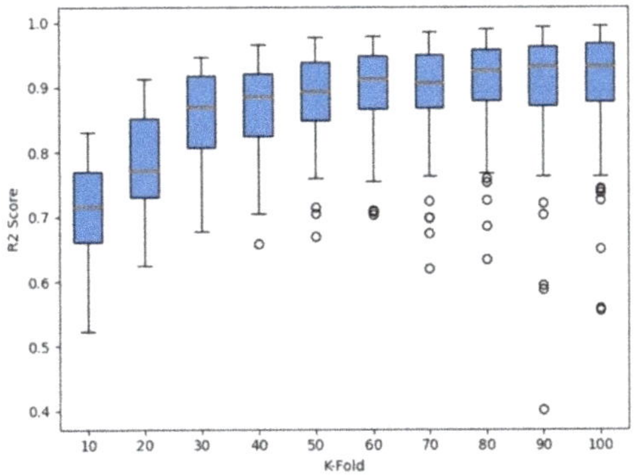

(a) Total street crime

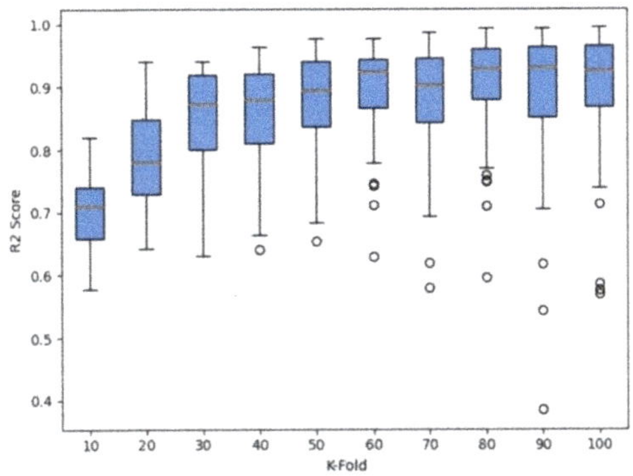

(b) Street property crime

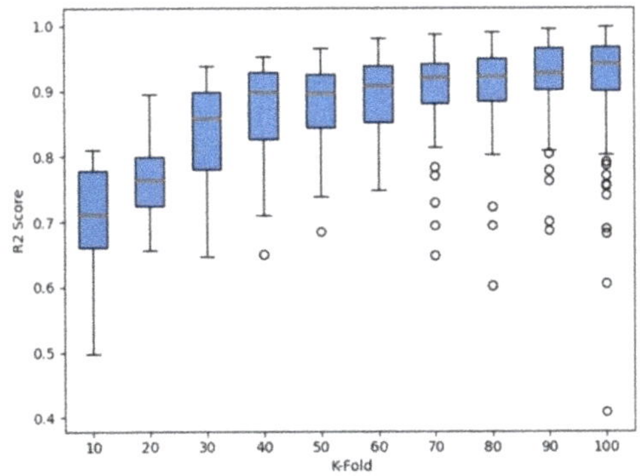

(c) Street violent crime

Figure 6. Cross-validation R^2 score of different fractions of training data for the regression models.

5. Discussion

Overall, the findings of this research are consistent with the previous literature. The number of people on the street had different effects on different types of crime. The total street crime and street property crime had significant positive associations with the number of people on the street, indicating that places with more people on the street have higher risks for total street crime and street property crime. Although the number of people on the street also positively influenced street violent crime, this effect was not significant statistically (p = 0.128). Therefore, street violent crime in a place would not witness a remarkable change in the number of people on the street. Such a disparity advocates for previous theoretical and empirical studies. First, both crime pattern theory and routine activity theory argue that offenders make rational decisions by balancing the potential benefits, costs, and risks when committing a crime. In general, offenders tend to commit crimes where suitable targets are present and capable guardians are absent. However, each type of crime has unique choice-structuring properties [66]. The influence of the presence of people in a given area is mainly dependent upon the nature of the crime; that is, the presence of people has diverse impacts on different types of crime [23]. Street property crimes, such as snatching, pickpocketing, and theft, have the nature of concealment and transience. Property crime offenders prefer to "fish in troubled waters" in that they usually commit crimes in crowded places when people are not paying attention and then flee the scene quickly. The whole process of committing a crime must be completed rapidly without being noticed. A thick crowd of people in a location not only offers a lot of targets and opportunities, but the presence of a dense population can also provide perfect cover for the whole process of a crime: looking for a target, committing a crime, and fleeing the scene after a crime. This makes streets with dense populations ideal places for property crime. This is proven by the fact that property crimes have high exposure rates and low detection rates [31]. However, street violent crimes such as robbery, intentional injury, and assault do not happen secretly but usually with sounds of a struggle, making it easy to be spotted and draw people's attention. Therefore, these types of crimes are unlikely to happen in crowded places. Social disorganization theory focuses on the static of the residential population rather than the environment or the simple magnitude of floating populations [21]. The theory highlights that the local characteristics in a location improve individual tendencies toward delinquent behaviors or hinder collective efforts to preserve public order.

Second, the inconsistent influences of people on the street on different types of crime were also found in previous empirical studies. For example, Vomfell et al. utilized Twitter and taxi data to help forecast crime [24]. They concluded that using these features can significantly improve the prediction accuracy of property crime. For violent crime, however, the spatiotemporal dimension of these features adds very little value. They further explained that long-term neighborhood structural conditions are the primary influences of violent crime. Social deprivation, for example, provides the context for violent behaviors. Therefore, violent crimes commonly take place in locations with poor social cohesion. As for property crime, it is local opportunities through anonymity that matter, rather than deprivation. Another study conducted by Malleson and Andresen in Leeds found similar results; although the study region had a large volume of violent crimes, there was no statistically significant elevation in the risk of violent criminal victimization when considering a theoretically informed population at risk [47].

The physical environment variables deduced from BSV images also had meaningful associations with street crime. The average proportion of trees had the most significant influence on all types of crime among the street view variables. Specifically, the impacts were all significantly negative. Therefore, places with higher eye-level street green spaces were associated with less crime. Green spaces have been proven to improve community cohesion, making people's desire to survey their surroundings and intervene in an ongoing crime stronger [67,68]. Well-maintained vegetation in a place can also act as a cue to care, indicating that inhabitants actively care about their territory and potentially suggesting that an intruder would be noticed and confronted [69]. Additionally, green spaces, including trees, parks, and other natural features, could play a relieving effect that can make human psychological and emotional states calm, improve cognitive functioning, and inhibit people from committing crimes [70,71]. A series of empirical research has proven green spaces to be inhabitable to crime [7,67–72].

The average proportion of paths was also negatively associated with all three types of crime, and the effects were statistically significant. This result supports the previous advocation of adopting design principles that facilitate walking and social interaction, because human activity could help promote a sense of community belonging, which is beneficial for crime prevention [73]. Places with high proportions of paths provide spaces for outdoor activity. People living here are more likely to go out and associate with others. Therefore, people's anonymity decreases, as there is an opportunity to gain mutual acquaintance with other residents, enabling social control of inhibiting crime because unfamiliar people will stand out as strangers in these neighborhoods.

The average proportion of buildings had significantly negative associations with all three types of crime. As shown in Figure 4e, communities with high average proportions of buildings are generally dispersed along arterial roads. Buildings extracted from GSVs in these areas are usually high-rise buildings fronting the street, indicating these districts are work areas with many employees. Jacobs assumed that such vibrant locations would have less crime given the presence of many guardians [74]. A similar study about the associations between the built environment and crime using GSV images demonstrated that the presence of buildings was generally unrelated to crime, except for robberies [10].

Other street view variables such as the average proportion of roads, the average proportion of walls, number of streetlamps, and number of traffic lights had no statistically significant relationships with any of the three types of crime. Therefore, we do not discuss these variables further.

The spatial lags of the dependent variables significantly impacted all types of crime. Specifically, according to the results of the full models, a one standard deviation increase in the spatial lag of the dependent variable was associated with an 85.1% (IRR = 1.851) increase in the number of total street crimes, an 88.2% (IRR = 1.882) increase in the number of street property crimes, and a 93.9% (IRR = 1.939) increase in the number of street violent crimes. These results reveal the widely existing spatial autocorrelation effect of geographic events. Modeling crime using a spatial regression model is thus necessary.

6. Conclusions

The presence and size of people on the street and physical streetscape characteristics have close associations with criminal activities [7,8,75]. However, large-scale environment conditions, especially fine-grained streetscape features, are difficult or expensive to obtain. The absence of precise quantitative data for street scenes leaves the relationship between crime and the visual characteristics of a streetscape unrevealed [76,77]. This study integrated BSV images and deep learning methods to retrieve detailed and rich information about the streetscape context. Controlling for the spatial autocorrelation effect, we constructed spatial lag negative binomial regression models to evaluate the influences of people on the street and the streetscape physical features on crime.

The results of this study are promising. First, the significant improvements in model performance after incorporating the street view variables demonstrate the necessity for accounting for the effect of the streetscape context when studying crime. Second, the number of people on the street had significant positive impacts on the total street crime and street property crime. However, no significant impacts were found on street violent crime. Therefore, the effect of human presence on crime is greatly determined by the nature of the crime. The phenomenon that the same street view variable had different effects on different types of crime reveals that the study of the general type of crime is insufficient. It ignores the different occurrence mechanisms of different types of crime. Third, regarding the physical streetscape features, the average proportion of paths, buildings, and trees had statistically significant and negative impacts on both the occurrence of street property crime and street violent crime.

This study is the first attempt at combining street view images and deep learning algorithms to extract both the on-street population and a series of eye-level physical streetscape features to investigate street crime. Previous studies could not distinguish between the street and indoor populations, and they were potentially biased by counting indoor people for their possible influence on street crime. This study provides evidence that streetscape features, including people on the street retrieved from street view images, can effectively explain street crime. The available street view images provide new opportunities for gathering large-scale quantitative streetscape characteristics which provide a basis for place-based crime research.

The presence of people on the street is a prerequisite for the occurrence of a street crime. Therefore, the on-street population is an essential factor of street crime. Traditional field observation methods used to collect the pedestrian volume are time-consuming, labor-intensive, and inefficient, while big data sources like mobile phone data, geotagged social media data, metro smart card data, and taxi and bicycle trajectory data are incapable of capturing the movement of pedestrians. Such limitations can be overcome by the method used in this study. Assessing the pedestrian volume automatically from street view images with deep learning techniques is a reliable method for determining the on-street population size. The availability of street view images offers broad geographic coverage. The methods utilized in this study could be applied to street crime research in other countries and regions. Researchers from other fields, such as public health, urban vitality, and street design, could also borrow ideas from this study, because the on-street population and streetscape features are important in these fields.

This study's findings not only validate criminology theories but also have implications for crime prevention and urban planning. Trees have a significantly negative impact on street crime. Therefore, urban designers may improve the environment by planting trees. Paths are also a design element found to have a crime deterring effect. Therefore, this design principle can be adopted to create spaces for outdoor activity. Police patrols should be deployed in targeted areas with high proportions of young people and renters. Furthermore, although communities with more POIs promote vitality and have other advantages, they may also have some unexpected drawbacks, such as street crime.

Several limitations of this research should be noted and addressed in the future. First, although the street view image is a valuable and accessible data source for determining the on-street population size, most street view images were captured in the daytime. Therefore, it cannot be known what the population size on the street was in the evening. Future studies could use satellite night light data to proxy on-street populations in the evening, as facilities are typically associated with lights at night [78]. Second, some objects such as litter, graffiti, broken windows, and property damage are not easily detected using existing methods. These objects are signs of physical incivilities, which have been demonstrated to attract crimes. Future studies could collect such fine-scale quantitative data using the environmental audit approach [5]. Third, apart from the physical environment, human visual perception of the urban environment can also affect the occurrence of crimes. Therefore, future research could account for perception [50].

Author Contributions: Conceptualization, Han Yue and Lin Liu; methodology, Han Yue, Huafang Xie and Jianguo Chen; writing—original draft preparation, Han Yue and Huafang Xie; writing—review and editing, Lin Liu and Jianguo Chen. All authors have read and agreed to the published version of the manuscript.

Funding: This research was partially supported by the National Key Research and Development Program of China No. 2018YFB0505500 and 2018YFB0505503.

Data Availability Statement: Not available.

Acknowledgments: The authors are very grateful for the comments of the reviewers and the editor, which have helped improve the article considerably.

Conflicts of Interest: The authors declare no conflict of interest.

References

1. Cohen, L.; Felson, M. Social change and crime rate trends: A routine activity approach. *Am. Sociol. Rev.* **1979**, *44*, 588–608. [CrossRef]
2. Brantingham, P.L.; Brantingham, P.J. A theoretical model of crime hot spot generation. *Stud. Crime Crime Prev.* **1999**, *8*, 7–26.

3. Weisburd, D. From criminals to criminal contexts: Reorienting criminal justice research and policy. *Adv. Criminol. Theory* **2002**, *10*, 197–216.
4. Cozens, P.; Davies, T. Crime and residential security shutters in an Australian suburb: Exploring perceptions of 'eyes on the street', social interaction and personal safety. *Crime Prev. Community Saf.* **2013**, *15*, 175–191. [CrossRef]
5. He, L.; Páez, A.; Liu, D. Built environment and violent crime: An environmental audit approach using Google Street View. *Comput. Environ. Urban Syst.* **2017**, *66*, 83–95. [CrossRef]
6. Rundle, A.G.; Bader, M.D.M.; Richards, C.A.; Neckerman, K.M.; Teitler, J.O. Using google street view to audit neighborhood environments. *Am. J. Prev. Med.* **2011**, *40*, 94–100. [CrossRef]
7. Wolfe, M.K.; Mennis, J. Does vegetation encourage or suppress urban crime? Evidence from Philadelphia, PA. *Landsc. Urban Plan.* **2012**, *108*, 112–122. [CrossRef]
8. Zhou, H.; Liu, L.; Lan, M.; Yang, B.; Wang, Z. Assessing the Impact of Nightlight Gradients on Street Robbery and Burglary in Cincinnati of Ohio State, USA. *Remote Sens.* **2019**, *11*, 1958. [CrossRef]
9. Patino, J.E.; Duque, J.C.; Pardo-Pascual, J.E.; Ruiz, L.A. Using remote sensing to assess the relationship between crime and the urban layout. *Appl. Geogr.* **2014**, *55*, 48–60. [CrossRef]
10. Hipp, J.R.; Lee, S.; Ki, D.; Kim, J.H. Measuring the Built Environment with Google Street View and Machine Learning: Consequences for Crime on Street Segments. *J. Quant. Criminol.* **2021**, 1–29. [CrossRef]
11. Amiruzzaman, M.; Curtis, A.; Zhao, Y.; Jamonnak, S.; Ye, X. Classifying crime places by neighborhood visual appearance and police geonarratives: A machine learning approach. *J. Comput. Soc. Sci.* **2021**, *4*, 813–837. [CrossRef] [PubMed]
12. Khorshidi, S.; Carter, J.; Mohler, G.; Tita, G. Explaining Crime Diversity with Google Street View. *J. Quant. Criminol.* **2021**, *37*, 361–391. [CrossRef]
13. Zhou, H.; Liu, L.; Lan, M.; Zhu, W.; Song, G.; Jing, F.; Zhong, Y.; Su, Z.; Gu, X. Using Google Street View imagery to capture micro built environment characteristics in drug places, compared with street robbery. *Comput. Environ. Urban Syst.* **2021**, *88*, 101631. [CrossRef]
14. Zhang, F.; Fan, Z.; Kang, Y.; Hu, Y.; Ratti, C. "Perception bias": Deciphering a mismatch between urban crime and perception of safety. *Landsc. Urban Plan.* **2021**, *207*, 104003. [CrossRef]
15. Yue, H.; Zhu, X. The influence of urban built environment on residential burglary in China: Testing the encounter and enclosure hypotheses. *Criminol. Crim. Justice* **2019**, *21*, 508–528. [CrossRef]
16. Clarke, R.V.; Felson, M. *Routine Activity and Rational Choice*; Transaction: New Brunswick, NJ, USA, 1993.
17. Roncek, D.W.; Bell, R. Bars, blocks, and crimes. *J. Environ. Syst.* **1981**, *11*, 35–47. [CrossRef]
18. McCord, E.S.; Ratcliffe, J.H. Intensity value analysis and the criminogenic effects of land use features on local crime patterns. *Crime Patterns Anal.* **2009**, *2*, 17–30.
19. Groff, E.; Mccord, E.S. The role of neighborhood parks as crime generators. *Secur. J.* **2012**, *25*, 1–24. [CrossRef]
20. Kubrin, C.E.; Squires, G.D.; Graves, S.M.; Ousey, G.C. Does fringe banking exacerbate neighborhood crime rates? *Criminol. Public Policy* **2011**, *10*, 437–466. [CrossRef]
21. Jung, Y.; Chun, Y.; Kim, K. Modeling Crime Density with Population Dynamics in Space and Time: An Application of Assault in Gangnam, South Korea. *Crime Delinq.* **2020**, *68*, 253–283. [CrossRef]
22. Caminha, C.; Furtado, V.; Pequeno, T.H.C.; Ponte, C.; Melo, H.P.M.; Oliveira, E.A.; Andrade, J.S. Human mobility in large cities as a proxy for crime. *PLoS ONE* **2017**, *12*, e0171609. [CrossRef] [PubMed]
23. Boivin, R. Routine activity, population(s) and crime: Spatial heterogeneity and conflicting Propositions about the neighborhood crime-population link. *Appl. Geogr.* **2018**, *95*, 79–87. [CrossRef]
24. Vomfell, L.; Härdle, W.K.; Lessmann, S. Improving crime count forecasts using Twitter and taxi data. *Decis. Support Syst.* **2018**, *113*, 73–85. [CrossRef]
25. Felson, M.; Eckert, M. *Crime and Everyday Life*, 5th ed.; Sage: Thousand Oaks, CA, USA, 2015.
26. Song, G.; Liu, L.; Bernasco, W.; Xiao, L.; Zhou, S.; Liao, W. Testing indicators of risk populations for theft from the person across space and time: The significance of mobility and outdoor activity. *Ann. Am. Assoc. Geogr.* **2018**, *108*, 1370–1388. [CrossRef]
27. Wilcox, P.; Land, K.C.; Hunt, S.A. *Criminal Circumstance: A Dynamic, Multi-Contextual Criminal Opportunity Theory*; Aldine de Gruyter: New York, NY, USA, 2003.
28. Brantingham, P.L.; Brantingham, P.J. Crime Pattern Theory. In *Environmental Criminology and Crime Analysis*; Wortley, R., Mazerolle, L., Eds.; Willan Publishing: Portland, OR, USA, 2008.
29. Nee, C.; Taylor, M. Examining burglars' target selection: Interview, experiment or ethnomethodology? *Psychol. Crime Law* **2000**, *6*, 45–59. [CrossRef]
30. Mayhew, P. Defensible space: The current status of a crime prevention theory. *Howard J. Crim. Justice* **1979**, *18*, 150–159. [CrossRef]
31. Yue, H.; Zhu, X.; Ye, X.; Guo, W. The Local Colocation Patterns of Crime and Land-Use Features in Wuhan, China. *ISPRS Int. J. Geo-Inf.* **2017**, *6*, 307. [CrossRef]
32. Jones, H.R. *Crime and the Urban Environment*; Avebury: Brookfield, VT, USA, 1993.
33. Shu, C.-F. Housing layout and crime vulnerability. *Urban Des. Int.* **2000**, *5*, 177–188.
34. Yue, H.; Zhu, X.; Ye, X.; Hu, T.; Kudva, S. Modelling the effects of street permeability on burglary in Wuhan, China. *Appl. Geogr.* **2018**, *98*, 177–183. [CrossRef]
35. Hillier, B. Can streets be made safe? *Urban Des. Int.* **2004**, *9*, 31–45. [CrossRef]

36. Skogan, W.G.; Maxfield, M.G. *Coping with Crime: Individual and Neighborhood Reactions*; Sage Publications: Beverley Hills, CA, USA, 1981.
37. Walker, B.B.; Schuurman, N. The pen or the sword: A situated spatial analysis of graffiti and violent injury in Vancouver, British Columbia. *Prof. Geogr.* **2015**, *67*, 608–619. [CrossRef]
38. Spelman, W. Abandoned buildings: Magnets for crime? *J. Crim. Justice* **1993**, *21*, 481–495. [CrossRef]
39. Russo, S.; Roccato, M.; Vieno, A. Criminal victimization and crime risk perception: A multilevel longitudinal study. *Soc. Indic. Res.* **2013**, *112*, 535–548. [CrossRef]
40. Andresen, M.A.; Jenion, G.W. Ambient populations and the calculation of crime rates and risk. *Secur. J.* **2008**, *23*, 114–133. [CrossRef]
41. Panigutti, C.; Tizzoni, M.; Bajardi, P.; Smoreda, Z.; Colizza, V. Assessing the use of mobile phone data to describe recurrent mobility patterns in spatial epidemic models. *R. Soc. Open Sci.* **2017**, *4*, 160950. [CrossRef]
42. Mburu, L.W.; Helbich, M. Crime Risk Estimation with a Commuter-Harmonized Ambient Population. *Ann. Am. Assoc. Geogr.* **2016**, *106*, 804–818. [CrossRef]
43. Hipp, J.R.; Bates, C.; Lichman, M.; Smyth, P. Using Social Media to Measure Temporal Ambient Population: Does it Help Explain Local Crime Rates? *Justice Q.* **2018**, *36*, 718–748. [CrossRef]
44. Tucker, R.; O'Brien, D.T.; Ciomek, A.; Castro, E.; Wang, Q.; Phillips, N.E. Who 'Tweets' Where and When, and How Does it Help Understand Crime Rates at Places? Measuring the Presence of Tourists and Commuters in Ambient Populations. *J. Quant. Criminol.* **2021**, *37*, 333–359. [CrossRef]
45. Johnson, P.; Andresen, M.A.; Malleson, N. Cell Towers and the Ambient Population: A Spatial Analysis of Disaggregated Property Crime. *Eur. J. Crim. Policy Res.* **2020**, *27*, 313–333. [CrossRef]
46. Hanaoka, K. New insights on relationships between street crimes and ambient population: Use of hourly population data estimated from mobile phone users' locations. *Environ. Plan. B Urban Anal. City Sci.* **2016**, *45*, 295–311. [CrossRef]
47. Malleson, N.; Andresen, M.A. The impact of using social media data in crime rate calculations: Shifting hot spots and changing spatial patterns. *Cartogr. Geogr. Inf. Sci.* **2014**, *42*, 112–121. [CrossRef]
48. Algahtany, M.; Kumar, L. A Method for Exploring the Link between Urban Area Expansion over Time and the Opportunity for Crime in Saudi Arabia. *Remote Sens.* **2016**, *8*, 863. [CrossRef]
49. Zhang, F.; Zhang, D.; Liu, Y.; Lin, H. Representing place locales using scene elements. *Comput. Environ. Urban Syst.* **2018**, *71*, 153–164. [CrossRef]
50. Naik, N.; Philipoom, J.; Raskar, R.; Hidalgo, C. Streetscore—Predicting the perceived safety of one million streetscapes. In Proceedings of the IEEE Conference on Computer Vision and Pattern Recognition Workshops, Columbus, OH, USA, 23–28 June 2014.
51. Jiang, Y.; Chen, L.; Grekousis, G.; Xiao, Y.; Ye, Y.; Lu, Y. Spatial disparity of individual and collective walking behaviors: A new theoretical framework. *Transp. Res. Part D Transp. Environ.* **2021**, *101*, 103096. [CrossRef]
52. Chen, L.; Lu, Y.; Sheng, Q.; Ye, Y.; Wang, R.; Liu, Y. Estimating pedestrian volume using Street View images: A large-scale validation test. *Comput. Environ. Urban Syst.* **2020**, *81*, 101481. [CrossRef]
53. Yin, L.; Cheng, Q.; Wang, Z.; Shao, Z. 'Big data' for pedestrian volume: Exploring the use of Google Street View images for pedestrian counts. *Appl. Geogr.* **2015**, *63*, 337–345. [CrossRef]
54. Sytsma, V.A.; Connealy, N.; Piza, E.L. Environmental Predictors of a Drug Offender Crime Script: A Systematic Social Observation of Google Street View Images and CCTV Footage. *Crime Delinq.* **2020**, *67*, 27–57. [CrossRef]
55. Ren, S.; He, K.; Girshick, R.; Sun, J. Faster r-cnn: Towards real-time object detection with region proposal networks. *Adv. Neural Inf. Processing Syst.* **2015**, *28*, 91–99. [CrossRef]
56. Zhao, H.; Shi, J.; Qi, X.; Wang, X.; Jia, J. Pyramid scene parsing network. In Proceedings of the IEEE Conference on Computer Vision and Pattern Recognition (CVPR), Honolulu, HI, USA, 21–26 July 2017; pp. 2881–2890.
57. Li, X.; Zhang, C.; Li, W.; Ricard, R.; Meng, Q.; Zhang, W. Assessing street-level urban greenery using Google Street View and a modified green view index. *Urban For. Urban Green.* **2015**, *14*, 675–685. [CrossRef]
58. Wu, L.; Liu, X.; Ye, X.; Leipnik, M.; Lee, J.; Zhu, X. Permeability, space syntax, and the patterning of residential burglaries in urban China. *Appl. Geogr.* **2015**, *60*, 261–265. [CrossRef]
59. Du, F.; Liu, L.; Jiang, C.; Long, D.; Lan, M. Discerning the Effects of Rural to Urban Migrants on Burglaries in ZG City with Structural Equation Modeling. *Sustainability* **2019**, *11*, 561. [CrossRef]
60. He, Q.; Li, J. The roles of built environment and social disadvantage on the geography of property crime. *Cities* **2022**, *121*, 103471. [CrossRef]
61. Kurtz, E.M.; Koons, B.A.; Taylor, R.B. Land use, physical deterioration, resident-based control, and calls for service on urban streetblocks. *Justice Q.* **2006**, *15*, 121–149. [CrossRef]
62. Kim, Y.-A.; Hipp, J.R. Density, diversity, and design: Three measures of the built environment and the spatial patterns of crime in street segments. *J. Crim. Justice* **2021**, *77*, 101864. [CrossRef]
63. Chen, J.G.; Liu, L.; Xiao, L.Z.; Xu, C.; Long, D.P. Integrative Analysis of Spatial Heterogeneity and Overdispersion of Crime with a Geographically Weighted Negative Binomial Model. *ISPRS Int. J. Geo-Inf.* **2020**, *9*, 60. [CrossRef]
64. Bernasco, W.; Block, R. Robberies in Chicago: A block-level analysis of the influence of crime generators, crime attractors, and offender anchor points. *J. Res. Crime Delinq.* **2011**, *48*, 33–57. [CrossRef]
65. Fotheringham, A.S.; Yue, H.; Li, Z. Examining the influences of air quality in China's cities using multi—Scale geographically weighted regression. *Trans. GIS* **2019**, *23*, 1444–1464. [CrossRef]

66. Cornish, D.B.; Clarke, R.V. Understanding crime displacement: An application of rational choice theory. *Criminology* **1987**, *25*, 933–947. [CrossRef]
67. Bogar, S.; Beyer, K.M. Green Space, Violence, and Crime: A Systematic Review. *Trauma Violence Abus.* **2016**, *17*, 160–171. [CrossRef]
68. Blair, L. *Community Gardens and Crime: Exploring the Roles of Criminal Opportunity and Informal Social Control*; University of Cincinnati: Cincinnati, OH, USA, 2014.
69. Kuo, F.E.; Sullivan, W.C. Environment and Crime in the Inner City. *Environ. Behav.* **2016**, *33*, 343–367. [CrossRef]
70. Lin, B.B.; Egerer, M.H.; Ossola, A. Urban Gardens as a Space to Engender Biophilia: Evidence and Ways Forward. *Front. Built Environ.* **2018**, *4*, 4. [CrossRef]
71. Roe, J.J.; Thompson, C.W.; Aspinall, P.A.; Brewer, M.J.; Duff, E.I.; Miller, D.; Mitchell, R.; Clow, A. Green space and stress: Evidence from cortisol measures in deprived urban communities. *Int. J. Environ. Res. Public Health* **2013**, *10*, 4086–4103. [CrossRef] [PubMed]
72. Garvin, E.C.; Cannuscio, C.C.; Branas, C.C. Greening vacant lots to reduce violent crime: A randomised controlled trial. *Inj. Prev.* **2013**, *19*, 198–203. [CrossRef] [PubMed]
73. Talen, E. *Charter of the New Urbanism*; McGraw Hill Education: New York, NY, USA, 2013.
74. Jacobs, J. *The Death and Life of Great American Cities*; Random House: New York, NY, USA, 1961.
75. Chalfin, A.; Hansen, B.; Lerner, J.; Parker, L. Reducing Crime Through Environmental Design: Evidence from a Randomized Experiment of Street Lighting in New York City. *J. Quant. Criminol.* **2021**, 1–31. [CrossRef]
76. Barnum, J.D.; Caplan, J.M.; Kennedy, L.W.; Piza, E.L. The crime kaleidoscope: A cross-jurisdictional analysis of place features and crime in three urban environments. *Appl. Geogr.* **2017**, *79*, 203–211. [CrossRef]
77. Lee, I.; Jung, S.; Lee, J.; Macdonald, E. Street crime prediction model based on the physical characteristics of a streetscape: Analysis of streets in low-rise housing areas in South Korea. *Environ. Plan. B Urban Anal. City Sci.* **2017**, *46*, 862–879. [CrossRef]
78. Liu, L.; Zhou, H.; Lan, M. Agglomerative Effects of Crime Attractors and Generators on Street Robbery? An Assessment by Luojia 1-01 Satellite Nightlight. *Ann. Am. Assoc. Geogr.* **2021**, *112*, 350–367. [CrossRef]

International Journal of
Geo-Information

Article

The Use of Machine Learning Algorithms in Urban Tree Species Classification

Zehra Cetin * and **Naci Yastikli**

Department of Geomatic Engineering, Faculty of Civil Engineering, Yildiz Technical University, Davutpasa Campus, Esenler, Istanbul 34220, Turkey; ynaci@yildiz.edu.tr
* Correspondence: zerisir@yildiz.edu.tr

Abstract: Trees are the key components of urban vegetation in cities. The timely and accurate identification of existing urban tree species with their location is the most important task for improving air, water, and land quality; reducing carbon accumulation; mitigating urban heat island effects; and protecting soil and water balance. Light detection and ranging (LiDAR) is frequently used for extracting high-resolution structural information regarding tree objects. LiDAR systems are a cost-effective alternative to the traditional ways of identifying tree species, such as field surveys and aerial photograph interpretation. The aim of this work was to assess the usage of machine learning algorithms for classifying the deciduous (broadleaf) and coniferous tree species from 3D raw LiDAR data on the Davutpasa Campus of Yildiz Technical University, Istanbul, Turkey. First, ground, building, and low, medium, and high vegetation classes were acquired from raw LiDAR data using a hierarchical-rule-based classification method. Next, individual tree crowns were segmented using a mean shift clustering algorithm from high vegetation points. A total of 25 spatial- and intensity-based features were utilized for support vector machine (SVM), random forest (RF), and multi-layer perceptron (MLP) classifiers to discriminate deciduous and coniferous tree species in the urban area. The machine learning-based classification's overall accuracies were 80%, 83.75%, and 73.75% for the SVM, RF, and MLP classifiers, respectively, in split 70/30 (training/testing). The SVM and RF algorithms generally gave better classification results than the MLP algorithm for identifying the urban tree species.

Keywords: machine learning; classification; LiDAR; 3D point cloud; urban trees

Citation: Cetin, Z.; Yastikli, N. The Use of Machine Learning Algorithms in Urban Tree Species Classification. *ISPRS Int. J. Geo-Inf.* **2022**, *11*, 226. https://doi.org/10.3390/ijgi11040226

Academic Editors: Wolfgang Kainz, Gloria Bordogna and Cristiano Fugazza

Received: 17 January 2022
Accepted: 25 March 2022
Published: 26 March 2022

1. Introduction

Urban areas have become one of the main habitats for human beings in recent years. Cities are suffering from various problems, such as air and water pollution, flood risk, and urban heat island effects, and urban life for citizens is becoming extremely difficult due to overpopulation and unplanned urbanization. Urban forests, especially trees, provide a sustainable solution to solve these ecological problems and help to improve the living conditions of the urban residents [1]. Urban trees are of major importance for the residents, offering various economic, environmental, health, and aesthetic benefits in urban environments [2]. Trees are crucial to improving the air, land, and water quality; absorbing and mitigating carbon dioxide (CO_2); lowering urban temperatures; reducing the storm water runoff, wind speed, and noise pollution; as well as supporting biodiversity and providing shelters for different animals [1,3–6]. In addition to these environmental and ecological benefits, urban trees have many social and psychological effects, such as improving physical/mental health, alleviating life stresses, encouraging residents to build stronger social relationships, potentially reducing crime, and making neighborhoods more attractive places [2,7,8]. As a main component of city structures, urban trees decorate parks, roads, and pavements, provide recreational areas, and create shade, as well as influencing real estate value [2,4]. Contrary to the numerous benefits for cities and residents, some urban trees have some adverse effects, such as causing allergic reactions [9] and environmental pollution, damaging historical texture, and obstructing the silhouette of cities [2]. Different tree species face different environmental stresses, so they have different benefits or disadvantages for the urban ecology [10,11]. Due to the mostly positive, and few negative, effects of urban trees, accurate information about individual tree species in cities is important to enable city planners and local administrators to understand the value of urban vegetation for ecosystem services. Thus, the detection and monitoring of urban tree species is necessary for urban planning and protection, disaster management, and sustainable development of urban areas, and require detailed up-to-date data sources [2,12].

To date, many tree species identification studies have focused on forest areas rather than urban areas [13]. Urban environments have a complex structure with many different objects, such as buildings with different types and heights, vegetation on top of buildings, power lines, temporary objects, paved roads, driveways, road signs, and parking lots, in comparison with forests, where the surrounding areas are comparatively homogeneous and the tree crowns are generally densely distributed [13,14]. In urban environments, trees spatially exist with other urban elements, and they can be in groups of trees or in irregular spatial designs, as well as being isolated or evenly spaced [15]. Urban trees generally have a large variation in structural characteristics, according to planting purposes. Urban areas are faced with specific challenges in tree species identification applications because of the above mentioned factors [13].

Aerial photo interpretation and field surveys are the two traditional methods used to identify urban tree species in farraginous urban environments [3]. These traditional ways are successful in local and small-scale studies, but in fact are labor-intensive, expensive, time-consuming, and are generally not appropriate for the entire coverage of large urban areas [3,6,16,17]. Based on specialized photo interpreters' experience, the large discrepancies between different interpreters are one of the disadvantages of the aerial photo interpretation technique. Today, the latest remote-sensing technologies offer a significant solution to the drawbacks of the traditional methods with their efficient, reliable, rapid, and repeatable methods for monitoring and analyzing urban tree species. Furthermore, they enable more cost-effective project budgets, especially for large-scale applications [6,8,11,18]. Over the last few decades, space-borne or airborne multispectral and hyperspectral images have been utilized for tree species identification [19–22]. However, multispectral and hyperspectral images have their own limitations, such as inadequate spectral resolution, shadowing, and obscuring impacts, which are caused by background features, spectral mixtures, etc. [21,23,24]. During multispectral and hyperspectral image acquisition, the lighting conditions change, both in time and in space [25]. In different parts of urban environments, the same urban tree species can have dissimilar spectral reflectance, or different urban tree species can have a similar spectral signature in the process of obtaining tree species using optical remote-sensing methods [21,26,27]. In addition, optical remote-sensing imagery is generally restricted to obtaining detailed information about the understory due to the heterogeneity of the urban environment [28]. Recently, active sensing light detection and ranging (LiDAR) systems have been successfully used in the detection of urban tree species, providing 3D information with high spatial resolution, returning multiple signals and high scanning speed [22,28,29]. Airborne light detecting and ranging systems can penetrate tree crowns, supply geometric and radiometric information, exhibit some of the inner structure of trees, and collect intensity data [30]. LiDAR capabilities can also be improved by increasing point density.

In tree species classification studies using LiDAR datasets, the test areas, used data types, number of classified tree species, and classification methods vary from application to application. Discrete-return LiDAR data are used in general, but full-waveform LiDAR data, which use a newer technology, have been used in recent applications for tree species classification [31–36]. While a LiDAR point cloud can be used alone in the classification of classic tree species, aerial photographs, satellite images, or hyperspectral images can be used additionally to a LiDAR point cloud as supplementary data for integrated tree species classification applications [37–45]. Raw LiDAR data or gridded LiDAR data, which enable the direct use of traditional image classification algorithms, can be used for tree species classification [46–49]. In some tree species classification studies, only deciduous and coniferous tree species are distinguished, while in others, a large number of different species are classified [50,51].

Nowadays, machine learning algorithms are used in many remote-sensing applications, such as change detection [52], monitoring of active wildfires [53], land usage identification [54], road edge detection [55], ground water potential assessment [56], marine oil spill detection, prediction, and vulnerability assessment [57], biomass and soil moisture retrievals [58], etc. The support vector machine (SVM), decision tree, random forest (RF), k-nearest neighbor (KNN), and artificial neural network (ANN) methods, which are traditional machine learning methods, have been mostly utilized for the classification of the LiDAR data to acquire tree species [17,59]. Zhang and Liu [47] aimed to analyze the applicability of LiDAR-derived geometric and intensity metrics to classify adjacent and dominant tree species using the support vector machine method at the individual tree level in their study area. Koma et al. [60] examined the object-based classification of urban trees using full-waveform three-dimensional LiDAR data in Vienna, Austria. The applicability of the geometric and radiometric features of deciduous trees and the coniferous pine species in an urban environment have been investigated using a random forest classification algorithm with the combined use of geometric and radiometric features. An overall accuracy of 87.5% was achieved as the most reliable classification by Koma et al. [60]. LiDAR data and hyperspectral imagery (0.5 m) were fused to differentiate eight common tree species in Dian et al. [61] by using vertical, spatial, and spectral features as an input to the classification of a support vector machine in Anyang, Henan, China, and a voting procedure was used to generate a tree species map. Shen and Cao [62] used hyperspectral images and LiDAR point cloud, which are acquired simultaneously, to classify five tree species with a RF classifier, aimed at detecting the most important variables for the classification, and an evaluation of the contribution of combined use of two different datasets. Shi et al. [21] used an RF classifier to obtain six different tree species and drew attention to the importance of using 37 different LiDAR features derived under leaf-on and leaf-off conditions. Kim et al. [63] analyzed the possible usage of LiDAR intensity data using a linear discriminant function to differentiate broadleaved and coniferous tree species and obtained overall classification results verifying their success. A methodology for coniferous and deciduous tree species classification in a forest area using both a k-means and expectation maximation (EM) classifier with full-waveform LiDAR features was proposed in Reitberger et al. [64], and a maximum overall accuracy of 85% in a leaf-on situation was reported. Ørka et al. [31] classified coniferous and deciduous tree species with linear discriminant analysis (LDA) in a boreal forest reserve in Norway using intensity and ALS-derived structural features and succeeded with 88% overall classification accuracy.

The primary aim for this research is to assess the potential usage of the support vector machine, random forest, and multi-layer perceptron machine learning algorithms for the classification of deciduous and coniferous tree species, using the Davutpasa Campus of Yildiz Technical University, Istanbul, Turkey as the urban study area.

The paper is subdivided into four main sections. Section 2 defines the materials and methods, including the study area, used dataset, general workflow, high vegetation classification, individual tree crown segmentation, feature extraction, urban tree species classification, and performance analysis subsections. The results and discussion in Section 3 presents an experimental evaluation and analysis of the machine learning-based urban tree classification process. Finally, Section 4 concludes the work presented in this paper.

2. Materials and Methods

2.1. Study Area

The Davutpasa Campus of Yildiz Technical University, located in the Istanbul province district in northwest Turkey (41°01′33″ N, 28°53′21″ E), was selected as the urban study area and includes buildings of several types and heights, vegetation of different types and heights, paved roads, driveways, road signs, and parking lots (Figure 1). The selected urban study area (approximately 4.7 ha) in the Davutpasa Campus of Yildiz Technical University (approximately 125 ha) was used for testing the performance of SVM, RF, and MLP machine learning algorithms to discriminate deciduous and coniferous tree species (Figure 2). In the study area, different types of deciduous trees, such as linden, cherry, sycamore, mulberry, quince, plum, apple, and locust, and coniferous trees, including red pine, stone pine, blue spruce, and Norway spruce were available.

Figure 1. Davutpasa Campus of Yildiz Technical University (**left**) and the urban study area (**right**).

Figure 2. Deciduous and coniferous tree species in the urban study area.

2.2. LiDAR Data and Field Data

The 3D LiDAR point-cloud data used in this study were acquired in September 2013 with a "Riegl LSM-Q680i" full-waveform laser scanner by the Istanbul Metropolitan Municipality. The LiDAR point cloud was provided in Log ASCII Standard (LAS format) with an average density of 16 points/m^2. The flying height and speed of the helicopter were approximately 600 m and 148 km/h, respectively. The used LiDAR dataset was acquired at up to 400,000 Hz pulse repetition frequency (at a near infrared laser wavelength) with a scanning angle of 60° (±30°). The data were recorded with a rotating polygon mirror in parallel scan lines with beam divergences of less than or equal to 0.5 mrad. The laser data were recorded with multiple returns (echoes) and 16-bit uncalibrated intensity information.

The ground-truth data for tree species were acquired by field investigation and photo interpretation within the study area. The species of each tree in the study area was determined accurately as deciduous or coniferous.

2.3. General Workflow

In this research, a tree species classification method from raw 3D LiDAR data based on SVM, RF, and MLP machine learning algorithms was developed to provide tree species identification with high efficiency in terms of time and cost for large-scale applications. As a first step, the ground, building, and low, medium, and high vegetation classes were acquired from a raw LiDAR point cloud with a hierarchical rule-based classification method; then, individual tree crowns were segmented using a mean shift clustering algorithm from high vegetation points. Feature extraction was conducted for each individual tree, and these features were used to classify the deciduous and coniferous tree species in the urban study area with SVM, RF, and MLP algorithms. An accuracy assessment for the classified tree species was carried out. A 10-fold cross-validation and feature importance process with Mean Decrease Gini (MDG) were also performed to evaluate the stability of the models and to analyze the effects of the classification features. TerraScan software and the Python programming language were utilized in this research. The flow chart of the proposed machine learning-based urban tree species classification process is shown in Figure 3.

Figure 3. Block diagram of proposed machine learning-based classification of urban tree species using SVM, RF, and MLP algorithms.

2.4. High Vegetation Classification

The proposed method in this study starts with a classification of the high vegetation points from raw LiDAR point cloud data. The 3D raw LiDAR point cloud was classified with a hierarchical rule-based classification method using spatial features. The ground, building, low vegetation, medium vegetation, high vegetation, low point, air point, and default classes were obtained based on the standard point classes of the American Society for Photogrammetry and Remote Sensing (ASPRS) [65]. In the rule-based classification algorithms, the information about the terrain surface is converted to a set of rules; thus each terrain class has its own characteristic rules, and the classification is carried out based on these rules [66,67]. In this study, each individual LiDAR point was classified into the appropriate classes with point-based rules using spatial features. The details of the used point-based classification of the LiDAR data with the hierarchical rule-based classification method to acquire the high vegetation class (using the TerraScan module of TerraSolid software) can be found in Yastikli and Cetin [68].

2.5. The Segmentation of Individual Tree Crowns

In the segmentation process, the high vegetation points are partitioned into subsets of neighboring points called "segments". Individual tree crowns were achieved as a consequence of the segmentation step. The segmentation process was carried out using the mean shift clustering algorithm, which was first proposed by Fukunaga and Hostitler in 1975 [69]. Mean shift is an iterative and non-parametric method that shifts each data point based on the local maxima of density function with Kernel Density Estimation (KDE) [70]. This method chooses a random point as the cluster center from the used dataset, updating the cluster centers on the condition that the mean of the candidate points should be in a certain region. This segmentation algorithm automatically sets the number of clusters according to the bandwidth that determines the size of the region to be searched. The bandwidth is the most important parameter that needs to be specified in the mean shift process [71]. The bandwidth can be estimated with manual iterations or by using the bandwidth function. In this study, the 2D point-based segmentation of high vegetation has been conducted based on the x and y Cartesian coordinate pair of each raw LiDAR point to acquire individual tree crowns. The bandwidth parameter was estimated with manual iterations, and the high vegetation points were segmented by the flat kernel with a proper bandwidth of 3. The segmentation processes with mean shift clustering were performed with the Python programming language (Python 3.6.4), along with scikit-learn library. The segmentation results could be categorized as correct detection, under-segmentation, over-segmentation, missed, and noise [72]. The generated segments in this study include both the under-segmentation, which means multiple crowns were segmented as a single crown, and over-segmentation, which means a single crown was segmented as multiple crowns.

2.6. Feature Extraction

The extraction of the features to be used in the classification model by applying statistical analysis is a critical step in the machine learning field [73]. In our proposed approach (see Figure 1), the classification features were calculated using the height and intensity information of the LiDAR points, and they were then used as input to the machine learning-based classification algorithms for the classification of urban tree species. The spatial- and intensity-based features were obtained from the raw 3D LiDAR point clouds to differentiate the urban tree species. The determined features were generated for each tree crown using LiDAR height and intensity information, including minimum and maximum values, and results of statistical analyses (mean, standard deviation, skewness, kurtosis, etc.) [11,21,74,75]. In Table 1, the 25 generated features for the classification of targeted urban tree species using SVM, RF, and MLP algorithms are given. Z indicates the Z coordinate of the LiDAR points in the national coordinate system, and intensity indicates the uncalibrated intensity value of the LiDAR points in Table 1.

Table 1. The generated spatial- and intensity-based features from LiDAR data.

LiDAR Data	
Spatial-Based Features	**Intensity-Based Features**
Number of points	–
Maximum Z	Maximum intensity
Minimum Z	Minimum intensity
Standard deviation of Z	Standard deviation of intensity
Mean Z	Mean intensity
Skewness of Z	Skewness of intensity
Kurtosis of Z	Kurtosis of intensity
Z range	Intensity range
5th percentile of Z	5th percentile of intensity
25th percentile of Z	25th percentile of intensity
50th percentile of Z	50th percentile of intensity
75th percentile of Z	75th percentile of intensity
90th percentile of Z	90th percentile of intensity

2.7. The Classification of Urban Tree Species

In the classification step, class labels were assigned to the obtained segments based on generated feature values. The generated segments on the study area were classified as deciduous or coniferous trees using machine learning algorithms. Machine learning is a type of automation as a branch of artificial intelligence, and it works on the function and structure of algorithms by constructing a data-driven model for estimations from sample inputs [76]. SVM, RF, and MLP are the chosen machine learning classifiers used in this study. Information about the used machine learning classifiers and performance analyses is given in the following subsection.

2.7.1. Support Vector Machine

Support vector machine (SVM) was developed in 1995 by Vladimir Naumovich Vapnik [77]. SVM is a non-parametric supervised machine learning algorithm that performs the classification process based on the statistical learning concept with adaptive computational learning [78]. SVMs are very popular kernel-based statistical machine learning algorithms [79]. Kernel-based methods, such as support vector machines, are the main subject of a study on classification, clustering, and regression problems [80]. A hyperplane or a set of hyperplanes were constructed between groups or observation classes by support vector machines in an infinite dimensional space to separate the samples [81]. The hyperplane is the decision surface for maximizing the distance to the neighboring data points in the classes [82,83]. The data points nearest to the obtained hyperplane are called support vectors [84]. Detecting the best separation hyperplane with the highest margin distance between the nearest points of the two classes is the objective of the SVM approach [85]. The linear hyperplane is only sufficient for linearly separable data. If the data are not linearly separable, the support vector machine method can map the data onto a higher dimensional space where they are linearly separable with a kernel function, such as sigmoid, polynomial, normalization, radial basis function (RBF), and Laplacian radial basis function kernel [79,83].

In the present work, the radial basis function (RBF) kernel was used in SVM algorithm. The most important C and gamma (γ) parameters were analyzed manually in detail, and the determined C value was 100,000,000, while the gamma value was $5.092462164188164 \times 10^{-10}$.

2.7.2. Random Forest

Random forest (RF) [86] is a widespread, powerful, non-parametric machine learning algorithm based on the bagging principle of decision tree classifiers [21]. The RF algorithm provides reliable classifications with the estimations acquired from an ensemble of classification and regression trees (CARTs) [72]. The random vectors in RF are used to develop individual trees consisting of root nodes, internal nodes, branches, and leaf nodes in the forest. In its simplest form, RF requires the number of trees (n) to constitute the "forest" and the number of features (m) to be used in each node in the trees [87]. In random forest, each tree votes for the most popular class at each input instance, and the final classification is defined by the majority votes of the entire forest trees [87,88]. The RF commonly uses the Gini index as a splitting criterion to determine which attribute to split during the learning phase of the tree [89,90]. The impurity level of the samples assigned to a node is measured with the Gini index [89]. A bootstrap sample, which is two-thirds of the original data, also known as an "in-of bag" sample, is used in the training of trees, and the one-third remainder data called as "out-of bag" samples are used in estimating the classification error and determining the importance of the classification features [6,91].

RF provides a variable importance measure (VIM), which is a key advantage according to alternative machine learning algorithms [92]. The Mean Decrease Accuracy (MDA) and Mean Decrease Gini (MDG) are two different VIMs in RF to identify the most relevant features or perform a feature selection procedure [90,93]. While MDA is the average of the difference between two out-of-bag test errors, MDG assesses the difference between the Gini index before and after classification [94].

In this research, n_estimators (the number of decision trees in the forest) and max_features (the number of features considered for the best split) parameters in the RF algorithm were analyzed for accurate classification. A total of 150 trees and 10 features at each split were used as n_estimators and max features, respectively. We also present the feature importance measures with MDG.

2.7.3. Multi-Layer Perceptron

Today, one of the most popular research topics in the machine learning and artificial intelligence fields is artificial neural network (ANNs) [95]. Multi-layer perceptron (MLP), a specific form of ANN, consists of the connection of neurons (the process elements) with each other in a given order, and each neuron is connected to another neuron in the next layer with connections that are named as weights [95,96]. Each neuron in a multi-layer perceptron structure receives an input array, and then generates a single output. MLP has one input layer, one or more hidden layer, and one output layer [76].

The information flows from the input to the output layer unidirectionally through the hidden layers, as MLP is a feed-forward neural network [97]. Each layer has a different role in a multi-layer perceptron algorithm. The first (input) layer indicates the inputs of the problem, and the last (output) layer represents the outputs of the problem. The main computational core of the multi-layer perceptron algorithm is the hidden layers [98]. The MLP structure used in this study is shown in Figure 4. As can be seen from Figure 4, the proposed MLP approach has a hidden layer with 20 neurons, as well as an input layer with 25 neurons and an output layer with 2 neurons.

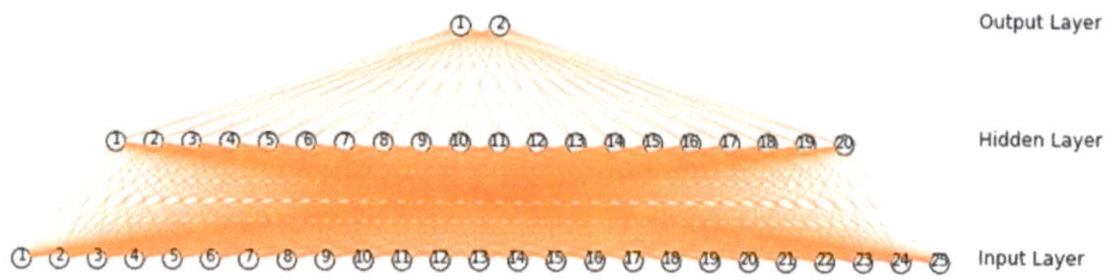

Figure 4. Structure of the proposed multi-layer perceptron algorithm.

2.8. Performance Analysis

Evaluation methods are needed to determine the classification success of a model [99]. The performance of the proposed machine learning-based classification model is evaluated by dividing the used dataset into training/testing samples. The accuracy of the classifications cannot be truly achieved, if the reference dataset is comparably small, and only a single split into training and test samples is used [18]. By applying iterative data-splitting, the cross-validation process allows us to validate the stability of the proposed classification technique [100]. In cross-validation, first the dataset is split into several different subsets, then a group is determined as a test set and the remaining groups are used as training sets, and the process is repeated for all possible training and test sets. Therefore, all the combinations are tested, and a performance value is acquired with the cross-validation by taking the average of each split result [101,102].

The classification performance of the used machine learning algorithms was quantitatively assessed with a statistical measurement of accuracy [73]. Equations (1)–(4) show the computed performance measures: the accuracy, recall, precision, and F1-score for tree species classification using a confusion matrix [103–105]. Accuracy indicates the ratio of correctly classified samples to all samples. Recall is used to assess the proportion of correctly predicted positive samples by the classification algorithm to the total number of samples that should be recognized as correct. Precision is used to estimate the ratio of correctly classified positive samples to the total predicted positive samples [106,107]. The F1-score is calculated with the harmonic mean of recall and precision measures [108]. The potential values of accuracy, recall, precision, and F1-score range from 0 to 1. The values closer to 1 describe a better classification performance, and the values closer to 0 express lower classification results [109]:

$$Accuracy = \frac{TP + TN}{TP + TN + FP + FN} \tag{1}$$

$$Recall = \frac{TP}{TP + FN} \tag{2}$$

$$Precision = \frac{TP}{TP + FP} \tag{3}$$

$$F1 - score = 2\frac{Precision \times Recall}{Precision + Recall} \tag{4}$$

TP, *TN*, *FP*, and *FN* in Equation (1) define the true positive, true negative, false positive, and false negative samples, respectively. *TP* refers to the entities classified correctly according to the ground-truth data, and *TN* defines the entities that were not acquired in the classification and also do not exist in the ground-truth data. *FP* represents the entities that have been obtained with the classification but do not exist in the ground-truth data, and *FN* refers to the entities defined as correct in the ground-truth data that were not acquired in the classification [2,110,111].

In this study, the machine learning-based classification models were formed with 70% training and 30% test samples. The classification results were interpreted using the accuracy, recall, precision, and F1-score values on the test dataset (30%). A 10-fold cross-validation was also used to evaluate the stability of the proposed machine learning-based SVM, RF, and MLP algorithms for tree species classification. All the segmentation processes of the high vegetation points, the urban tree species classification, and the performance analysis in this research were performed in the Jupyter Notebook environment with the Python programming language (Python 3.6.4), along with scikit-learn library.

3. Results and Discussion

As the first step in the proposed methodology, the high vegetation classification results acquired with a hierarchical rule-based classification of a three-dimensional raw LiDAR point cloud using spatial (geometric) features are given in Figure 5. When the point-based classification results were analyzed, it was clearly seen that the points of high vegetation were obtained successfully; therefore, almost all the tree points in the study area were accurately assigned to the high vegetation class.

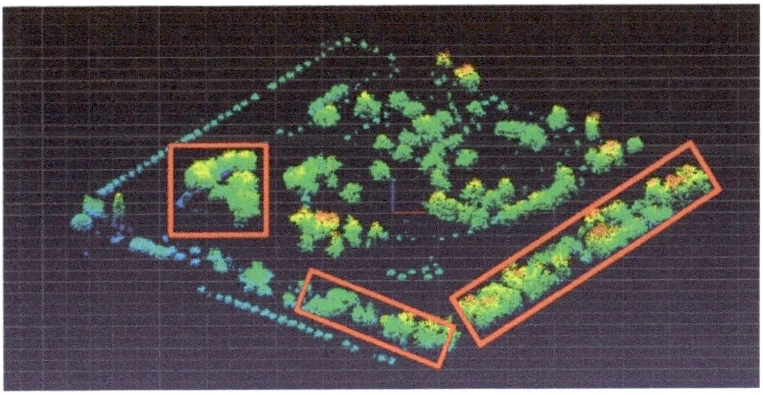

Figure 5. Classified high vegetation points in the study area (dense and mixed tree points in red rectangles).

Individual tree crown segmentation is the second step of the proposed machine learning-based urban tree species classification after the high vegetation classification process. A total of 346 tree crowns were obtained using mean shift segmentation (2D kernel with a proper bandwidth of 3) in the study area as the primary results of 2D point-based tree crown segmentation. Under- and over-segmented urban trees existed in the study area, especially in areas where the trees were dense and the tree crowns were mixed with each other (see the red rectangles in Figure 5). The segmentation accuracy of the tree crowns was 77% based on ground-truth data. As we were focused on individual trees, the under-segmented and over-segmented tree crowns were removed from the output of mean shift segmentation. As a result, 265 individual tree crowns were obtained for the classification process of tree species (Figure 6). The ground-truth data, which were required for training and testing the machine learning-based classification models, were created by manually labeling the obtained individual tree segments as deciduous or coniferous trees based on visual interpretation and field investigation (Figure 7).

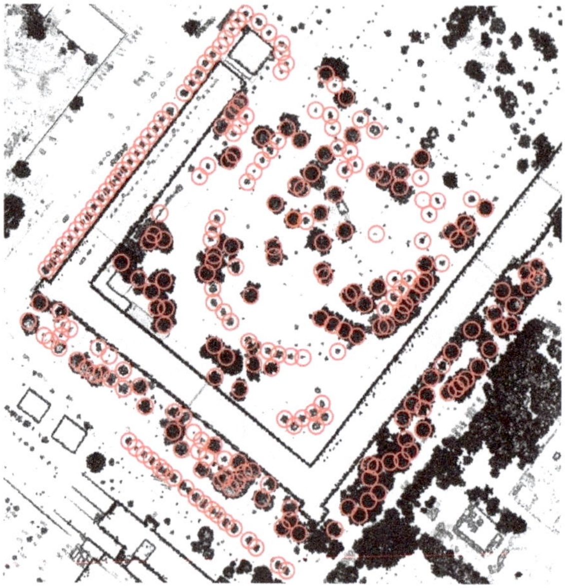

Figure 6. A total of 265 individual tree crowns (red circles) acquired with mean shift segmentation.

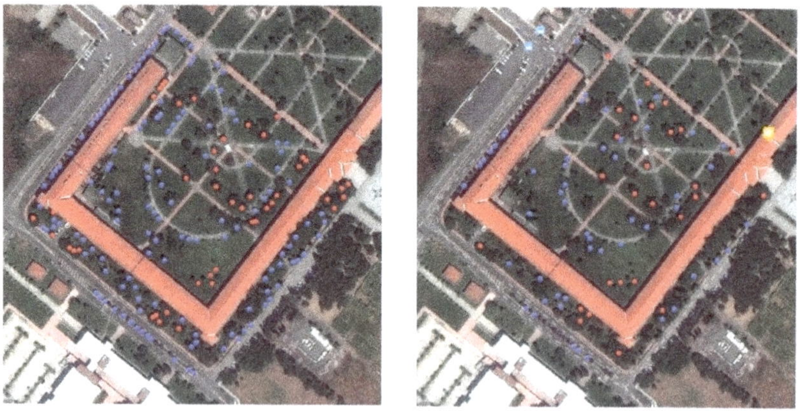

(a) (b)

Figure 7. Training and test samples in the study area: (**a**) training sample set; (**b**) test sample set.

The segmented individual trees were classified into deciduous or coniferous tree species based on the spatial- and intensity-related features (see Table 1) of the LiDAR data with support vector machine, random forest, and multi-layer perceptron machine learning classification methods using determined parameters and 25 generated features. In the 3D point-based classification process, 193 deciduous and 72 coniferous trees, in total 265 individual tree crowns, were used. Each classifier (SVM, RF, and MLP) was trained with the same randomly determined training sample set (185 trees, which was 70% of the total segmented 265 tree crowns), and the remaining test samples (80 trees, which was 30% of the total segmented 265 tree crowns) were used to validate the classification performance of the models (Figure 7).

The classification results for urban deciduous and coniferous trees using SVM, RF, and MLP classification models are shown in Figure 8. As mentioned earlier, the classification results were interpreted by means of accuracy, recall, precision, and F1-score on the test dataset (30%). The overall accuracies for the proposed machine learning-based classification of urban tree species with SVM, RF, and MLP classification models are given in Table 2. The urban tree species classification results obtained from our study area show that the machine learning methods were able to classify the urban tree species using the 3D raw airborne LiDAR data with sufficient accuracy. According to the obtained overall classification accuracies of urban tree species, the RF classifier had the best classification accuracy with 83.75%, while SVM had a 3.75% lower classification accuracy and MLP had a 10% lower classification accuracy than RF.

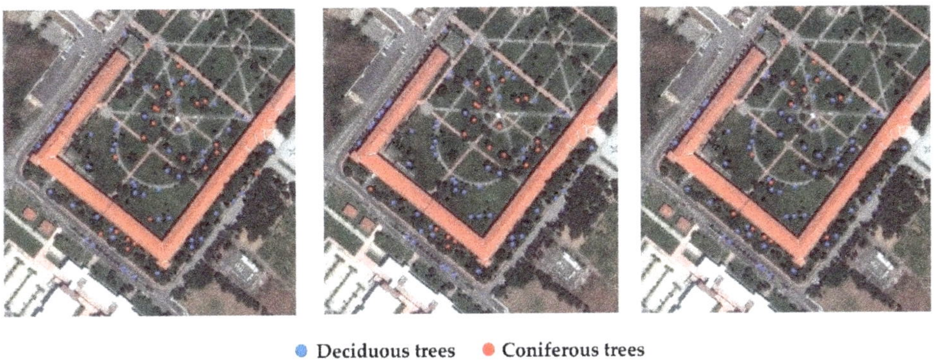

(a) (b) (c)

Figure 8. The classification results for urban deciduous and coniferous trees: (**a**) SVM; (**b**) RF; (**c**) MLP.

Table 2. Overall accuracy values of the proposed three classification algorithms.

Classifier	Overall Accuracy
SVM	80.00%
RF	83.75%
MLP	73.75%

The recall, precision, and F1-score values of each tree species (deciduous and coniferous trees) were also calculated and are shown in Figure 9. According to the recall values, RF was the best classification method for deciduous trees with a 0.943 recall value in the study area. The recall values of the SVM and MLP classifiers were also high, similar to the RF method, and were 0.887 and 0.925, respectively for deciduous trees. Lower recall values of the SVM, RF, and MLP algorithms were obtained for coniferous trees compared with deciduous trees in the study area. While the SVM and RF values were quite similar, the recall of the MLP algorithm was the worst with 0.370. Therefore, MLP is an insufficient method for the coniferous tree species. Similar precision values (0.825 and 0.833) were obtained for deciduous trees with the SVM and RF classifiers. MLP was the least successful classification method for deciduous trees according to the precision values (0.742). For the coniferous tree species, the RF classifier was the best method with 0.850 precision values. The values of 0.739 and 0.714 are the lower precision values for the coniferous tree species using the SVM and MLP methods, respectively, in the study area. Regarding the F1-scores, the coniferous tree species were classified worse than the deciduous tree species, and all the classification methods (SVM, RF, and MLP) were the successful for the deciduous tree species (with values of 0.855, 0.885, and 0.824). For the coniferous tree species, the MLP classifier was the worst classification method according to the 0.488 F1-score value. RF was the best classification method, and SVM was a relatively good algorithm for deciduous trees with regard to F1-scores (0.723 and 0.680, respectively). In general, the recall, precision, and F1-score values for the deciduous tree species were higher than for the coniferous tree species in the study area. Consequently, the SVM, RF, and MLP machine learning classifiers produced more successful classification results for the deciduous trees species compared with the coniferous tree species.

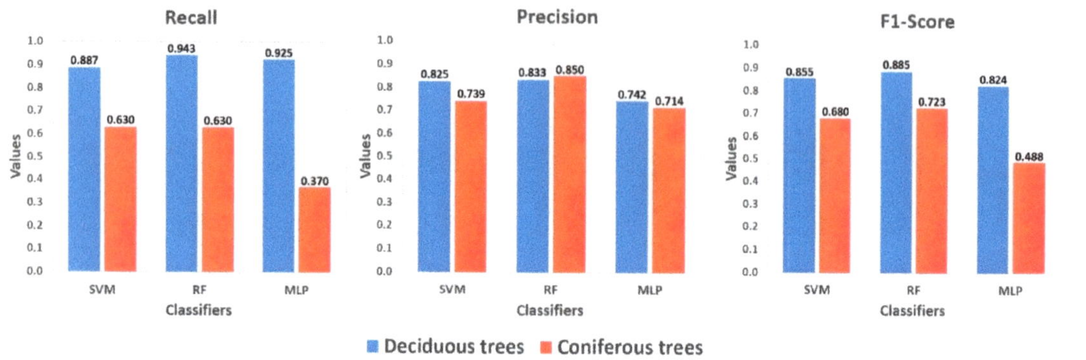

Figure 9. Recall, precision, and F1-score values of SVM, RF, and MLP classification algorithms for each deciduous and coniferous tree species.

In this study, the classification success of the SVM, RF, and MLP classification models were tested using cross-validation. A 10-fold cross-validation was performed for the classification models in order to obtain better confidence classification results, evaluate the stability of the models, and avoid overfitting, as can be seen in Table 3. The average RF classification accuracy was the highest (81.54%) with the 10-fold cross-validation, which was 0.44% higher than SVM and 17.82% higher than the average MLP classification accuracy (see Table 3). The MLP classifier had the worst average classification accuracy. While the accuracy of the two urban tree species classified with SVM and RF algorithms was similar, the accuracy acquired with the MLP algorithm was relatively low compared to these.

Table 3. The 10-fold cross-validation results of the proposed three classification methods.

Classifier	10-Fold Cross-Validation Average Accuracy
SVM	81.10%
RF	81.54%
MLP	63.72%

A feature importance process was also conducted in this study to analyze the impact of the used 25 spatial- and intensity-based features of the tree species classification with RF. The Mean Decrease Gini (MDG) was used to indicate the feature importance in the RF classification (Figure 10). While the most important classification feature among the 25 features is "90th percentile of Z", the least important classification feature was the "5th percentile of Z". "Minimum Z", "Number of points", "Z range", "75th percentile of Z", and "Standard deviation of Z" were the next five most important features following "90th percentile of Z". Generally, the spatial-based features had a higher importance than the intensity-based features for the proposed 3D point-based classification according to MDG in the RF classification.

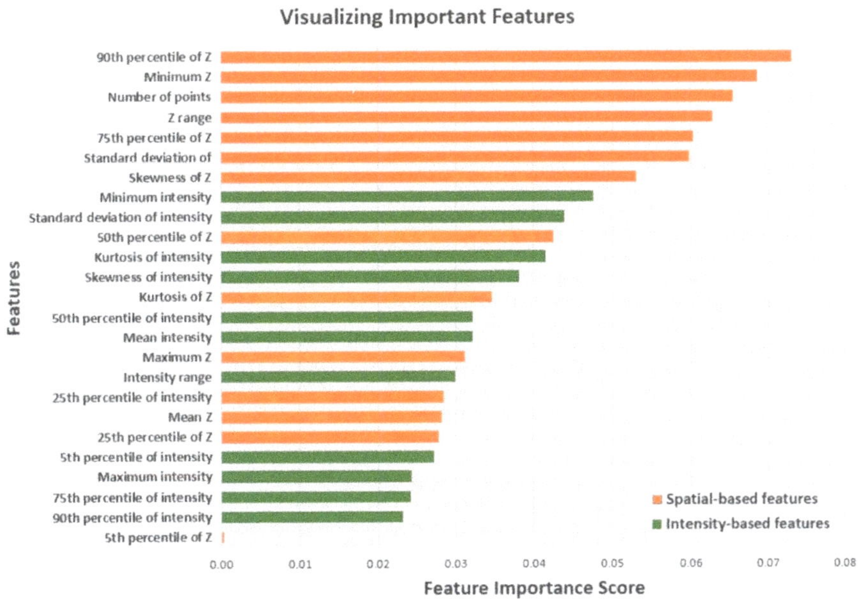

Figure 10. The feature importance scores according to MDG in the RF classification.

The three widely-used machine learning algorithms (SVM, RF, and MLP) were selected to assess their usage for urban tree species classification. The results of this study provide preliminary findings for deciduous and coniferous tree species classification using 3D raw LiDAR data in an urban area. When the performances of our machine learning-based classification algorithms were analyzed, it was noted that the support vector machine, and particularly the random forest classifier, discovered reliable solutions, even for the urban tree objects that showed similar geometrical and textural properties, whereas the multi-layer perceptron classifier solutions were not competitive (Tables 2 and 3). The success of the RF classifier against the SVM and MLP classifiers can be explained by the fact that the RF classifier is suitable for handling unbalanced samples and adds additional randomness to the classification model during the growing trees as well as searching for only the best features among a random subset of features in the splitting process of each node [112–115]. Compared with the SVM and RF classification models, the MLP models can have numerous weights for optimizing in each iteration [114,116,117]. In addition, the MLP needs more training data and more parameter tuning in the training stage [117]. Considering these reasons, the MLP classifier is less successful than the SVM and RF classifiers for the classification of urban tree species.

Most of the tree species classification studies in the literature have used a canopy height model (CHM) produced from LiDAR points, as well as features derived from full-waveform LiDAR data for deciduous and coniferous tree species [47,49,118,119]. In this study, commonly used spatial- and intensity-based features were computed from a traditional LiDAR dataset without using any reference terrain surface model or any CHM [11,75,120,121]. The features derived from full-waveform LiDAR data were not used since the full-waveform information was not available in our LiDAR dataset [21,64]. Based on deciduous and coniferous tree species classification, the maximum overall classification accuracy of 85% in Reitberger et al. [64], and the overall classification accuracy of 88% in Ørka et al. [31] are comparable with our maximum overall classification accuracy of 84% with random forest. However, the k-means and the expectation maximation (EM) classifier used in Reitberger et al. [64] and the linear discriminant analysis (LDA) classifier used in Ørka et al. [31] are different from our SVM, RF, and MLP machine learning-based classifiers. The full-waveform LiDAR features used in Reitberger et al. [64] and the structural and echo-based features used in Ørka et al. [31] were not included in our classification features. The study areas in Reitberger et al. [64] and in Ørka et al. [31] were selected as forest areas that were not comparable with our urban study area. Considering the obtained overall accuracy metrics in our urban study area, the proposed machine learning-based classification algorithms in tree species classification using the 3D raw LiDAR data were successful.

The number of training and test samples in the study area was relatively small in comparison with similar studies such as Yu et al. [121] and Nguyen et al. [122], but the obtained accuracies of the machine learning-based classifications are comparable. The classification results obtained from our study could serve as the basis for a pilot study in future deciduous and coniferous tree species classification studies using machine learning algorithms in large-scale urban applications.

4. Conclusions

In this paper, a machine learning-based 3D LiDAR point cloud classification algorithm was proposed to classify urban tree species as deciduous or coniferous using SVM, RF, and MLP. The experimental results indicate that SVM and RF classifications generally outperform MLP classification. The obtained results could be improved by extending the size of the used training and test samples, in addition to using full-waveform LiDAR to a produce larger number of spatial- and intensity-based features for discriminating deciduous and coniferous tree species in urban environments.

The classification results are encouraging with respect to the difficult study area, which included heterogeneous urban structures with dense trees in different sizes, ages, and species. This study offers insights for urban authorities regarding the potential use of machine learning algorithms for classifying deciduous (broadleaf) and coniferous tree species from 3D raw LiDAR data in urban environments using spatial- and intensity-based LiDAR features. Compared with traditional field surveys and aerial photograph interpretation methods for tree species classification, the proposed approach has essential benefits, such as automation, especially in terms of reducing the manpower and field study requirements. Many activities, such as the management, planning, and maintenance of urban trees, as well as the identification of endemic tree species, can be easily carried out in urban environments with the proposed machine learning-based 3D raw LiDAR point classification approach.

Author Contributions: Conceptualization, Zehra Cetin and Naci Yastikli; methodology, Zehra Cetin and Naci Yastikli; software, Zehra Cetin and Naci Yastikli; validation, Zehra Cetin and Naci Yastikli; formal analysis, Zehra Cetin and Naci Yastikli; investigation, Zehra Cetin and Naci Yastikli; resources, Zehra Cetin and Naci Yastikli; data curation, Zehra Cetin and Naci Yastikli; writing—original draft preparation, Zehra Cetin and Naci Yastikli; writing—review and editing, Zehra Cetin and Naci Yastikli; visualization, Zehra Cetin and Naci Yastikli; supervision, Naci Yastikli. All authors have read and agreed to the published version of the manuscript.

Funding: This research received no external funding.

Institutional Review Board Statement: Not applicable.

Informed Consent Statement: Not applicable.

Data Availability Statement: Not applicable.

Acknowledgments: The authors would like to thank the Metropolitan Municipality of Istanbul for supplying the LiDAR dataset in the study area of Davutpasa, Istanbul, Turkey.

Conflicts of Interest: The authors declare no conflict of interest.

References

1. Wang, K.; Wang, T.; Liu, X. A Review: Individual Tree Species Classification Using Integrated Airborne LiDAR and Optical Imagery with a Focus on the Urban Environment. *Forests* **2019**, *10*, 1. [CrossRef]
2. Yastikli, N.; Cetin, Z. Detection of Individual Trees in Urban Areas Using the Point Cloud Produced by Dense Image Matching Algorithms. In Proceedings of the FIG Working Week 2020, Amsterdam, The Netherlands, 10–14 May 2020.
3. Pu, R.; Landry, S. A comparative analysis of high spatial resolution IKONOS and WorldView-2 imagery for mapping urban tree species. *Remote Sens. Environ.* **2012**, *124*, 516–533. [CrossRef]
4. Ciesielski, M.; Sterenczak, K. Accuracy of determining specific parameters of the urban forest using remote sensing. *Iforest Biogeosci. For.* **2019**, *12*, 498–510. [CrossRef]
5. Feng, X.; Li, P. A Tree Species Mapping Method from UAV Images over Urban Area Using Similarity in Tree-Crown Object Histograms. *Remote Sens.* **2019**, *11*, 1982. [CrossRef]
6. Jombo, S.; Adam, E.; Byrne, M.J.; Newete, S.W. Evaluating the capability of Worldview-2 imagery for mapping alien tree species in a heterogeneous urban environment. *Cogent Soc. Sci.* **2020**, *6*, 1754146. [CrossRef]
7. Strunk, J.L.; Mills, J.R.; Ries, P.; Temesgen, H.; Jeroue, L. An urban forest-inventory-and-analysis investigation in Oregon and Washington. *Urban For. Urban Green.* **2016**, *18*, 100–109. [CrossRef]
8. Liu, L.; Coops, N.C.; Aven, N.W.; Pang, Y. Mapping urban tree species using integrated airborne hyperspectral and LiDAR remote sensing data. *Remote Sens. Environ.* **2017**, *200*, 170–182. [CrossRef]
9. Xu, J.; Cai, Z.; Wang, T.; Liu, G.; Tang, P.; Ye, X. Exploring Spatial Distribution of Pollen Allergenic Risk Zones in Urban China. *Sustainability* **2016**, *8*, 978. [CrossRef]
10. Dawe, G. *The Routledge Handbook of Urban Ecology*, 1st ed.; Douglas, I., Goode, D., Houck, M., Wang, R., Eds.; Routledge: New York, NY, USA, 2011; ISBN 9781138824423.
11. Roffey, M.; Wang, J. Evaluation of Features Derived from High-Resolution Multispectral Imagery and LiDAR Data for Object-Based Support Vector Machine Classification of Tree Species. *Can. J. Remote Sens.* **2020**, *46*, 473–488. [CrossRef]
12. Yastikli, N.; Cetin, Z. Detection of Individual Trees in Urban Areas Using Raw LiDAR Data. In Proceedings of the International Symposium on Applied Geoinformatics ISAG2019, Istanbul, Turkey, 7–9 November 2019.
13. Li, D.; Ke, Y.; Gong, H.; Li, X. Object-Based Urban Tree Species Classification Using Bi-Temporal WorldView-2 and WorldView-3 Images. *Remote Sens.* **2015**, *7*, 16917–16937. [CrossRef]
14. Höfle, B.; Hollaus, M.; Hagenauer, J. Urban vegetation detection using radiometrically calibrated small-footprint full-waveform airborne LiDAR data. *ISPRS J. Photogramm. Remote Sens.* **2012**, *67*, 134–147. [CrossRef]
15. Ardila, J.P. Object-Based Methods for Mapping and Monitoring of Urban Trees with Multitemporal Image Analyses. Ph.D. Thesis, University of Twente Faculty of Geo-Information and Earth Observation (ITC), Enschede, The Netherlands, 2012.
16. Alonzo, M.; Bookhagen, B.; Roberts, D.A. Urban tree species mapping using hyperspectral and lidar data fusion. *Remote Sens. Environ.* **2014**, *148*, 70–83. [CrossRef]
17. Yan, S.; Jing, L.; Wang, H. A New Individual Tree Species Recognition Method Based on a Convolutional Neural Network and High-Spatial Resolution Remote Sensing Imagery. *Remote Sens.* **2021**, *13*, 479. [CrossRef]
18. Fassnacht, F.E.; Latifi, H.; Stereńczak, K.; Modzelewska, A.; Lefsky, M.; Waser, L.T.; Straub, C.; Ghosh, A. Review of studies on tree species classification from remotely sensed data. *Remote Sens. Environ.* **2016**, *186*, 64–87. [CrossRef]
19. Mustafa, Y.; Habeeb, H.; Stein, A.; Sulaiman, F. Identification and Mapping of Tree Species in Urban Areas Using Worldview-2 Imagery. *ISPRS Ann. Photogramm. Remote Sens. Spat. Inf. Sci.* **2015**, *II-2/W2*, 175–181. [CrossRef]
20. Shojanoori, R.; Shafri, H.Z.M.; Mansor, S.; Ismail, M.H. The use of worldview-2 satellite data in urban tree species mapping by object-based image analysis technique. *Sains Malays.* **2016**, *45*, 1025–1034.
21. Shi, Y.; Skidmore, A.; Heurich, M. Important LiDAR metrics for discriminating forest tree species in Central Europe. *ISPRS J. Photogramm. Remote Sens.* **2018**, *137*, 163–174. [CrossRef]
22. Jombo, S.; Adam, E.; Odindi, J. Classification of tree species in a heterogeneous urban environment using object-based ensemble analysis and World View-2 satellite imagery. *Appl. Geomat.* **2021**, *13*, 373–387. [CrossRef]
23. Wang, J.; Lindenbergh, R.; Menenti, M. Scalable individual tree delineation in 3d point clouds. *Photogramm. Rec.* **2018**, *33*, 315–340. [CrossRef]
24. Man, Q.; Dong, P.; Yang, X.; Wu, Q.; Han, R. Automatic Extraction of Grasses and Individual Trees in Urban Areas Based on Airborne Hyperspectral and LiDAR Data. *Remote Sens.* **2020**, *12*, 2725. [CrossRef]
25. Pu, R. Broadleaf species recognition with in situ hyperspectral data. *Int. J Remote Sens.* **2009**, *30*, 2759–2779. [CrossRef]
26. Ghiyamat, A.; Shafri, H.Z.M. A review on hyperspectral remote sensing for homogeneous and heterogeneous forest biodiversity assessment. *Int. J. Remote Sens.* **2010**, *31*, 1837–1856. [CrossRef]
27. Immitzer, M.; Atzberger, C.; Koukal, T. Tree Species Classification with Random Forest Using Very High Spatial Resolution 8-Band WorldView-2 Satellite Data. *Remote Sens.* **2012**, *4*, 2661–2693. [CrossRef]
28. Wang, Y.; Wang, J.; Chang, S.; Sun, L.; An, L.; Chen, Y.; Xu, J. Classification of Street Tree Species Using UAV Tilt Photogrammetry. *Remote Sens.* **2021**, *13*, 216. [CrossRef]
29. Hartling, S.; Sagan, V.; Sidike, P.; Maimaitijiang, M.; Carron, J. Urban Tree Species Classification Using a WorldView-2/3 and LiDAR Data Fusion Approach and Deep Learning. *Sensors* **2019**, *19*, 1284. [CrossRef]

30. Moradi, A.; Satari, M.; Momeni, M. Individual Tree of Urban Forest Extraction from Very High Density LiDAR Data. *Int. Arch. Photogramm. Remote Sens. Spatial Inf. Sci.* **2016**, *XLI-B3*, 337–343. [CrossRef]
31. Ørka, H.O.; Næsset, E.; Bollandsås, O.M. Classifying species of individual trees by intensity and structure features derived from airborne laser scanner data. *Remote Sens. Environ.* **2009**, *113*, 1163–1174. [CrossRef]
32. Vaughn, N.R.; Moskal, L.M.; Turnblom, E.C. Fourier transformation of waveform LiDAR for species recognition. *Remote Sens. Lett.* **2011**, *2*, 347–356. [CrossRef]
33. Ko, C.; Remmel, T.K.; Sohn, G. Mapping tree genera using discrete LiDAR and geometric tree metrics. *Bosque* **2012**, *33*, 313–319. [CrossRef]
34. Vaughn, N.R.; Moskal, L.M.; Turnblom, E.C. Tree species detection accuracies using discrete point lidar and airborne waveform lidar. *Remote Sens.* **2012**, *4*, 377–403. [CrossRef]
35. Lindberg, E.; Eysn, L.; Hollaus, M.; Holmgren, J.; Pfeifer, N. Delineation of tree crowns and tree species classification from full-waveform airborne laser scanning data using 3-D ellipsoidal clustering. *IEEE J. Sel. Top. Appl. Earth Obs. Remote Sens.* **2014**, *7*, 3174–3181. [CrossRef]
36. Yu, X.; Litkey, P.; Hyyppä, J.; Holopainen, M.; Vastaranta, M. Assessment of low density full-waveform airborne laser scanning for individual tree detection and tree species classification. *Forest* **2014**, *5*, 1011–1031. [CrossRef]
37. Brandtberg, T. Classifying individual tree species under leaf-off and leaf-on conditions using airborne lidar. *ISPRS J. Photogramm. Remote Sens.* **2007**, *61*, 325–340. [CrossRef]
38. Heinzel, J.N.; Weinacker, H.; Koch, B. Full automatic detection of tree species based on delineated single tree crowns—A data fusion approach for airborne laser scanning data and aerial photographs. In *SilviLaser 2008, Proceedings of the 8th International Conference on LiDAR Applications in Forest Assessment and Inventory*; Hill, R.A., Rosette, J., Suárez, J., Eds.; SilviLaser: Edinburgh, UK, 2008; pp. 76–85.
39. Holmgren, J.; Persson, A.; Söderman, U. Species identification of individual trees by combining high resolution LiDAR data with multi-spectral images. *Int. J. Remote Sens.* **2008**, *29*, 1537–1552. [CrossRef]
40. Jones, T.G.; Coops, N.C.; Sharma, T. Assessing the utility of airborne hyperspectral and LiDAR data for species distribution mapping in the coastal Pacific Northwest, Canada. *Remote Sens. Environ.* **2010**, *114*, 2841–2852. [CrossRef]
41. Kim, S.; Hinckley, T.; Briggs, D. Classifying individual tree genera using stepwise cluster analysis based on height and intensity metrics derived from airborne laser scanner data. *Remote Sens. Environ.* **2011**, *115*, 3329–3342. [CrossRef]
42. Li, J. Individual Tree Delineation and Species Identification in Deciduous and Mixed Canadian Forests Using High Spatial Resolution Airborne Lidar And Image Data. Ph.D. Thesis, Graduate Program in Earth and Space Science York University, Toronto, ON, Canada, 2013.
43. Ghosh, A.; Fassnacht, F.E.; Joshi, P.K.; Koch, B. A framework for mapping tree species combining hyperspectral and LiDAR data: Role of selected classifiers and sensor across three spatial scales. *Int. J. Appl. Earth Obs. Geoinf.* **2014**, *26*, 49–63. [CrossRef]
44. Sommer, C.; Holzwarth, S.; Heiden, U.; Heurich, M.; Mueller, J.; Mauser, W. Feature-based treee species classification using airborne hyperspectral and lidar data in the Bavarian Forest National Park. *EARSeL Eproc.* **2015**, *14*, 49–70. [CrossRef]
45. Alonzo, M.; McFadden, J.P.; Nowak, D.J.; Roberts, D.A. Mapping urban forest structure and function using hyperspectral imagery and lidar data. *Urban For. Urban Green.* **2016**, *17*, 135–147. [CrossRef]
46. Sasaki, T.; Imanishi, J.; Ioki, K.; Morimoto, Y.; Kitada, K. Object-based classification of land cover and tree species by integrating airborne LiDAR and high spatial resolution imagery data. *Landsc. Ecol. Eng.* **2012**, *8*, 157–171. [CrossRef]
47. Zhang, Z.; Liu, X. Support Vector Machines for Tree Species Identification Using Lidar Derived Structure and Intensity Variables. *Geocarto Int.* **2013**, *28*, 364–378. [CrossRef]
48. Schumacher, J.; Nord-Larsen, T. Wall-to-Wall Tree Type Classification Using Airborne Lidar Data and CIR Images. *Int. J. Remote Sens.* **2014**, *35*, 3057–3073. [CrossRef]
49. Hovi, A.; Korhonen, L.; Vauhkonen, J.; Korpela, I. LiDAR waveform features for tree species classification and their sensitivity to tree and acquisition related parameters. *Remote Sens. Environ.* **2016**, *173*, 224–237. [CrossRef]
50. Kim, S.; Schreuder, G.; McGaughey, R.J.; Andersen, H.-E. Individual tree species identification using LIDAR intensity data. In Proceedings of the ASPRS 2008 Annual Conference, Portland, OR, USA, 28 April–2 May 2008; pp. 382–393.
51. Cho, M.A.; Mathieu, R.; Asner, G.P.; Naidoo, L.; Van Aardt, J.A.; Ramoelo, A.; Debba, P.; Wessels, K.; Main, R.; Smit, I.P.; et al. Mapping tree species composition in South African savannas using an integrated airborne spectral and LiDAR system. *Remote Sens. Environ.* **2012**, *125*, 214–226. [CrossRef]
52. Pati, C.; Panda, K.A.; Tripathy, A.K.; Pradhan, K.S.; Patnaik, S. A novel hybrid machine learning approach for change detection in remote sensing images. *Eng. Sci. Technol. Int. J.* **2020**, *23*, 973–981. [CrossRef]
53. McCarthy, N.F.; Tohidi, A.; Valero, M.M.; Dennie, M.; Aziz, Y.; Hu, N. A Machine Learning Solution for Operational Remote Sensing of Active Wildfires. In Proceedings of the IGARSS 2020-2020 IEEE International Geoscience and Remote Sensing Symposium, Waikoloa, HI, USA, 26 September–2 October 2020; pp. 6802–6805. [CrossRef]
54. Thepade, S.D.; Dindorkar, M.Y. Fusing deep convolutional neural network features with Thepade's SBTC for land usage identification. *Eng. Sci. Technol. Int. J.* **2022**, *27*, 101014. [CrossRef]
55. Senchuri, R.; Kuras, A.; Burud, I. Machine Learning Methods for Road Edge Detection on Fused Airborne Hyperspectral and LIDAR Data. In Proceedings of the 11th Workshop on Hyperspectral Imaging and Signal Processing: Evolution in Remote Sensing (WHISPERS), Amsterdam, The Netherlands, 24–26 March 2021; pp. 1–5. [CrossRef]

56. Kamali Maskooni, E.; Naghibi, S.A.; Hashemi, H.; Berndtsson, R. Application of Advanced Machine Learning Algorithms to Assess Groundwater Potential Using Remote Sensing-Derived Data. *Remote Sens.* **2020**, *12*, 2742. [CrossRef]
57. Temitope Yekeen, S.; Balogun, A.-L. Advances in Remote Sensing Technology, Machine Learning and Deep Learning for Marine Oil Spill Detection, Prediction and Vulnerability Assessment. *Remote Sens.* **2020**, *12*, 3416. [CrossRef]
58. Ali, I.; Greifeneder, F.; Stamenkovic, J.; Neumann, M.; Notarnicola, C. Review of Machine Learning Approaches for Biomass and Soil Moisture Retrievals from Remote Sensing Data. *Remote Sens.* **2015**, *7*, 16398–16421. [CrossRef]
59. Cao, L.; Coops, N.; Innes, J.L.; Dai, J.; Ruan, H.; She, G. Tree species classification in subtropical forests using small-footprint full-waveform LiDAR data. *Int. J. Appl. Earth Obs.* **2016**, *49*, 39–51. [CrossRef]
60. Koma, Z.; Koenig, K.; Höfle, B. Urban Tree Classification Using Full-Waveform Airborne Laser Scanning. *ISPRS Ann. Photogramm. Remote Sens. Spat. Inf. Sci.* **2016**, *III-3*, 185–192. [CrossRef]
61. Dian, Y.; Pang, Y.; Dong, Y.; Li, Z. Urban Tree Species Mapping Using Airborne LiDAR and Hyperspectral Data. *J. Indian Soc. Remote Sens.* **2016**, *44*, 595–603. [CrossRef]
62. Shen, X.; Cao, L. Tree-Species Classification in Subtropical Forests Using Airborne Hyperspectral and LiDAR Data. *Remote Sens.* **2017**, *9*, 1180. [CrossRef]
63. Kim, S.; McGaughey, R.J.; Andersen, H.E.; Schreuder, G. Tree species differentiation using intensity data derived from leaf-on and leaf-off airborne laser scanner data. *Remote Sens. Environ.* **2009**, *113*, 1575–1586. [CrossRef]
64. Reitberger, J.; Krzystek, P.; Stilla, U. Analysis of full waveform LIDAR data for the classification of deciduous and coniferous trees. *Int. J. Remote Sens.* **2008**, *29*, 1407–1431. [CrossRef]
65. *LAS Specification 1.4–R15*; The American Society for Photogrammetry & Remote Sensing: Bethesda, ML, USA, 2019.
66. Mehta, A.; Dikshit, O.; Venkataramani, K. Integration of high-resolution imagery and LiDAR data for object-based classification of urban area. *Geocarto Int.* **2014**, *29*, 418–432. [CrossRef]
67. Gevaert, C.M.; Persello, C.; Nex, F.C.; Vosselman, G. A Deep Learning Approach to DTM Extraction from Imagery Using Rule-Based Training Labels. *ISPRS J. Photogramm. Remote Sens.* **2018**, *142*, 106–123. [CrossRef]
68. Yastikli, N.; Cetin, Z. Classification of raw LiDAR point cloud using point-based methods with spatial features for 3D building reconstruction. *Arab. J. Geosci.* **2021**, *14*, 146. [CrossRef]
69. Fukunaga, K.; Hostetler, L. The estimation of the gradient of a density function, with applications in pattern recognition. *IEEE Trans. Inf. Theory* **1975**, *21*, 32–40. [CrossRef]
70. Wen, Z.Q.; Cai, Z.X. Mean shift algorithm and its application in tracking of objects. In Proceedings of the 5th International Conference on Machine Learning and Cybernetics, Dalian, China, 13–16 August 2006; pp. 4024–4028. [CrossRef]
71. Chen, W.; Hu, X.; Chen, W.; Hong, Y.; Yang, M. Airborne LiDAR Remote Sensing for Individual Tree Forest Inventory Using Trunk Detection-Aided Mean Shift Clustering Techniques. *Remote Sens.* **2018**, *10*, 1078. [CrossRef]
72. Le Louarn, M.; Clergeau, P.; Briche, E.; Deschamps-Cottin, M. "Kill Two Birds with One Stone": Urban Tree Species Classification Using Bi-Temporal Pléiades Images to Study Nesting Preferences of an Invasive Bird. *Remote Sens.* **2017**, *9*, 916. [CrossRef]
73. Aydin, F.; Aslan, Z. Recognizing Parkinson's disease gait patterns by vibes algorithm and Hilbert-Huang transform. *Eng. Sci. Technol. Int. J.* **2021**, *24*, 112–125. [CrossRef]
74. Chi, D.; Degerickx, J.; Yu, K.; Somers, B. Urban Tree Health Classification Across Tree Species by Combining Airborne Laser Scanning and Imaging Spectroscopy. *Remote Sens.* **2020**, *12*, 2435. [CrossRef]
75. Michałowska, M.; Rapiński, J. A Review of Tree Species Classification Based on Airborne LiDAR Data and Applied Classifiers. *Remote Sens.* **2021**, *13*, 353. [CrossRef]
76. Kececi, A.; Yildirak, A.; Ozyazici, K.; Ayluctarhan, G.; Agbulut, O.; Zincir, I. Implementation of machine learning algorithms for gait recognition. *Eng. Sci. Technol. Int. J.* **2020**, *23*, 931–937. [CrossRef]
77. Vapnik, V.N. *The Nature of Statistical Learning Theory*; Springer: New York, NY, USA, 1995; ISBN 978-1-4757-3264-1.
78. Ray, P.; Mishra, D.P. Support vector machine based fault classification and location of a long transmission line. *Eng. Sci. Technol. Int. J.* **2016**, *19*, 1368–1380. [CrossRef]
79. Mallet, C.; Bretar, F.; Roux, M.; Soergel, U.; Heipke, C. Relevance Assessment of Full-Waveform Lidar Data for Urban Area Classification. *ISPRS J. Photogramm. Remote Sens.* **2011**, *66*, S71–S84. [CrossRef]
80. Lodha, S.K.; Kreps, E.J.; Helmbold, D.P.; Fitzpatirck, D. Aerial lidar data classification using support vector machines (SVM). In Proceedings of the Third International Symposium on 3D Data Processing, Visualization and Transmission, Chapel Hill, NC, USA, 14–16 June 2006. [CrossRef]
81. Nicolas, P.R. *Scala for Machine Learning*; Packt Publishing Ltd.: Birmingham, UK; ProQuest Ebook Central: Morrisville, NC, USA, 2014; ISBN 9781787122383.
82. Petropoulos, G.P.; Arvanitis, K.; Sigrimis, N. Hyperion hyperspectral imagery analysis combined with machine learning classifiers for land use/cover mapping. *Expert Syst. Appl.* **2012**, *39*, 3800–3809. [CrossRef]
83. Liu, H.J.; Wu, C.S. Crown-level tree species classification from AISA hyperspectral imagery using an innovative pixel-weighting approach. *Int. J. Appl. Earth Obs. Geoinf.* **2018**, *68*, 298–307. [CrossRef]
84. Awad, M.; Khanna, R. *Efficient Learning Machines: Theories, Concepts, and Applications for Engineers and System Designers*; Apress: Berkeley, CA, USA, 2015; ISBN 978-1-4302-5989-3.
85. Thome, A.C.G. SVM Classifiers—Concepts and Applications to Character Recognition. In *Advances in Character Recognition*; IntechOpen Book Series; Ding, X., Ed.; InTech: Rijeka, Croatia, 2012. [CrossRef]

86. Breiman, L. Random Forests. *Mach. Learn.* **2001**, *45*, 5–32. [CrossRef]
87. Sothe, C.; De Almeida, C.M.; Schimalski, M.B.; La Rosa, L.E.C.; Castro, J.D.B.; Feitosa, R.Q.; Dalponte, M.; Lima, C.L.; Liesenberg, V.; Miyoshi, G.T.; et al. Comparative performance of convolutional neural network, weighted and conventional support vector machine and random forest for classifying tree species using hyperspectral and photogrammetric data. *GISci. Remote Sens.* **2020**, *57*, 369–394. [CrossRef]
88. Chehata, N.; Guo, L.; Mallet, C. Airborne LiDAR feature selection for urban classification using random forests. *Int. Arch. Photogramm. Remote Sens. Spat. Inf. Sci.* **2009**, *39*, 207–212.
89. Qi, Y. Random Forest for Bioinformatics. In *Ensemble Machine Learning;* Zhang, C., Ma, Y., Eds.; Springer: New York, NY, USA, 2012; pp. 307–323. [CrossRef]
90. Nembrini, S.; König, I.R.; Wright, M.N. The revival of the Gini importance? *Bioinformatics* **2018**, *34*, 3711–3718. [CrossRef] [PubMed]
91. Belgiu, M.; Dragut, L. Random Forest in Remote Sensing: A Review of Applications and Future Directions. *ISPRS J. Photogramm. Remote Sens.* **2016**, *114*, 24–31. [CrossRef]
92. Han, H.; Guo, X.; Yu, H. Variable selection using Mean Decrease Accuracy and Mean Decrease Gini based on Random Forest. In Proceedings of the 7th IEEE International Conference on Software Engineering and Service Science (ICSESS), Beijing, China, 26–28 August 2016; pp. 219–224. [CrossRef]
93. Wang, H.; Yang, F.; Luo, Z. An experimental study of the intrinsic stability of random forest variable importance measures. *BMC Bioinform.* **2016**, *17*, 60. [CrossRef]
94. García Moreno, A.I.; Alvarado Orozco, J.M.; Ibarra-Medina, J.R.; Martinez Franco, E. Image-based porosity classification in Al-alloys by laser metal deposition using random forests. *Int. J. Adv. Manuf. Technol.* **2020**, *110*, 2827–2845. [CrossRef]
95. Turkoglu, B.; Kaya, E. Training multi-layer perceptron with artificial algae algorithm. *Eng. Sci. Technol. Int. J.* **2020**, *23*, 1342–1350. [CrossRef]
96. Shoaib, M.; Shamseldin, A.Y.; Melville, B.W. Comparative study of different wavelet based neural network models for rainfall–runoff modeling. *J. Hydrol.* **2014**, *515*, 47–58. [CrossRef]
97. Lek, S.; Guégan, J.-F. Artificial neural networks as a tool in ecological modelling, an introduction. *Ecol. Model.* **1999**, *120*, 65–73. [CrossRef]
98. Nezami, S.; Khoramshahi, E.; Nevalainen, O.; Pölönen, I.; Honkavaara, E. Tree Species Classification of Drone Hyperspectral and RGB Imagery with Deep Learning Convolutional Neural Networks. *Remote Sens.* **2020**, *12*, 1070. [CrossRef]
99. Topic, A.; Russo, M. Emotion recognition based on EEG feature maps through deep learning network. *Eng. Sci. Technol. Int. J.* **2021**, *24*, 1442–1454. [CrossRef]
100. Yibre, A.M.; Kocer, B. Semen quality predictive model using Feed Forwarded Neural Networktrained by Learning-Based Artificial Algae Algorithm. *Eng. Sci. Technol. Int. J.* **2021**, *24*, 310–318. [CrossRef]
101. Karagul Yildiz, T.; Yurtay, N.; Onec, B. Classifying anemia types using artificial learning methods. *Eng. Sci. Technol. Int. J.* **2021**, *24*, 50–70. [CrossRef]
102. Zhang, Z.; Kazakova, A.; Moskal, L.M.; Styers, D.M. Object-Based Tree Species Classification in Urban Ecosystems Using LiDAR and Hyperspectral Data. *Forests* **2016**, *7*, 122. [CrossRef]
103. Congalton, R.G.; Green, K. *Assessing the Accuracy of Remotely Sensed Data: Principles and Practises*, 2nd ed.; CRC Press Taylor and Francis Group: New York, NY, USA, 2009. [CrossRef]
104. Ni, H.; Lin, X.; Zhang, J. Classification of ALS Point Cloud with Improved Point Cloud Segmentation and Random Forests. *Remote Sens.* **2017**, *9*, 288. [CrossRef]
105. Atik, M.E.; Duran, Z.; Seker, D.Z. Machine Learning-Based Supervised Classification of Point Clouds Using Multiscale Geometric Features. *ISPRS Int. J. Geo Inf.* **2021**, *10*, 187. [CrossRef]
106. Vo, A.-V.; Truong-Hong, L.; Laefer, D.F.; Bertolotto, M. Octree-based region growing for point cloud segmentation. *ISPRS J. Photogramm. Remote Sens.* **2015**, *104*, 88–100. [CrossRef]
107. Pan, Y.; Dong, Y.; Wang, D.; Chen, A.; Ye, Z. Three-Dimensional Reconstruction of Structural Surface Model of Heritage Bridges Using UAV-Based Photogrammetric Point Clouds. *Remote Sens.* **2019**, *11*, 1204. [CrossRef]
108. Dai, C.; Zhang, Z.; Lin, D. An Object-Based Bidirectional Method for Integrated Building Extraction and Change Detection between Multimodal Point Clouds. *Remote Sens.* **2020**, *12*, 1680. [CrossRef]
109. Yancho, M.; Coops, N.; Tompalski, P.; Goodbody, T.; Plowright, A. Fine-Scale Spatial and Spectral Clustering of UAV-Acquired Digital Aerial Photogrammetric (DAP) Point Clouds for Individual Tree Crown Detection and Segmentation. *IEEE J. Sel. Top. Appl. Earth Obs. Remote Sens.* **2019**, *12*, 4131–4148. [CrossRef]
110. Rutzinger, M.; Rottensteiner, F.; Pfeifer, N. A comparison of evaluation techniques for building extraction from airborne laser scanning. *IEEE J. Sel. Top. Appl. Earth Obs. Remote Sens.* **2009**, *1*, 11–20. [CrossRef]
111. Uzar, M.; Yastikli, N. Automatic building extraction using LiDAR and aerial photographs. *Bol. Ciênc. Geod.* **2013**, *19*, 153–171. [CrossRef]
112. Díaz-Uriarte, R.; Alvarez de Andrés, S. Gene selection and classification of microarray data using random forest. *BMC Bioinform.* **2006**, *7*, 3–15. [CrossRef] [PubMed]
113. Azar, A.T.; Elshazly, H.I.; Hassanien, A.E.; Elkorany, A.M. A random forest classifier for lymph diseases. *Comput. Methods Programs Biomed.* **2014**, *113*, 465–473. [CrossRef]

114. You, H.; Ma, Z.; Tang, Y.; Wang, Y.; Yan, J.; Ni, M.; Cen, K.; Huang, Q. Comparison of ANN (MLP), ANFIS, SVM, and RF models for the online classification of heating value of burning municipal solid waste in circulating fluidized bed incinerators. *Waste Manag.* **2017**, *68*, 186–197. [CrossRef]
115. Indira, B.; Valarmathi, K. A perspective of the machine learning approach for the packet classification in the software defined network. *Intell. Autom. Soft Comput.* **2020**, *26*, 795–805. [CrossRef]
116. Zhang, J.; Zhang, J.; Lok, T.-M.; Lyu, M.R. A hybrid particle swarm optimization-back-propagation algorithm for feedforward neural network training. *Appl. Soft Comput.* **2007**, *185*, 1026–1037. [CrossRef]
117. Jozdani, S.E.; Johnson, B.A.; Chen, D. Comparing Deep Neural Networks, Ensemble Classifiers, and Support Vector Machine Algorithms for Object-Based Urban Land Use/Land Cover Classification. *Remote Sens.* **2019**, *11*, 1713. [CrossRef]
118. Li, J.; Hu, B. Exploring high-density airborne light detection and ranging data for classification of mature coniferous and deciduous trees in complex Canadian forests. *J. Appl. Remote Sens.* **2012**, *6*, 063536. [CrossRef]
119. Shi, Y.; Skidmore, A.; Holzwarth, S.; Heiden, U.; Pinnel, N.; Zhu, X.; Heurich, M. Tree species classification using plant functional traits from LiDAR and hyperspectral data. *Int. J. Appl. Earth Obs. Geoinf.* **2018**, *73*, 207–219. [CrossRef]
120. Lin, Y.; Hyyppä, J. A comprehensive but efficient framework of proposing and validating feature parameters from airborne LiDAR data for tree species classification. *Int. J. Appl. Earth Obs. Geoinf.* **2016**, *46*, 45–55. [CrossRef]
121. Yu, X.; Hyyppä, J.; Litkey, P.; Kaartinen, H.; Vastaranta, M.; Holopainen, M. Single-sensor solution to tree species classification using multispectral airborne laser scanning. *Remote Sens.* **2017**, *9*, 108. [CrossRef]
122. Nguyen, H.M.; Demir, B.; Dalponte, M. Weighted Support Vector Machines for Tree Species Classification Using Lidar Data. In Proceedings of the IGARSS 2019—2019 IEEE International Geoscience and Remote Sensing Symposium, Yokohama, Japan, 28 July–2 August 2019; pp. 6740–6743. [CrossRef]

International Journal of
Geo-Information

Article

Multi-Resolution Transformer Network for Building and Road Segmentation of Remote Sensing Image

Zhongyu Sun [1,2], Wangping Zhou [1,2,*], Chen Ding [2] and Min Xia [1,2]

1 Jiangsu Key Laboratory of Big Data Analysis Technology, Nanjing University of Information Science and Technology, Nanjing 210044, China; 201913360009@nuist.edu.cn (Z.S.); xiamin@nuist.edu.cn (M.X.)
2 Jiangsu Collaborative Innovation Center on Atmospheric Environment and Equipment Technology, Nanjing University of Information Science and Technology, Nanjing 210044, China; 20181223015@nuist.edu.cn
* Correspondence: wpzhou@nuist.edu.cn

Abstract: Extracting buildings and roads from remote sensing images is very important in the area of land cover monitoring, which is of great help to urban planning. Currently, a deep learning method is used by the majority of building and road extraction algorithms. However, for existing semantic segmentation, it has a limitation on the receptive field of high-resolution remote sensing images, which means that it can not show the long-distance scene well during pixel classification, and the image features is compressed during down-sampling, meaning that the detailed information is lost. In order to address these issues, Hybrid Multi-resolution and Transformer semantic extraction Network (HMRT) is proposed in this paper, by which a global receptive field for each pixel can be provided, a small receptive field of convolutional neural networks (CNN) can be overcome, and the ability of scene understanding can be enhanced well. Firstly, we blend the features by branches of different resolutions to keep the high-resolution and multi-resolution during down-sampling and fully retain feature information. Secondly, we introduce the Transformer sequence feature extraction network and use encoding and decoding to realize that each pixel has the global receptive field. The recall, F1, OA and MIoU of HMPR obtain 85.32%, 84.88%, 85.99% and 74.19%, respectively, in the main experiment and reach 91.29%, 90.41%, 91.32% and 84.00%, respectively, in the generalization experiment, which prove that the method proposed is better than existing methods.

Keywords: segmentation; high resolution; transformer; deep learning

Citation: Sun, Z.; Zhou, W.; Ding, C.; Xia, M. Multi-Resolution Transformer Network for Building and Road Segmentation of Remote Sensing Image . *ISPRS Int. J. Geo-Inf.* **2022**, *11*, 165. https://doi.org/10.3390/ijgi11030165

Academic Editors: Gloria Bordogna, Cristiano Fugazza and Wolfgang Kainz

Received: 9 January 2022
Accepted: 23 February 2022
Published: 25 February 2022

1. Introduction

Land resources have the following qualities as carriers of human existence and development: nonrenewable resources, fixed location, and imbalanced distribution [1]. With the population and economy expanding at such a rapid pace, the amount of available land resources is gradually diminishing. In modern society, buildings and roads are the basic components of urban layout, and accurately extracting buildings and roads from remote sensing satellite images helps to realize the macro-planning of the city [2]. The research methods for remote sensing images can be divided into two parts: traditional theoretical calculation methods and artificial intelligence big data analysis methods. The traditional theoretical calculation method is to extract image texture features through theoretical calculation of each pixel of the image, so as to realize remote sensing image segmentation and target extraction. Although the traditional method has reached a certain standard in terms of segmentation accuracy, it needs to manually set the calculation parameters, which consumes human resources and material resources and lacks in calculation efficiency. On the contrary, the deep learning method in artificial intelligence can complete the end-to-end remote sensing image segmentation [3,4] and realize the high-accuracy automatic image segmentation function without manual intervention [5,6]. The dividends brought by the big data era have made a qualitative leap in computing efficiency, which results in that the computing efficiency has been greatly improved compared to traditional methods. Therefore, it is of great importance to use the deep learning method in artificial intelligence to achieve semantic segmentation of remote sensing images.

In the past few years, there have been many works on the semantic segmentation of remote sensing images. For example, Yuan et al. [7] calculated the texture features and spectrum of the image using local spectral histograms, linearly combined representative features using each local spectral histograms, classified the pixel by weight estimation, and finally realized the segmentation of images. This method could greatly reduce the feature dimension of the network through subspace projection, and realize that the input dimension of the network could be selected adaptively. However, the disadvantage is that only spectral information is used in the calculation process. Li et al. [8] proposed a watershed algorithm for edge embedding markers, which was used for the segmentation of high-resolution remote sensing images. The method results in improvements in the two key steps of segmentation (one is label extraction and the other one is pixel labeling), which could improve the accuracy of edge segmentation in high-resolution images. Furthermore, this method used an edge embedding detector to extract edge information with confidence, which was usually used in situations with weak boundaries and improved the positioning accuracy of target boundaries. Although the accuracy of the segmentation boundary had been improved, there are also problems that the detailed feature information is complex and the interference factors make it difficult to obtain information. Fan et al. [9] found that these remote sensing segmentation methods rarely use prior information. As a result, he proposed a new approach based on prior information. A single-point iterative weighted fuzzy C-means clustering algorithm was used in this method, which solved the data distribution and the effect of random initialization of cluster centers on the quality of clustering. The above feature segmentation method can divide remote sensing images effectively, but there are also problems that exist, such as weak noise resistance, slow speed of segmentation, manual parameter design, etc., which cannot be used for the task of automatically segmenting large amounts of data.

Current deep learning is still under development in the area of building and road extractions. Panboonyuen et al. [10] proposed an enhanced deep convolutional encoding and decoding network for road segmentation of remote sensing images, combining ELU activation function [11] and SegNet network [12] to form an end-to-end segmentation network, and finally through optimizing indicators and removing false road objects to further improve the overall effect. However, for this method, fewer applications of continuous feature information are used when extracting roads from remote sensing images, which led to the interference map and fracture area. Aiming at the problem of loss of detailed information during the down-sampling process, Sun et al. [13] offered a new feature fusion strategy based on a full convolutional network with ultra-high resolution image segmentation, which maximized the fusion of deep-level semantic features and shallow-level detail information. Combining with this model, the effective digital surface model was proposed, and the information of high-resolution remote sensing images was extracted, which improved the accurate segmentation of the full convolutional network. However, in remote sensing image segmentation, there were problems that the scale of the target segmentation was inconsistent and the scale of information had not been mined. In order to solve the problem, Liu et al. [14] proposed a multi-channel deep convolutional neural network to alleviate the loss of spatial and scale features of segmented targets in images. Qi et al. [15] proposed a multi-scale convolution and attention mechanism based on a segmentation model. The attention mechanism, on the other hand, could only capture the local receptive field. Li et al. [16] proposed to use a two-way attention mechanism network for semantic segmentation of remote sensing images. One is focused on the spatial semantic information in the feature map, and the other is on the associated information between channels. Combining the two-way attention information could effectively improve the accuracy of segmentation. Lan et al. [17] proposed a global context road automatic segmentation neural network for road segmentation under complex background and field of view occlusion. In the network, a residual cavity convolutional network was used to provide a wide receptive field. Although with the multi-scale information of a larger receptive field, for the network, the relevance of the middle layer could not be ignored. He et al. [18] proposed a hybrid first-order and second-order attention network to enhance the relevance of feature information in the middle of the network.

In summary, for the above remote sensing image semantic segmentation methods [7–18] in deep learning, satisfactory results have been achieved. At present, in most semantic segmentation networks, down-sampling of the convolutional neural network (CNN) is used to extract features. Feature maps are compressed many times during extraction, which results in the loss of details. Up-sampling by feature maps with missing detailed features makes it difficult to restore feature maps with high-resolution and classify resolution accurately. In the process of feature extraction by deep convolutional networks, the receptive field is limited. Although a larger receptive field can be obtained by the use of hollow convolution and feature pyramids, the receptive field is still local, understanding of long-distance scenes cannot be achieved, and pixels cannot be classified precisely. To solve these problems, a Hybrid Multi-resolution and Transformer semantic extraction Network (HMRT) is proposed in this study. In general, this work has made three contributions: (1) The multi-resolution semantic extraction branch is constructed. In this structure, branches of different resolutions conduct feature fusion, which not only ensures that high and multi-resolution are kept during the down-sampling process but also ensures that feature information is retained. (2) The Transformer sequence feature extraction network is introduced. In this network, each pixel with a global receptive field is realized by the use of encoding and decoding, and in the meantime, the location information of the pixel is overlain. The small receptive field of the convolutional neural network can be overcome and the understanding of a long-distance scene can be improved. (3) Feature Channels Maximum Element is proposed to strengthen the class location information, which can effectively improve the accuracy of segmentation.

2. Methodology

In order to solve the two problems of the loss of details caused by scale compression in the down-sampling process and the lack of long-distance understanding due to the limitation of the receptive field, this paper proposes a Hybrid Multi-resolution and Transformer semantic extraction Network (HMRT). The overall framework of the HMRT is shown in Figure 1. The overall framework of the HMRT proposed in this work is divided into two parallel branches. The first branch provides the network with different resolution feature maps. The feature maps of different resolutions are divided into 2 times down-sampling, 4 times down-sampling, and 8 times down-sampling. There are 3 different stages in the down-sampling process of each resolution feature map. Each stage will map the channel of the feature map and increase the dimension, and the number of channels are 64, 128, and 256, respectively. Finally, the three feature maps of different resolutions are cross-fused. In the process of feature map cross-fusion, the feature maps of different scales are sampled and restored to the input image size, so that the second branch can use these feature maps directly when the features are merged. The second branch mainly uses a combination of convolutional neural networks and transformers to extract semantic information from the global receptive field of the feature map. First, it extracts the local feature map information of the input image through the convolutional neural network and obtains the 16 times down-sampled feature map. Next, unlike the current semantic segmentation network, it uses the Transformer method to continue to encode and decode the 16 times down-sampled feature map. The advantage of the Transformer encoding and decoding method is to make the entire feature map perform a global receptive field, which overcomes the limitation of the receptive field caused by the small convolution kernel of the convolutional neural network. In addition, the position information of each pixel in the feature map is introduced in the encoding process, so that each pixel adds semantic information in the position dimension. In the decoding process, the category high-dimensional mapping matrix is used as the query matrix, and the key-value matrix and the numerical matrix are from the output of the Transformer encoding layer. The structure diagram of the HMRT is shown in Figure 1, and there are two different dimensional transformations in the network layer. The dimension of the 3-dimensional feature map indicates the quantity of channels, the height and width of the feature map. In the 2-dimensional feature map, N represents the number of categories, and D represents the hidden layer mapping dimension of the Transformer encoding and decoding network. The upper half of the figure is the multi-resolution semantic extraction branch, and the lower half is the Transformer semantic extraction branch.

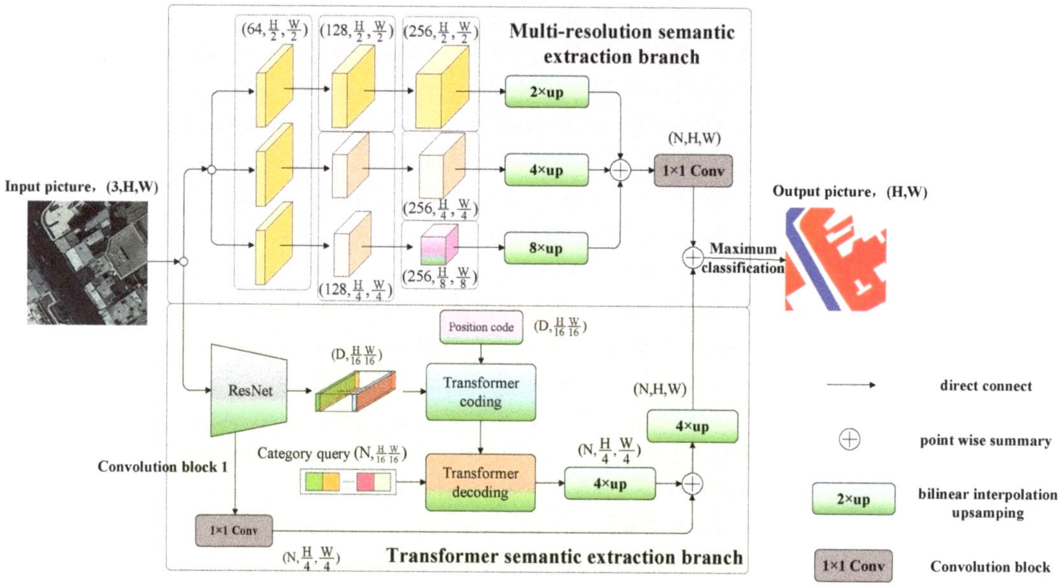

Figure 1. Hybrid multi-resolution and Transformer semantic extraction network framework.

2.1. Multi-Resolution Semantic Extraction Branch

The significance of the multi-resolution semantic extraction branch is that in the down-sampling feature extraction process of the convolutional neural network, the maximum pooling layer or the 3×3 convolution kernel with a sliding convolution stride of 2 is usually used for feature map length and width compression. Although rich semantic information is obtained in this way, it is inevitable that a lot of detailed information is discarded. In order to intuitively understand the differences in the feature map compression process, we show three feature compression ways, they are: (1) convolution with a 3×3 convolution kernel, a stride of 1, and a padding of 1. (2) Convolution with a 3×3 convolution kernel, a stride of 2, and a padding of 1. (3) The maximum pooling with a window of 2×2 size and a stride of 2. The visualization effects of the three feature compression ways are shown in Figure 2, Figure 2a shows the convolution with a stride of 1, Figure 2b shows the convolution with a stride of 2, Figure 2c shows the maximum pooling with a stride of 2. The size of the input picture demonstrated in Figure 2 is 512×512.

As is shown in Figure 2, the convolution with a stride of 1 is the most sufficient for feature extraction, almost retaining all the feature details in the original image, and the output feature map size is consistent with the input image specification (512×512). In the feature extraction process of the convolution operation with a stride of 2, the size of the output feature is 256×256, which is compressed to half of the original input image. When comparing Figure 2b with Figure 2a, it is not difficult to see that Figure 2b has significantly less feature information than Figure 2a. Then, compare the output result of the maximum pooling feature compression in Figure 2c with Figure 2a,b, the image in Figure 2c appears jagged and the loss of feature information is more serious than that of Figure 2b. Finally, we conclude that the semantic information expression capabilities of feature compression are as follows: the convolution with a stride of 1 is greater than the convolution with a stride of 2, and the convolution with a stride of 2 is greater than the maximum pooling operation with a stride of 2.

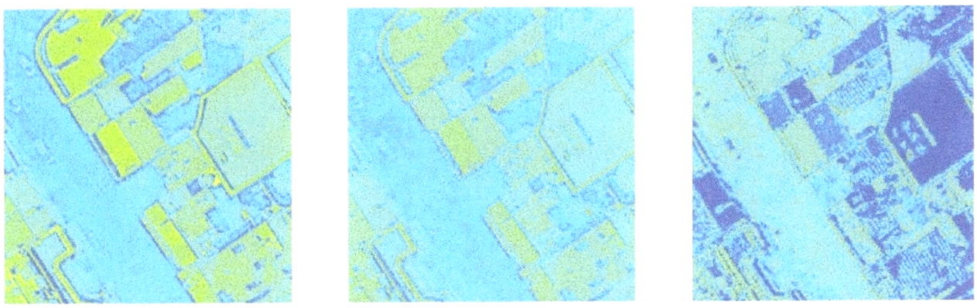

(a) (b) (c)

Figure 2. Different feature map compression visualization effects: (**a**) convolution with a stride of 1 (**b**) convolution with a stride of 1 (**c**) maximum pooling with a stride of 2.

At present, most of the current convolutional neural networks need to use the maximum pooling operation with a stride of 2 and use the convolution operations with a stride of 2 repeatedly in the feature extraction process, which will lead to the loss of detailed information in the feature map [19,20]. In order to overcome this difficulty, this section proposes a multi-resolution semantic extraction branch to provide rich multi-resolution feature maps for the network. The multi-resolution semantic extraction branch proposed in this section is divided into three branches. The three branches use the same input, but the input image is down-sampled at different multiples to obtain feature maps of different resolutions. The three feature maps with different resolutions are 2 times down-sampling, 4 times down-sampling and 8 times down-sampling.

The branch structure of multi-resolution semantic extraction is shown in Figure 3, the input of the whole branch is a picture of $3 \times H \times W$ size. The first branch in Figure 3 is 2 times down-sampling, consisting of a residual module with a stride of 2 and two residual modules with a stride of 1. The second branch in Figure 3 is 4 times down-sampling, consisting of two residual modules with a stride of 2 and a residual module with a stride of 1. The third branch in Figure 3 is 8 times down-sampling, consisting of three residual modules with a stride of 2. Each branch is composed of three residual modules, and the three residual modules will gradually increase the number of channel mapping during the down-sampling process. The number of channel mapping is 64, 128 and 256, respectively. After the three branches pass through the residual modules with different strides, the down-sampling feature maps of 2 times, 4 times and 8 times are obtained, respectively. Next, the feature maps need to be gathered and fused. However, the resolution of the feature maps is inconsistent, and it needs to be standardized to the same level. Therefore, the feature maps need to be up-sampled, and the multiple of the up-sampling is the inverse transform of the down-sampling multiple, which are 2 times up-sampling, 4 times up-sampling, and 8 times up-sampling, respectively. After sampling on the three branches, feature maps are restored to the size of the input image and then correspondingly added and fused in the channel dimension to obtain a feature map of $256 \times H \times W$ size. Finally, 1×1 convolution is used to map the number of channels of semantic information feature maps containing multiple resolutions to the number of categories N that can be learned by the model. As a result, the multi-resolution semantic extraction feature map ($N \times H \times W$) of category information is obtained.

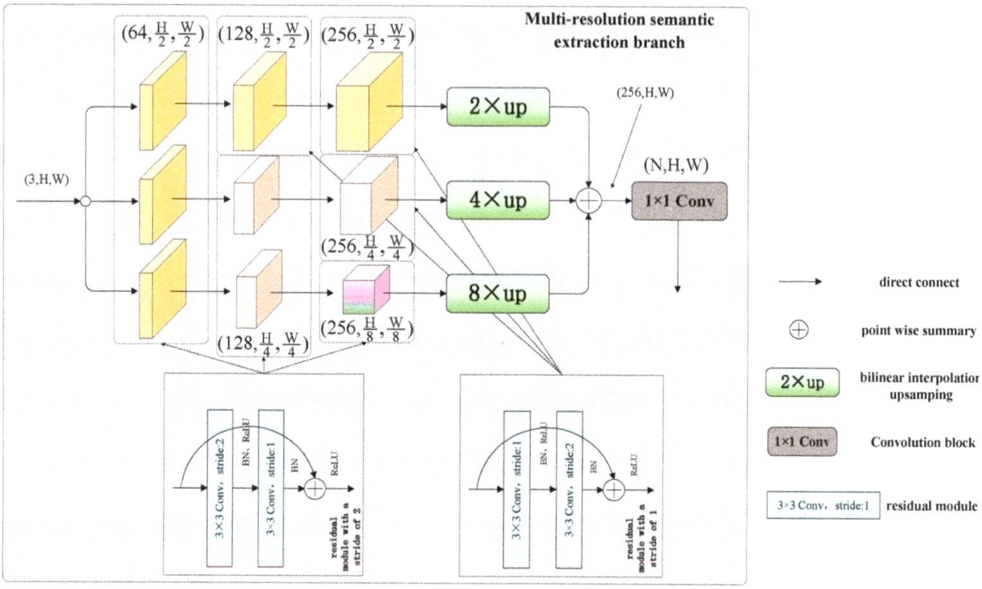

Figure 3. Structure diagram of multi-resolution semantic extraction branch.

Many excellent CNNs have appeared in recent years, including ResNet [21], VGG [22] and GoogLeNet [23]. After considering the amount of parameters and network accuracy, this work adopts the ResNet-18 basic module as the residual block in the three branches. The structure of the two residual modules in Figure 3 is shown in Figure 4. This work uses residual modules with two kinds of strides, the forward propagation of the residual module with a stride of 1 is shown in Equation (1).

$$X_{out} = \sigma(\beta(Conv_{3\times3}(\beta(\sigma(Conv_{3\times3}(X))))) + X), \tag{1}$$

where $Conv_{3\times3}$ is a 3×3 convolution with a stride of 1, β is BN, σ is a ReLU activation function.

The forward propagation of the residual module with a stride of 2 is shown in Equation (2).

$$X_{out} = \sigma(\beta(Conv'_{3\times3}(\beta(\sigma(Conv_{3\times3}(X))))) + X), \tag{2}$$

where $Conv_{3\times3}$ is a 3×3 convolution with a stride of 2, β is BN, σ is a ReLU activation function, $Conv'_{3\times3}$ is a 3×3 convolution with a stride of 1.

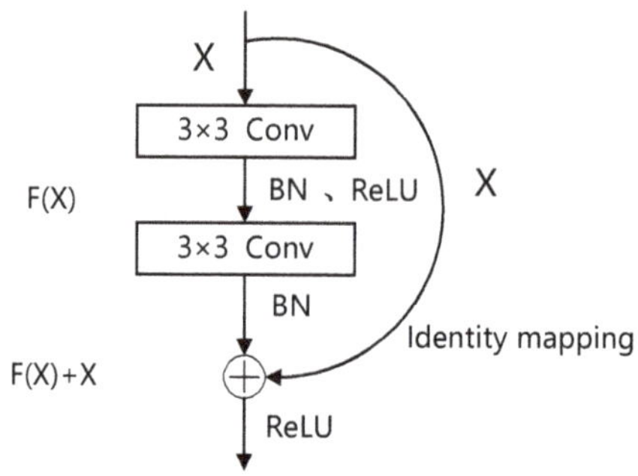

Figure 4. Structure diagram of ResNet-18 residual module.

The specific structure of the multi-resolution semantic extraction branch is shown in Table 1. The table shows the parameter settings of the six stages (input, first stage, second stage, third stage, feature convergence and output) of the branch.

Table 1. The specific structure of multi-resolution semantic extraction branch.

Stages	Number of Channels	2 Times Down-Sampling Branch	4 Times Down-Sampling Branch	8 Times Down-Sampling Branch
Input	3		$3 \times H \times W$	
The first stage	64	Convolution stride: 2	Convolution stride: 2	Convolution stride: 2
The second stage	128	Convolution stride: 1	Convolution stride: 2	Convolution stride: 2
The third stage	256	Convolution stride: 1	Convolution stride: 1	Convolution stride: 2
Feature convergence	256	2 times up-sampling	4 times up-sampling	8 times up-sampling
	N		1 × 1 convolution changes the number of channels	
Output	N		Accumulate to obtain the feature maps of $N \times H \times W$ size	

2.2. Transformer Semantic Extraction Branch

The significance of the Transformer semantic extraction branch is that the small receptive field of ordinary convolutional neural networks causes a lack of long-distance scene understanding. Although there have been many methods to improve the problem of small perception fields, such as enlarging the convolution kernel and using atrous convolution, they all have certain drawbacks. On the one hand, after enlarging the convolution kernel, the amount of model parameters and model calculations will increase, which will increase a lot of computational overhead. It is not a good choice in the case of lack of time and limited computing resources. On the other hand, although the use of atrous convolution can expand the receptive field of the original convolution kernel without adding additional calculations, the atrous convolution is filled with 0 when the convolution kernel is expanded, resulting in loss of internal details. The atrous convolution is more friendly to the extraction of large target objects and can capture the long-distance dependence of large target objects, but the advantages of small target objects are not obvious enough. Since the 0 padding used by the convolution kernel affects the continuity of the convolution kernel in the feature extraction process, small target objects are split or ignored, which affects the effectiveness of small target object extraction. The change in the size of the receptive field plays a big role in the feature extraction process. A convolution kernel with a large receptive field can extract the long-distance dependence of a large target, while a convolution kernel with a small receptive field can extract the complete features of a small target object. Taking the dimension of the convolution kernel width as an example, the derivation process of the receptive field is shown in Equation (3).

$$K' = K + (K - 1) \times (d - 1), \tag{3}$$

where K is the size and width of the convolution kernel, d is dilation rate, K' is the size of the receptive field.

In order to reflect the difference of the receptive fields conveniently and intuitively, this work designs different sizes of convolution kernels and different sizes of dilation rates for visualization. The comparison of the receptive fields with different dilation rates and convolution kernels is shown in Figure 5.

It can be seen that the current semantic segmentation networks have limitations in the receptive field, so this work combines the Transformer method with global receptive fields [24–28] to deeply mine the semantic information of the feature maps. On this basis, a hybrid convolutional neural network (ResNet-18) and a Transformer semantic extraction branch based on Transformer encoding and decoding are constructed.

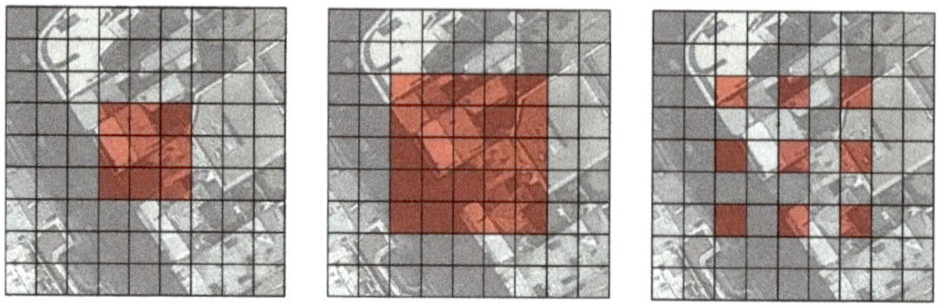

(a) (b) (c)

Figure 5. Comparison of different void rates and convolution kernel receptive fields. (**a**) Receptive field of 3 × 3 convolution kernel; (**b**) receptive field of 5 × 5 convolution kernel; (**c**) receptive field of 5 × 5 convolution kernel with a dilation rate of 2.

2.2.1. The Overall Framework of Transformer Semantic Extraction Branch

The overall framework of the Transformer semantic extraction branch is composed of a hybrid convolutional neural network (backbone network) and Transformer encoding and decoding modules. The structure of this branch is shown in Figure 6, *D* in the figure is the mapping dimension, and *N* is the number of categories. Firstly, the backbone network adopts ResNet-18 with a sliding window to extract features. Secondly, the backbone network performs 16 times down-sampling, flattening the feature map in the width and height dimensions to obtain a feature map with a dimension of (D, H/16, W/16). Next, we add the obtained feature map to the position coding matrix of the same size and input the result into the Transformer coding module for global coding. The number of times the encoding module repeats the encoding is set to 6. Corresponding Transformer decoding is performed after encoding, and a total of 6 decoding layers are set. The query matrix of the first decoding layer is provided by the category matrix and the query matrix after the second layer is the output of the previous decoding layer. In addition, the key-value matrix and the numerical matrix are also the output of the decoding matrix of the previous layer. After Transformer encoding and decoding, the feature map with a dimension of (H, H/16, W/16) is outputted. Then, the last dimension of the feature map is flattened into two dimensions to obtain a feature map (H, H/16, W/16) with the number of channels of category N. Next, the new feature map is up-sampled 4 times to obtain the feature map with a dimension of (H, H/4, W/4). The reason for using 4 times up-sampling is to make the size of the feature map consistent with the convolution block 1 of the backbone network. Since the feature map of the convolution block 1 is used to enrich position information, the fusion of the feature map can help position restoration. Finally, the feature map needs to be restored to the input image size to achieve pixel-level classification. In the restoration process, the feature map is up-sampled 4 times to obtain the feature map.

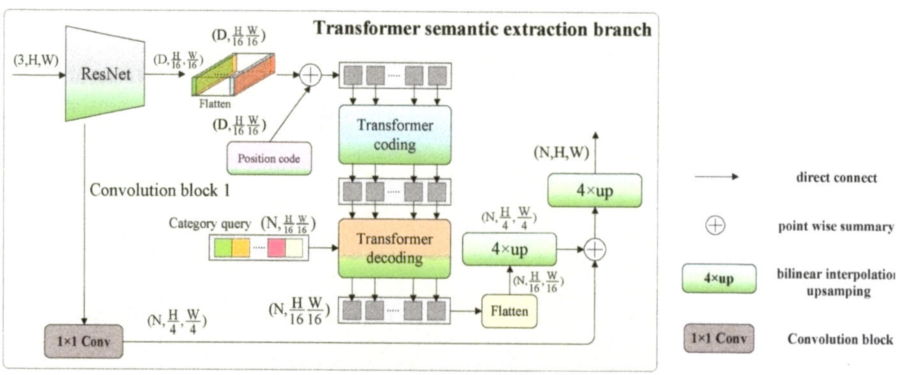

Figure 6. Structure diagram of Transformer semantic extraction branch.

2.2.2. The Feature Extraction of Backbone Network

Transformer semantic extraction branch proposed in this section uses a convolutional neural network as the backbone network for feature extraction. ResNet-18 is used to extract the deep semantic information of the image, but the structure of ResNet-18 used in this paper is slightly different from the structure in the original paper.

As shown in Table 2, after the picture of 512 × 512 size passes through the convolution block 1, convolution block 2, convolution block 3, and convolution block 4 that are consistent with the original paper, a feature map of 32 × 32 size can be obtained. Next, the network maintains this resolution and uses convolution to deepen the extracted features. Convolution deepening is completed by convolution block 5 and convolution block 5 is a 3 × 3 convolution with a stride of 1 and 256 channels. The final output feature map is 1/16 of the size of the input image, which is twice as large as the output size of the original backbone network. This maintains a large resolution feature map, which is conducive to the extraction of richer feature information from Transformer global features. In addition, the number of channels of the output feature map is also reduced by half. The reason is that the Transformer connected to the backbone network still has a high-dimensional mapping channel when it continues to encode and decode. The reduction in the number of channels in the backbone network can reduce a certain amount of parameters.

Table 2. The specific structure of the backbone network ResNet-18.

Modules	The Size of the Feature Map	ResNet-18
Input	512 × 512	-
convolution block 1	256 × 256	7 × 7, Number of channels: 64, Stride: 2, padding 3 × 3, Maximum pooling layer, Stride: 2
convolution block 2	128 × 128	3 × 3, Number of channels: 64, Stride: 2 3 × 3, Number of channels: 64, Stride: 1
convolution block 3	64 × 64	3 × 3, Number of channels: 128, Stride: 2 3 × 3, Number of channels: 128, Stride: 1
convolution block 4	32 × 32	3 × 3, Number of channels: 256, Stride: 2 3 × 3, Number of channels: 128, Stride: 1
convolution block 5	32 × 32	3 × 3, Number of channels: 256, Stride: 1 3 × 3, Number of channels: 128, Stride: 1

2.2.3. Transformer Encoding and Decoding

Transformer encoding and decoding was first proposed by Vaswani et al. [29] for natural language processing. This method extracts global information from the input feature map. Inspired by this innovation, this paper transplants and fine-tunes the Transformer to the semantic segmentation task to make up for the limitations of the convolutional neural network's receptive field when performing semantic segmentation. The overall structure of the improved Transformer in this paper is composed of encoding and decoding, and the encoding and decoding modules are spliced from the self-attention mechanism into a multi-head attention mechanism. The structure of the Transformer is shown in Figure 7.

Firstly, the input feature map is the feature map (D, H/16, W/16) extracted by the backbone network. Secondly, the flattening function maps the last two dimensions into a one-dimensional vector to obtain a new feature map (D, H/16, W/16). Then the position coding matrix $p \in R^{(D,H/16,W/16)}$ of each pixel in the feature map and the feature map are superimposed as the input of the coding layer. The coding layer first undergoes the feature normalization layer to normalize the channel dimensions, and then passes through different matrix mappings to obtain the query matrix $q \in R^{(H/16,W/16,D)}$, the key-value matrix $k \in R^{(H/16,W/16,D)}$, and the numerical matrix $v \in R^{(H/16,W/16,D)}$. The calculation process is shown in Equations (4)–(7). X_{input} is the input of the coding layer. Δ represents feature normalization layer. K_q is an encoding layer query matrix mapping function. K_k is the coding layer key-value matrix mapping function. K_v is the coding layer numerical matrix mapping function.

$$X = \Delta(X_{input}), \tag{4}$$

$$q = K_q(X), \tag{5}$$

$$k = K_k(X), \tag{6}$$

$$v = K_v(X), \tag{7}$$

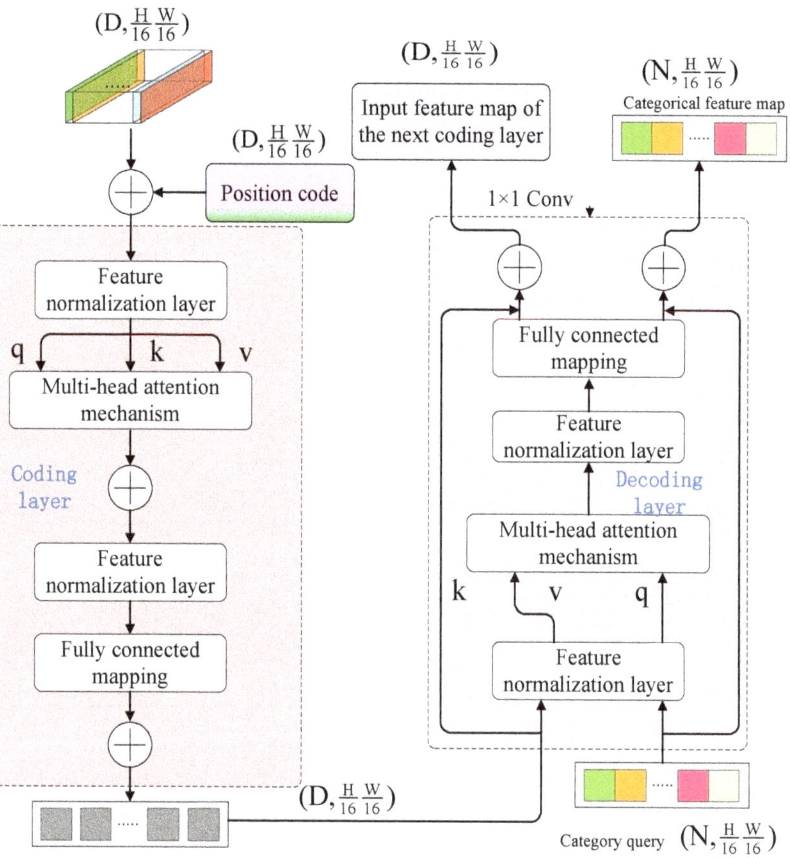

Figure 7. The structure diagram of Transformer encoding and decoding.

After obtaining the query matrix, the key-value matrix and the numerical matrix, the three matrices are input into the multi-head attention mechanism module for attention calculation. The multi-head attention mechanism module is obtained by splicing multiple self-attention mechanism modules (this paper sets the number of heads of the multi-head attention mechanism to 4). The advantage of having multi-headed attention is that feature information can be obtained from different branches to enrich semantic information, and different branches can independently extract features and then merge them, which can increase the diversity of feature extraction. After the multi-head attention mechanism module, the output feature map and the feature map before the feature normalization layer aggregate the original feature map information through jump connection. As is shown in Equation (8).

$$X_{atten} = \Gamma(q, k, v) + X_{input}, \tag{8}$$

where Γ represents multi-head attention mechanism module, X_{atten} is the output of the multi-head attention mechanism, q, k, v represent the query matrix, key-value matrix, and numeric matrix of the coding layer, respectively. X_{input} is the input of the coding layer.

Then through the feature normalization layer and the fully connected layer, the feature dimension is mapped to the high dimension. The fully connected mapping is a 4 times mapping. Finally, the original feature map before entering the full connection is also aggregated by jump connection to obtain the output feature map of the coding layer (HW,D). The calculation process is shown in Equation (9).

$$X_{encoder} = \Pi(\Delta(X_{atten})), \tag{9}$$

where $X_{encoder}$ is the output of the coding layer, Π represents fully connected mapping, Δ represents the feature normalization layer, and X_{atten} is the output of the multi-head attention mechanism.

The input of the decoding layer is composed of two parts: the output feature map of the coding layer and category query feature map created based on the number of categories. The feature layer is normalized before entering the multi-head attention mechanism of the decoding layer, and then the output feature map of the coding layer is decomposed into a key-value matrix $k' \in R^{(H/16,W/16,N)}$ and a numerical matrix $v' \in R^{(H/16,W/16,N)}$ through different matrix mappings. Among them, N is the number of categories. The calculation process is shown in Equations (10)–(12).

$$X'_{encoder} = \Delta(X_{encoder}), \tag{10}$$

$$k' = K'_k(X'_{encoder}), \tag{11}$$

$$v' = K'_v(X'_{encoder}), \tag{12}$$

where $X'_{encoder}$ is the output of the feature normalization layer, Δ represents feature normalization layer, $X_{encoder}$ is the output of the coding layer, K'_k represents key-value matrix mapping function in the decoding layer, K'_v represents the numerical matrix mapping function in the decoding layer.

The decoding layer query matrix $q' \in R^{(H/16,W/16,N)}$ is obtained according to the category initialization, and the key-value matrix $k' \in R^{(H/16,W/16,N)}$ and the numerical matrix $v' \in R^{(H/16,W/16,N)}$ are obtained through matrix mapping. Next, we input them into the multi-head attention mechanism for decoding at the same time, and the decoded result goes through the feature normalization layer and fully connected mapping. The fully connected mapping is a 4 times mapping. Then, the first feature of the decoding layer and the feature map before the feature normalization layer are correspondingly fused. Finally, the output category feature map of the decoding layer and the input feature map of the next coding layer are obtained. The calculation process is shown in Equation (13).

$$X_{decoder} = \Pi(\Delta(\Gamma(q', k', v'))) + q', \tag{13}$$

where $X_{decoder}$ is the output of the decoding layer, Π represents fully connected mapping, Δ represents the feature normalization layer, Γ represents the multi-head attention mechanism module, q', k', v' are the query matrix, key-value matrix and numerical matrix of the decoding layer

The structure of the multi-head attention mechanism module is shown in Figure 8. As shown in Figure 8, query matrix q, key-value matrix k, and numerical matrix v are inputed. The calculation process of a single attention mechanism is as follows: Firstly, the key-value matrix and the transpose of the query matrix are multiplied. Secondly, Softmax is performed at the last of the results obtained. Finally, the result of Softmax and numerical matrix are multiplied to obtain the attention result. In addition, the multi-head attention mechanism proposed in this paper splices the single-head attention mechanisms to obtain the attention mechanism information of different branches.

The decomposition process of a single self-attention mechanism module is shown in Figure 8. In the case of inputting three targets, the three targets are calculated by their corresponding query matrix mapping matrix W_{query}, key-value mapping matrix $W_{key-value}$, and numerical mapping matrix $W_{numericlvalue}$ to obtain their respective query matrix, key-value matrix, and numerical matrix ($q_1, q_2, q_3, k_1, k_2, k_3, v_1, v_2, v_3$). Next, Softmax calculates the weights of the target itself and all targets. The calculation process of measure weights is shown in Equation (14), and the decomposition structure of the calculation process of the self-attention mechanism is shown in Figure 9.

$$Weights_i = \frac{q_i * k_i^T}{\sum_i (q_i * k_i^T)}, \quad (14)$$

where $Weights_i$ is the measure weights of the i-th goal, q_i represents the query matrix of the i-th target, k_i represents the key-value matrix of the i-th target.

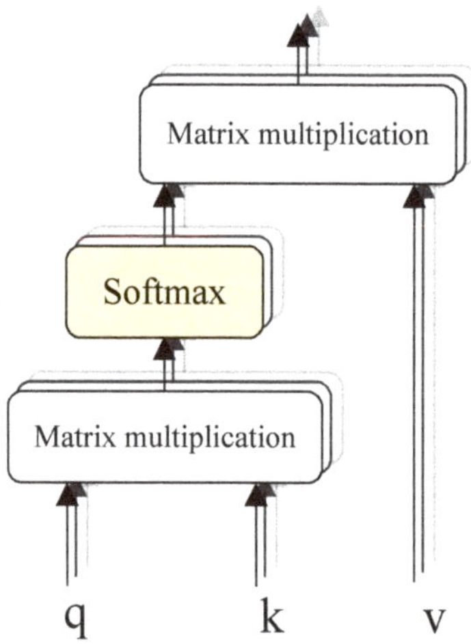

Figure 8. The structure diagram of multi-head attention mechanism.

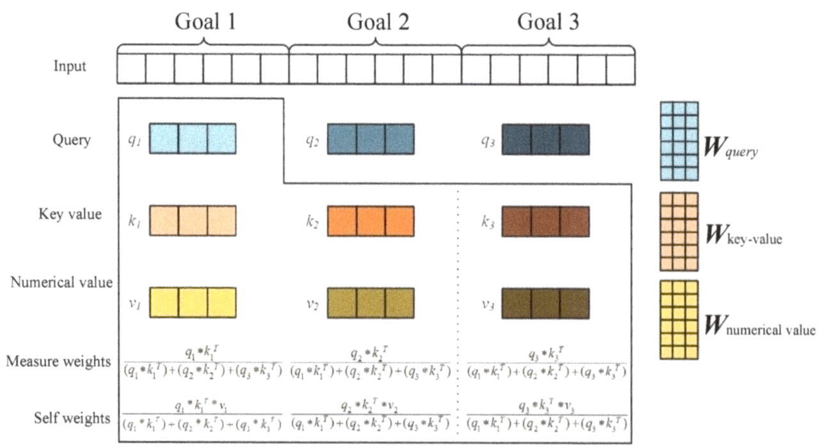

Figure 9. Decomposition structure diagram of the calculation process of the self-attention mechanism.

Finally, the self weights are obtained by multiplying the weights of each target and the corresponding numerical matrix. The calculation process is shown in Equation (15).

$$Attention_i = Weights_i * v_i, \tag{15}$$

where $Attention_i$ is the self weights (Attention information) of the i-th goal, $Weights_i$ is the measure weights of the i-th goal, v_i represents the numerical matrix of the i-th target.

3. Experiment and Result Analysis

In this chapter, experiments were carried out on the Aerial Image Segmentation Dataset (AISD) [30] and ISPRS 2D Semantic Labeling Contest (ISPRS) [31]; the HMRT model was compared with many of the best models currently available FCN-8S [32], U-Net [33], PSPNet [34] and DeeplabV3+ [35]; the overall accuracy rate (OA), recall rate (Recall), F1-Score and mean intersection over union (MIoU) are used as the quantitative analysis indicators of the experiment. The results show that HMRT is better than the comparison model in various assessment indicators.

3.1. Datasets

3.1.1. AISD Dataset

The original images of the AISD dataset were collected from OpenStreetMap online remote sensing image data, and the semantic segmentation dataset of high resolution remote sensing images were constructed by manual annotation. AISD included image data from six regions: Berlin, Chicago, Paris, Potsdam, and Zurich. This paper chose Potsdam regional data to conduct the experiment, and we named the data set as Potsdam-A. There are 24 original images and labels with an average size of 3000 × 3000 in Potsdam-A. The example of the original image and label is shown in Figure 10.

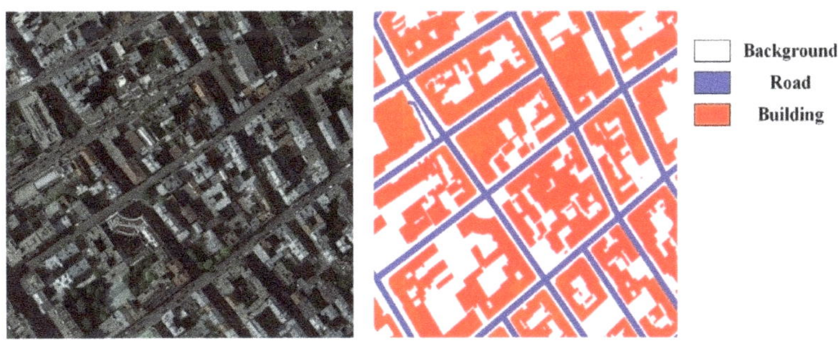

(a) (b)

Figure 10. Original image and label example from Potsdam-A. In (**b**), the red is building; the blue is road background; the white is background. (**a**) image; (**b**) label.

Because the original image was too large to be directly input to the model training, we used Python to crop the picture of 3000 × 3000 into the picture of 512 × 512 and obtained a total of 1728 pictures finally. In the case of a small amount of data, the generalization effect was poor and the feature learning ability of the model was weak. Therefore, data enhancement was needed to ensure that the model has reliable learning capability. The original data set was randomly flipped horizontally, vertically and rotated by 90 degrees to expand to 4307 pictures.

3.1.2. ISPRS Dataset

The ISPRS 2D Semantic Labeling Contest dataset is a high-resolution aerial image dataset with complete Semantic Labeling published by the International Society for Photogram-metry and Remote Sensing (ISPRS). Similarly, we selected the Potsdam region in ISPRS to verify the generalization performance of the model, and name the data set Potsdam-B. Potsdam-B was made up of 38 accurately labeled images and five foregrounds: impervious surfaces, buildings, low vegetation, tree and car. The example of the original image and label is shown in Figure 11.

We adopted the same cropping strategy of the Potsdam-A dataset on the Potsdam-B dataset to obtain 5184 pictures of 512 × 512 size.

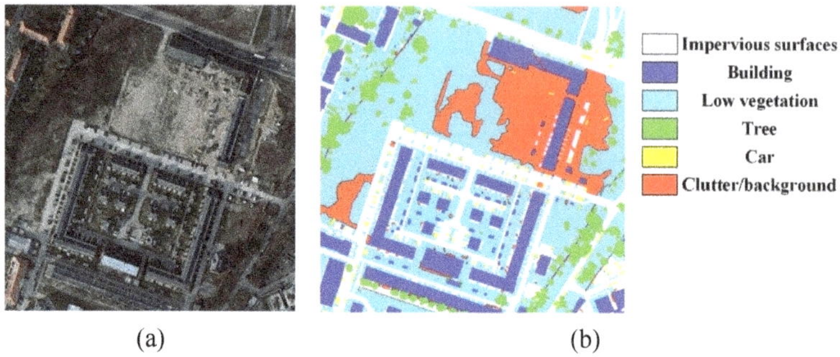

Figure 11. Original image and label example from Potsdam-B; (**a**) image; (**b**) label.

3.2. Implementation Details

In this experiment, we selected five evaluation indicators, including overall accuracy rate (*OA*), recall rate (*Recall*), F_1-Score and intersection over union (*IoU*). They are as follows:

$$OA = \frac{TP + TN}{P + N}, \quad (16)$$

$$Recall = \frac{TP}{TP + FN'} \quad (17)$$

$$F_1 = 2 \times \frac{Precision \times Recall}{Precision + Recall'}, \quad (18)$$

$$Precision = \frac{TP}{TP + FP'} \quad (19)$$

$$IoU = \frac{TP}{TP + FP + FN}. \quad (20)$$

The cross-entropy loss function (CEloss) was applied to calculate the difference value between the true value and the predicted value. The model performed backpropagation and learned the best parameters under the guidance of the difference value. The derivation process of CE_{loss} is shown in Equation (21):

$$CE_{loss}(p, q) = -\frac{1}{m} \sum_{i=1}^{m} \sum_{j=1}^{n} p(x_{ij}) log(q(x_{ij})), \quad (21)$$

where m is the quantity of samples, n represents the quantity of categories, $p(x_{ij})$ is a variable (if the category j is the same as the sample i, it is 1, otherwise it is 0), $q(x_{ij})$ is the probability sample, i is predicted to be class j.

All experiments were carried out on Ubuntu16.04 LTS with a Intel(R)Core(TM)i7-8750F CPU @2.20 GHz, 16 G of memory (RAM), and a NVIDIA GeForce RTX1060(8 GB). Python 3.8 was used, and the model was built using Pytorch1.0.1. All models were trained for 300 epochs with a batch size of 4, and the initial learning rate was 0.001.

This paper improves the post-processing of model prediction. The model prediction method adds multi-scale and sliding window splicing, which can significantly improve prediction results. The execution method of the multi-scale strategy is to enlarge the picture by 1.0, 1.25, 1.5, 1.75, 2.0 times on the basis of the original predicted picture and then predict.

After the prediction result is obtained, the size of the picture is reduced to the original picture size and added to obtain the final prediction result. The sliding window splicing strategy is to set the stride sliding window to predict the picture according to the rule from left to right and top to bottom in the upper left corner of the predicted picture. The schematic diagram of the sliding window splicing prediction strategy is shown in Figure 12. Figure 12a represents the size of the prediction window, Figure 12b represents the stride for sliding right, Figure 12c represents the stride for sliding down. In the panning process, in order to ensure that the entire picture can be predicted by the model and obtain the output, we set the panning stride to be less than or equal to the prediction window size. If the stride is smaller than the prediction window, repeated prediction parts will appear. Therefore, we use the overall summation method to realize the prediction for the repeated prediction parts.

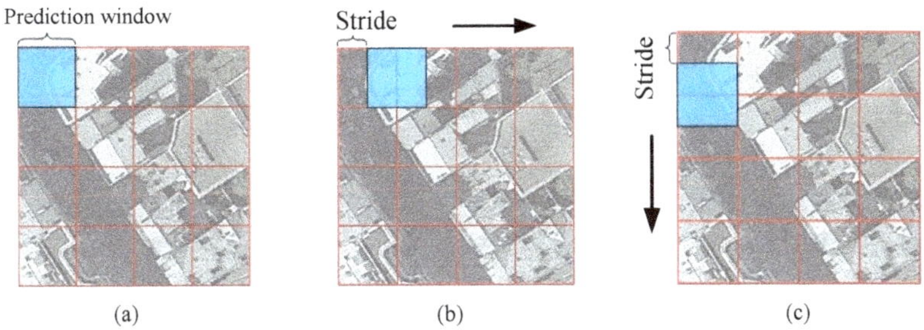

Figure 12. Schematic diagram of the sliding window splicing prediction strategy; (**a**) initial window postion; (**b**) stride for sliding right. (**c**) stride for sliding down.

3.3. Analysis of Results

3.3.1. Evaluation Metrics and Prediction Effect

(1) Main experiment

In order to test the proposed HMRT module, comprehensive experiments were carried out on the Potsdam-A dataset. The evaluation metrics are shown in Table 3, and a comparison of the prediction results is shown in Figure 13. Moreover, ablation experiments were carried out to verify the effectiveness of the multi-resolution semantic extraction branch. The network without the multi-resolution semantic extraction branch module was tested and named HMRT-1.

In Table 3, recall, F1, OA and MIoU of the HMRT obtained 85.32%, 84.88%, 85.99% and 74.19%, respectively. All four indicators were better than the comparison networks [32–35]. Among them, OA reached 85.99%, which was 0.92 higher than DeeplabV3+, and MioU reached 74.19, which was 1.37 higher than DeeplabV3+.

Table 3. The evaluation metrics on Potsdam-A test set.

Methods	Backbone	Recall (%) ↑	F1 (%) ↑	OA (%) ↑	MIoU (%) ↑
FCN-8S	VGG16	79.99	80.93	83.09	68.55
U-Net	-	82.63	82.94	84.49	71.28
PSPNet	ResNet-50	83.02	83.57	84.54	72.09
DeeplabV3+	ResNet-50	83.90	84.05	85.07	72.82
HMRT-1	-	85.17	85.14	85.80	73.75
HMRT	-	85.32	84.88	85.99	74.19

The IOU of each model on the Potsdam-A test set are illustrated in Table 4. The IOU indexes of the HMRT were 65.21%, 73.15% and 84.21%, respectively, exceeding the four comparison networks [32–35]. The IOU results showed that the HMRT has absolute advantages in segmentation accuracy.

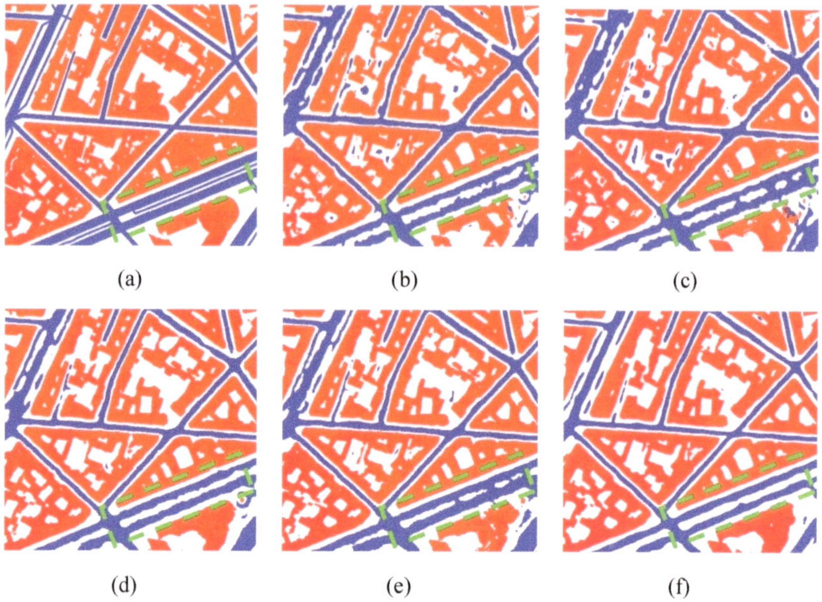

Figure 13. Comparison of the prediction results on Potsdam-A; (a) the superposition of the image and the label; (b) FCN-8S; (c) U-Net; (d) PSPNet; (e) DeeplabV3+; (f) HMRT.

Table 4. The IoU results on Potsdam-A test set.

Methods	Backbone	Background (%) ↑	Road (%) ↑	Building (%) ↑	MIoU (%) ↑
FCN-8S	VGG16	59.18	64.10	82.36	68.55
U-Net	-	62.29	68.64	82.91	71.28
PSPNet	ResNet-50	63.34	71.73	81.21	72.09
DeeplabV3+	ResNet-50	64.15	71.65	82.65	72.82
HMRT-1	-	65.00	72.54	83.71	73.75
HMRT	-	65.21	73.15	84.21	74.19

(2) Generalization experimental

To verify the generalization performance of the models proposed in this paper, the Potsdam-B data set was used for further experiment. The evaluation metrics are shown in Table 5. In Table 5, the recall, F1, OA and MIoU of HMRT reached 91.29%, 90.41%, 91.32% and 84.00%, respectively. All indicators were at their greatest levels, which could prove that the model proposed in this paper is not only effective, but has good generalization performance.

Table 5. The evaluation metrics on Potsdam-B test set.

Methods	Backbone	Recall (%) ↑	F1 (%) ↑	OA (%) ↑	MIoU(%) ↑
FCN-8S	VGG16	86.31	85.48	86.86	78.43
U-Net	-	87.74	87.41	88.51	80.87
PSPNet	ResNet-50	88.74	88.34	88.59	81.02
DeeplabV3+	ResNet-50	88.95	87.73	88.48	81.55
HMRT	-	91.29	90.41	91.32	84.00

Moreover, this paper visualizes the prediction results of each model. The comparison of the prediction results is shown in Figure 14. Figure 14a is the real label, Figure 14b–f correspond to the prediction results of FCN-8S, U-Net, PSPNet, DeeplabV3+ and HMRT, respectively. Through comparison, we can find that the HMRT model proposed in this paper has a global receptive field, and the segmentation accuracy is higher than that of the comparison model. The dashed box in the figure highlights the area where the segmentation effect is obvious.

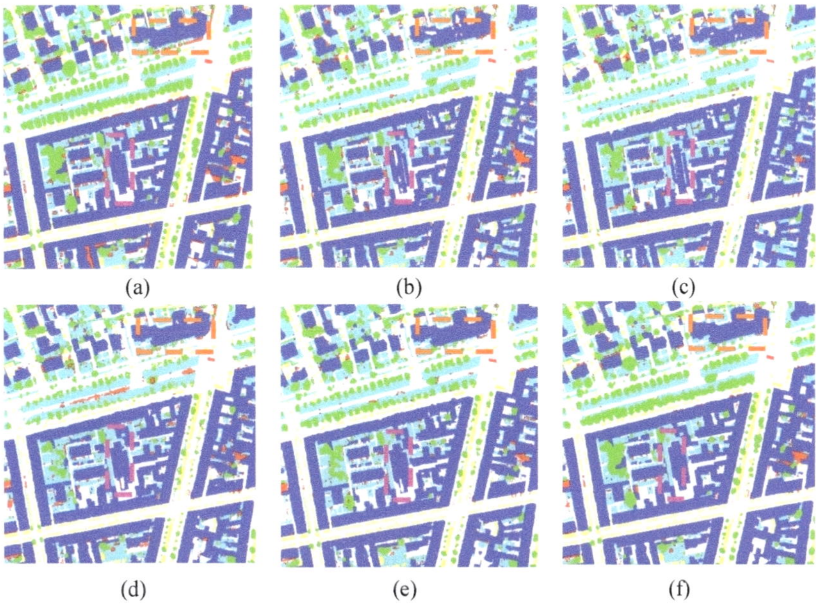

Figure 14. Comparison of the prediction results on Potsdam-B; (**a**) the superposition of the image and the label; (**b**) FCN-8S; (**c**) U-Net; (**d**) PSPNet; (**e**) DeeplabV3+; (**f**) HMRT.

3.3.2. Quantitative Analysis of Model Prediction Results Promotion Strategy

This work adopted two post-processing methods, multi-scale fusion and sliding stitching, to improve the prediction accuracy. The multi-scale fusion strategy is to adapt to different size targets in remote sensing images, and the experimental parameters are that the predicted picture is magnified by 1.0, 1.25, 1.5, 1.75, 2.0 times, respectively. The sliding stitching strategy is to alleviate the problem of jagged edges when the pictures are directly stitched, the experimental parameter is that the step size is half of the sliding window (512 × 512). Finally, we used the controlled variable method to experiment on the two post-processing strategies, and each network obtained 4 sets of experimental results. The results of the quantitative analysis experiment are shown in Table 6.

It is concluded from Table 6 that the two strategies of both multi-scale fusion and sliding splicing can improve the prediction accuracy to a certain extent, and the prediction accuracy reaches the highest when the two strategies of multi-scale fusion and sliding splicing are used at the same time. In order to visually demonstrate the effectiveness of the post-processing strategy, Figure 15 shows the comparison between the predicted result without using post-processing strategies and predicted the result using two post-processing strategies. From the comparison of Figure 15b,c, it can be seen that the post-processing strategies reduce the stitching traces of splicing pictures, and the outlines of the foreground target buildings and roads in the picture are clearer.

Table 6. Comparison of multi-scale fusion and sliding splicing prediction quantitative analysis. √ means this post-processing method is adopted, ↑ means the higher, the better.

Methods	Multi-Scale	Sliding Stitching (%) ↑	F1 (%) ↑	OA (%) ↑	MIoU(%) ↑
FCN-8S	-	-	80.93	83.09	68.55
	√	-	81.05	83.15	68.69
	-	√	81.13	83.18	68.97
	√	√	81.41	83.35	69.22
U-Net	-	-	82.94	84.54	71.28
	√	-	83.06	84.63	71.44
	-	√	83.15	84.71	71.51
	√	√	83.25	84.90	71.73
PSPNet	-	-	83.57	84.49	72.09
	√	-	83.62	84.53	72.18
	-	√	83.71	84.61	72.25
	√	√	84.04	85.04	72.78
DeeplabV3+	-	-	84.05	85.17	72.82
	√	-	84.12	85.21	72.91
	-	√	84.18	85.25	72.97
	√	√	84.30	85.36	73.18
HMRT	-	-	84.88	85.58	74.19
	√	-	85.40	86.55	74.82
	-	√	85.48	86.31	74.90
	√	√	85.79	86.85	75.39

(a) (b) (c)

Figure 15. Comparison of prediction results before and after post-processing; (**a**) label image; (**b**) the predicted result without using post-processing strategies; (**c**) predicted result using two post-processing strategies.

4. Conclusions

This paper proposes the HMRT to extract buildings and roads from high resolution remote sensing images. In comparison with the current networks, HMRT has three advantages: (1) The multi-resolution semantic extraction branch is constructed to use branches with different resolutions for feature fusion, which ensures that high resolution and multi-resolution can always be maintained during the down-sampling process, and feature information is fully retained. It solves the problem that feature map compression leads to loss of details and the convolutional neural network lacks long-distance scene understanding when the current semantic segmentation algorithm uses a convolutional neural network (CNN) to extract image features. (2) The Transformer sequence feature extraction network is introduced through which the global receptive field of the feature map can be obtained, the long-distance dependence of the segmentation target is improved, and the issue of reduced resolution is solved, which is caused by feature map compression during the use of convolutional feature extraction. (3) The model has the following advantages, such as the highest accuracy index, absolute superiority in segmentation accuracy, and sufficient robust performance.

However, there are still some shortcomings in the segmentation of buildings and roads: (1) The use of the Transformer global receptive field to extract features is still in the development stage, so there is development space in the accuracy of the edge segmentation of buildings and roads and the structure of the model. (2) The complexity of the parameters of the Transformer encoder and decoder is high. (3) When the remote sensing image contains a lot of noise, the accuracy of segmentation will decrease. As a result, we will optimize HMRT to improve the segmentation accuracy and overcome the problem of a decrease in segmentation accuracy in case of a lot of noise in remote sensing images.

Author Contributions: Zhongyu Sun: conceptualization, methodology, writing—original draft. Wangping Zhou: conceptualization, supervision, software, writing—review and editing. Min Xia: funding acquisition, writing—review and editing. Chen Ding: formal analysis, validation. All authors have read and agreed to the published version of the manuscript.

Funding: This work is supported by the National Natural Science Foundation of PR China (42075130).

Institutional Review Board Statement: Not applicable.

Informed Consent Statement: Not applicable.

Data Availability Statement: The data and the code of this study are available from the corresponding author upon request.

Conflicts of Interest: The authors declare no conflict of interest.

References

1. Pham, H.M.; Yamaguchi, Y.; Bui, T.Q. A case study on the relation between city planning and urban growth using remote sensing and spatial metrics. *Landsc. Urban Plan.* **2011**, *100*, 223–230. [CrossRef]
2. Song, L.; Xia, M.; Jin, J.; Qian, M.; Zhang, Y. SUACDNet: Attentional change detection network based on siamese U-shaped structure. *Int. J. Appl. Earth Obs. Geoinf.* **2021**, *105*, 102597. [CrossRef]
3. Xia, M.; Qu, Y.; Lin, H. PADANet: Parallel asymmetric double attention network for clouds and its shadow detection. *J. Appl. Remote Sens.* **2021**, *15*, 046512. [CrossRef]
4. Wen, Q.; Jiang, K.; Wang, W.; Liu, Q.; Guo, Q.; Li, L.; Wang, P. Automatic building extraction from google earth images under complex backgrounds based on deep instance segmentation network. *Sensors* **2019**, *19*, 333. [CrossRef] [PubMed]
5. Behera, M.D.; Gupta, A.K.; Barik, S.K.; Das, P.; Panda, R.M. Use of satellite remote sensing as a monitoring tool for land and water resources development activities in an Indian tropical site. *Environ. Monit. Assess.* **2018**, *190*, 401. [CrossRef]
6. Qu, Y.; Xia,M.; Zhang, Y. Strip pooling channel spatial attention network for the segmentation of cloud and cloud shadow. *Comput. Geosci.* **2021**, *157*, 104940. [CrossRef]
7. Yuan, J.; Wang, D.; Li, R. Remote sensing image segmentation by combining spectral and texture features. *IEEE Trans. Geosci. Remote Sens.* **2013**, *52*, 16–24. [CrossRef]
8. Li, D.; Zhang, G.; Wu, Z.; Yi, L. An edge embedded marker-based watershed algorithm for high spatial resolution remote sensing image segmentation. *IEEE Trans. Image Process.* **2010**, *19*, 2781–2787. [PubMed]
9. Fan, J.; Han, M.; Wang, J. Single point iterative weighted fuzzy C-means clustering algorithm for remote sensing image segmentation. *Pattern Recognit.* **2009**, *42*, 2527–2540. [CrossRef]
10. Panboonyuen, T.; Vateekul, P.; Jitkajornwanich, K.; Lawawirojwong, S. An enhanced deep convolutional encoder-decoder network for road segmentation on aerial imagery. In Proceedings of the International Conference on Computing and Information Technology 2017, Helsinki, Finland, 21–23 August 2017; pp. 191–201. [CrossRef]
11. Clevert, D.A.; Unterthiner, T.; Hochreiter, S. Fast and accurate deep network learning by exponential linear units (elus). *arXiv* **2015**, arXiv:1511.07289.
12. Badrinarayanan, V.; Kendall, A.; Cipolla, R. Segnet: A deep convolutional encoder-decoder architecture for image segmentation. *IEEE Trans. Pattern Anal. Mach. Intell.* **2017**, *39*, 2481–2495. [CrossRef] [PubMed]
13. Sun, W.; Wang, R. Fully convolutional networks for semantic segmentation of very high resolution remotely sensed images combined with DSM. *IEEE Geosci. Remote Sens. Lett.* **2018**, *15*, 474–478. [CrossRef]
14. Liu, W.; Zhang, Y.; Fan, H.; Zou, Y.; Cui, Z. A New Multi-Channel Deep Convolutional Neural Network for Semantic Segmentation of Remote Sensing Image. *IEEE Access* **2020**, *8*, 131814–131825. [CrossRef]
15. Qi, X.; Li, K.; Liu, P.; Zhou, X.; Sun, M. Deep Attention and Multi-Scale Networks for Accurate Remote Sensing Image Segmentation. *IEEE Access* **2020**, *8*, 146627–146639. [CrossRef]
16. Li, J.; Xiu, J.; Yang, Z.; Liu, C. Dual Path Attention Net for Remote Sensing Semantic Image Segmentation. *ISPRS Int. J. Geo-Inf.* **2020**, *9*, 571. [CrossRef]
17. Lan, M.; Zhang, Y.; Zhang, L.; Du, B. Global Context based Automatic Road Segmentation via Dilated Convolutional Neural Network. *Inf. Sci.* **2020**, *535*, 156–171. [CrossRef]

18. He, N.; Fang, L.; Plaza, A. Hybrid first and second order attention Unet for building segmentation in remote sensing images. *Inf. Sci.* **2020**, *63*, 140305. [CrossRef]
19. Xia, M.; Zhang, X.; Liu, W.; Weng, L.; Xu, Y. Multi-stage Feature Constraints Learning for Age Estimation. *IEEE Trans. Inf. Forensics Secur.* **2020**, *15*, 2417–2428. [CrossRef]
20. Wang, P.; Chen, P.; Yuan, Y.; Liu, D.; Huang, Z.; Hou, X.; Cottrell, G. Understanding convolution for semantic segmentation. In Proceedings of the 2018 IEEE Winter Conference on Applications of Computer Vision (WACV), Lake Tahoe, NV, USA, 12–15 March 2018; pp. 1451–1460. [CrossRef]
21. He, K.; Zhang, X.; Ren, S.; Sun, J. Deep residual learning for image recognition. In Proceedings of the IEEE Conference on Computer Vision and Pattern Recognition (CVPR), Las Vegas, NV, USA, 27–30 June 2016; pp. 770–778.
22. Simonyan, K.; Zisserman, A. Very deep convolutional networks for large-scale image recognition. *arXiv* **2014**, arXiv:1409.1556.
23. Szegedy, C.; Liu, W.; Jia, Y.; Sermanet, P.; Reed, S.; Anguelov, D.; Erhan, D.; Vanhoucke, V.; Rabinovich, A. Going deeper with convolutions. In Proceedings of the IEEE Conference on Computer Vision and Pattern Recognition (CVPR), Boston, MA, USA, 7–12 June 2015; pp. 1–9.
24. Xia, M.; Wang, K.; Song, W.; Chen, C.; Li, Y. Non-intrusive load disaggregation based on composite deep long short-term memory network. *Expert Syst. Appl.* **2020**, *160*, 113669. [CrossRef]
25. Xie, E.; Wang, W.; Wang, W.; Sun, P.; Xu, H.; Liang, D.; Luo, P. Segmenting transparent object in the wild with transformer. *arXiv* **2021**, arXiv:2101.08461.
26. Dosovitskiy, A.; Beyer, L.; Kolesnikov, A.; Weissenborn, D.; Zhai, X.; Unterthiner, T.; Dehghani, M.; Minderer, M.; Heigold, G.; Gelly, S.; et al. An image is worth 16x16 words: Transformers for image recognition at scale. *arXiv* **2020**, arXiv:2010.11929.
27. Carion, N.; Massa, F.; Synnaeve, G.; Usunier, N.; Kirillov, A.; Zagoruyko, S. End-to-end object detection with transformers. In Proceedings of the European Conference on Computer Vision, Glasgow, UK, 23–28 August 2020; pp. 213–229. [CrossRef]
28. Zheng, S.; Lu, J.; Zhao, H.; Zhu, X.; Luo, Z.; Wang, Y.; Fu, Y.; Feng, J.; Xiang, T.; Torr, P.H.S.; et al. Rethinking Semantic Segmentation from a Sequence-to-Sequence Perspective with Transformers. *arXiv* **2020**, arXiv:2012.15840.
29. Vaswani, A.; Shazeer, N.; Parmar, N. Attention is all you need. In *Advances in Neural Information Processing Systems*; Curran Associates Inc.: Red Hook, NY, USA, 2017; pp. 5998–6008.
30. Kaiser, P.; Wegner, J.D.; Lucchi, A.; Jaggi, M.; Hofmann, T.; Schindler, K. Learning aerial image segmentation from online maps. *IEEE Trans. Geosci. Remote Sens.* **2017**, *55*, 6054–6068. [CrossRef]
31. Rottensteiner, F.; Sohn, G.; Gerke, M.; Wegner, J.D. ISPRS Semantic Labeling Contest. *ISPRS* **2014**, *1*, 4.
32. Long, J.; Shelhamer, E.; Darrell, T. Fully convolutional networks for semantic segmentation. In Proceedings of the IEEE Conference on Computer Vision and Pattern Recognition (CVPR), Boston, MA, USA, 7–12 June 2015; pp. 3431–3440.
33. Ronneberger, O.; Fischer, P.; Brox, T. U-net: Convolutional networks for biomedical image segmentation. In Proceedings of the International Conference on Medical Image Computing and Computer-Assisted Intervention, Munich, Germany, 5–9 October 2015; Springer: Berlin/Heidelberg, Germany, 2015; pp. 234–241.
34. Zhao, H.; Shi, J.; Qi, X.; Wang, X.; Jia, J. Pyramid scene parsing network. In Proceedings of the IEEE Conference on Computer Vision and Pattern Recognition(ECCV), Honolulu, HI, USA, 21–26 July 2017; pp. 2881–2890.
35. Chen, L.C.; Zhu, Y.; Papandreou, G.; Schroff, F.; Adam, H. Encoder-decoder with atrous separable convolution for semantic image segmentation. In Proceedings of the European Conference on Computer Vision (ECCV), Munich, Germany, 8–14 September 2018; pp. 801–818.

International Journal of
Geo-Information

Article

RepDarkNet: A Multi-Branched Detector for Small-Target Detection in Remote Sensing Images

Liming Zhou [1,2], Chang Zheng [1,2], Haoxin Yan [1,2], Xianyu Zuo [1,2,*], Yang Liu [1,2,3], Baojun Qiao [1,2] and Yong Yang [4]

1 Henan Key Laboratory of Big Data Analysis and Processing, Henan University, Kaifeng 475000, China; lmzhou@henu.edu.cn (L.Z.); czheng@henu.edu.cn (C.Z.); 104754190933@henu.edu.cn (H.Y.); sea@vip.henu.edu.cn (Y.L.); qbj@henu.edu.cn (B.Q.)
2 School of Computer and Information Engineering, Henan University, Kaifeng 475000, China
3 Henan Province Engineering Research Center of Spatial Information Processing, Henan University, Kaifeng 475004, China
4 Institute of Plant Stress Biology, State Key Laboratory of Cotton Biology, Department of Biology, Henan University, Kaifeng 475000, China; 104752160097@vip.henu.edu.cn
* Correspondence: xianyu_zuo@henu.edu.cn

Abstract: Recent years have seen rapid progress in target-detection missions, whereas small targets, dense target distribution, and shadow occlusion continue to hinder progress in the detection of small targets, such as cars, in remote sensing images. To address this shortcoming, we propose herein a backbone feature-extraction network called "RepDarkNet" that adds several convolutional layers to CSPDarkNet53. RepDarkNet considerably improves the overall network accuracy with almost no increase in inference time. In addition, we propose a multi-scale cross-layer detector that significantly improves the capability of the network to detect small targets. Finally, a feature fusion network is proposed to further improve the performance of the algorithm in the *AP*@0.75 case. Experiments show that the proposed method dramatically improves detection accuracy, achieving *AP* = 75.53% for the Dior-vehicle dataset and *mAP* = 84.3% for the Dior dataset, both of which exceed the state-of-the-art level. Finally, we present a series of improvement strategies that justifies our improvement measures.

Keywords: deep learning; convolutional neural network; backbone network; target detection; remote sensing images

Citation: Zhou, L.; Zheng, C.; Yan, H.; Zuo, X.; Liu, Y.; Qiao, B.; Yang, Y. RepDarkNet: A Multi-Branched Detector for Small-Target Detection in Remote Sensing Images. *ISPRS Int. J. Geo-Inf.* **2022**, *11*, 158. https://doi.org/10.3390/ijgi11030158

Academic Editors: Gloria Bordogna and Cristiano Fugazza

Received: 7 January 2022
Accepted: 20 February 2022
Published: 22 February 2022

1. Introduction

A basic task in computer vision is target detection, which frames the region of interest in the input image in a bounding box. As space technology continues to develop, super-resolution remote sensing images are becoming increasingly important. However, target detection in remote sensing imaging remains a huge challenge and is thus receiving increasing research attention. According to different imaging bands, remote sensing can be divided into optical remote sensing, infrared remote sensing, SAR (Synthetic Aperture Radar), and other categories. Target detection in optical remote sensing images plays an important role in a wide range of applications such as environmental monitoring, geological hazard detection, land-use/land-cover (LULC) mapping, geographic information system (GIS) updating, precision agriculture, and urban planning [1].

In optical remote sensing imaging, targets usually have various orientations because of the differences in overhead views. Targets in optical remote sensing images thus consume a small number of target pixels distributed widely across different directions, making them easily affected by the surrounding environment. For example, vehicles in optical remote sensing images are often obscured by shadows, as shown in Figure 1. These factors make the difficult task of target detection in optical remote sensing images a research priority.

(a) (b)

Figure 1. The vehicles in optical remote sensing imagery. Vehicles in (**a**) are generally below 10 pixels, and in (**b**) are obscured by shadows. Image in (**a**) is from the Dior dataset, and (**b**) from the HRRSD dataset.

In the last few years, the rise of deep learning has contributed greatly to the development of target detection. Good results have been achieved from a series of powerful target-detection algorithms, such as RCNN [2], Fast RCNN [3], Faster RCNN [4], SSD [5], YOLO [6,7], and ResNet [8]. However, the performance of these algorithms when applied to aerial-imaging tasks remains unsatisfactory.

In many convolutional network models, the backbone network is often used to extract target features, such as color, texture, and scale. The backbone network can provide several combinations of sensory field sizes and center steps to meet the target-detection requirements of different scales and classes [9]. Some commonly used backbone networks for various computer-vision tasks are VGG [10], ResNet [8], MobileNet [11–13], and DarkNet [6,7].

To detect small targets, the backbone network plays an important role. A good backbone network extracts more feature information, but feature information is more difficult to extract from small targets. However, in optical remote sensing imaging, car targets may consume fewer than 10 pixels because of the angle of the shot. To detect those targets in optical remote sensing images, we made the following propositions:

1. Influenced by RepVgg [14], we proposed a backbone network named "RepDarkNet" that combines training accuracy and detection speed. Experiments show that RepDarkNet performs better when applied to the Dior dataset than do YOLOv3 and YOLOv4 with DarkNet style as the backbone.
2. We proposed a cross-layer converged network for small targets in optical remote sensing images. The network contains multi-scale cross-layer detection and feature fusion networks.
3. Besides, larger input and GIoU [15] were used to improve very-small-target detection and tested separately by applying them to the Dior-vehicle dataset.

The remainder of this paper is organized as follows: Section 1 introduces related research on detection in optical remote sensing images. Section 2 describes the proposed approach to detect small targets. Section 3 describes the experimental design and experimental details, and Section 4 describes the experimental results and analysis. Finally, we conclude this study in Section 5.

2. Related Work

Usually, the horizontal bounding box (HBB) target-detection method is used for general scenes and for optical remote sensing target detection. Rabbi et al. [9] used GAN

(Generative Adversarial Networks) to convert images into super-resolution images and extract feature information from the images to improve small-target detection. Zhang et al. [16] proposed a hierarchical robust convolutional neural network (CNN) and built a large high-resolution target-detection dataset called HRSSD. Adam et al. [17] proposed YOLT, which evaluates satellite images of arbitrary size at a rate exceeding 0.5 km^2/s and locates small targets (5 pixels) at high resolution.

Optical remote sensing imagery contains mostly densely distributed targets of arbitrary orientations; therefore, HBB-based targets may overlap significantly. Oriented bracketing boxes are often used to detect rotating targets when characterizing targets in aerial images. Ding et al. [18] proposed a rotational region-of-interest learner to convert horizontal regions of interest to rotational regions of interest. Based on the rotation-sensitive regions, a rotation position-sensitive region alignment (RPS-RoI-Align) module was proposed to extract rotation-invariant features from rotation-sensitive regions and thereby facilitate subsequent classification and regression. Yang et al. [19] proposed a multi-class rotational detector SCRDet for small, randomly oriented, and intensively distributed targets in aerial imagery, which improves sensitivity to small targets through a sampling fusion network that fuses multilayer features with anchored sampling. Finally, Zhou et al. [20] proposed an anchorless polar-coordinate optical remote sensing target detector (P-RSDet) that provides competitive detection accuracy by using a simpler target representation model and fewer regression parameters.

Concerning vehicles in optical remote sensing imagery, numerous researchers have investigated the characteristics of car targets. Audebert et al. [21] trained a deep, fully convolutional network on ISPRS Potsdam and the NZAM/ONERA Christchurch datasets, and used the learned semantic graph to extract the exact vehicle segments, thereby detecting vehicles by simply extracting connected parts. Zhang et al. [22] proposed a YOLOv3-based deeply separable attention-guided network for the real-time detection of small vehicle targets in optical remote sensing imagery. Shi et al. [23] proposed a single-stage anchorless detection method to detect vehicles in arbitrary directions. This method transforms vehicle detection into a multitask learning problem that requires the use of a full convolutional network to directly predict high-level vehicle features.

3. Materials and Methods

This section introduces the backbone network RepDarkNet, which is a multi-scale cross-layer detector and feature fusion network. In addition, we tested some common methods to improve the detection rate of small targets.

3.1. Reasons for Choosing DarkNet Style

We chose the DarkNet structure because it combines accuracy and speed. Currently, backbone networks are carefully designed to include a multi-branch structure, such as MobileNet [11–13], Inception [24], or DenseNet [25]. They have common characteristics; for instance, DenseNet makes the topology more complex by connecting lower layers to many higher layers, and while it offers a high-performance conversion network, it does so at the cost of a mass of GPUs.

Darknet53 is a simple two-branch network that adds branches in a top-down convolution process. It is mainly composed of a series of fully designed residual blocks. As can be seen in Figure 2, the residual block loads the output of the two convolutions plus the output of a skip connection together as the input information for the next layer of convolution. The elaborate network layer of DarkNet produces surprisingly fast detection.

In a multi-branch network structure, the paths between branches are not strongly interdependent, so the structure can be seen as a collection of many paths [26]. On a small scale, the DarkNet-style network is composed of multiple residual block results, whereas on a larger scale, it can be seen as five layers, as in Figure 2, where we divide the backbone network into a five-block structure ($R1$–$R5$). In the field of target detection, deeper network structures tend to be stronger than shallower structures, although building multiple branching structures in each residual layer increases training time and detection time and wastes significant computational resources in return for poor accuracy. Therefore, implementing such network structures in one big module is a good way to improve them. YOLOv4 thus makes a breakthrough in accuracy, so we improve on the backbone network CSPDarkNet.

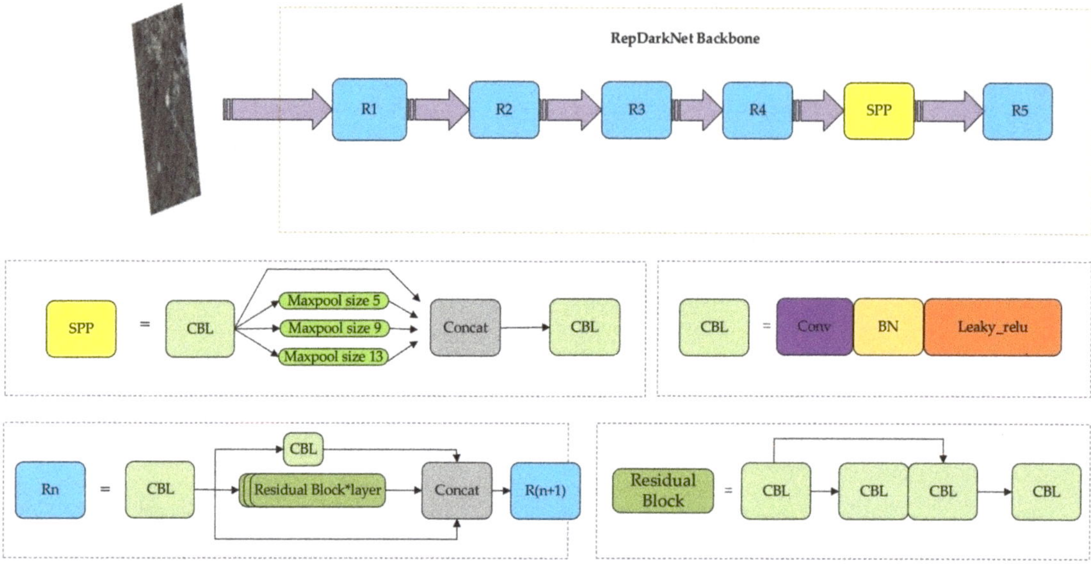

Figure 2. Details of backbone network. In Rn, when $n = 1$, layer = 1; when $n = 2$, layer = 2; when $n = 3$ or 4, layer = 8; and when $n = 5$, layer = 4. The purple arrow represents the connection between R_n and R_{n+1} blocks, the output information of R_n is the input information of R_{n+1}.

3.2. Overview of DarkNet-Style Backbone Network

The proposed RepVGG (re-parameterization VGG) [14] makes VGG shine again in classification tasks. It is a simple, yet powerful, neural network structure. For target detection, DarkNet is an excellent and fast framework, but accuracy is what many researchers now focus on. In this work, we continue the simple structure of DarkNet and design a muti-branch backbone network RepDarkNet (re-plan the branching structure of DarkNet), the details of which are given in Figure 2, where R_n represents a four-branch module in RepDarkNet. In the SPP module, the three maxpool layers are of sizes 5, 9, and 13, respectively.

CSPDarkNet is one of the backbone networks of the CSP(Cross-Stage-Partial-connections) module [7]. If CSPDarkNet is a two-branch structure, RepDarkNet is a simple, but powerful, three-branch structure. Compared with CSPDarkNet, it adds only one layer of branches per layer and constitutes a module with the information $y = f(x) + g(x) + x$, where $f(x)$ can be expressed as:

$$f_i(x) = W_i \cdot \delta(B(x)), \tag{1}$$

where x is the output of the previous layer, W is the weight information, $\cdot$ indicates convolution, B is the batch normalization operation, and $\delta(x) = \max(x, 0)$ is the Relu function. To explain $g(x)$, we take as an example R2, which contains two residual modules, each of which can be expressed as:

$$y_i = f_i(y_{i-1}) + y_{i-1}. \tag{2}$$

Given an input value y_0, then the first residual output is:

$$y_3 = y_2 + f_3(y_2) \tag{3}$$

$$= [y_1 + f_2(y_1)] + f_3(y_1 + f_2(y_1)) \tag{4}$$

$$= [y_0 + f_1(y_0) + f_2(y_0 + f_1(y_0))] + f_3(y_0 + f_1(y_0) + f_2(y_0 + f_1(y_0))). \tag{5}$$

Similarly, the second convolution result is $g(x) = y_6$, so:

$$y_6 = y_5 + f_6(y_5) \tag{6}$$

$$= [y_4 + f_5(y_4)] + f_6(y_4 + f_5(y_4)) \tag{7}$$

$$= [y_3 + f_4(y_3) + f_5(y_3 + f_4(y_3))] + f_6(y_3 + f_4(y_3) + f_5(y_3 + f_4(y_3))). \tag{8}$$

In the process of convolution, CNNs gradually lose the feature information of small targets, whereas the multi-branch module is built to send the upper-layer information directly to the deep layer, enriching the feature information of the deep layer network, especially for small targets. The experimental results also show that RepDarkNet detects small targets better than other algorithms do.

3.3. Cross-Layer Fusion Network

The cross-layer fusion network consists of two parts: a multi-scale cross-layer detector and a neck feature fusion network.

In a YOLOv4 network, the YOLO head usually splits the image into 19 × 19, 38 × 38, and 76 × 76 grids. However, in optical remote sensing images, small targets are usually smaller than 30 or even 20 pixels; therefore, adding a detection head at a shallow level is a good way to improve the detection accuracy of small targets, and studies show that this is feasible [27]. To detect small targets, as shown in Figure 3, N3 does not work well. Segmenting a 1080 × 1080 pixel image into a 76 × 76 grid makes for difficult detection based on the existing feature information. Therefore, a multi-scale cross-layer detector was designed that takes into account the size of the network and the inference time. The detector uses only layers N1, N2, and N4 and divides the image into three scales of 19 × 19, 38 × 38, and 152 × 152 to detect large and small targets in the image during inference. Experiments show that the multi-scale cross-layer detector is much more effective, especially in detecting small targets. Figure 4 shows its network structure.

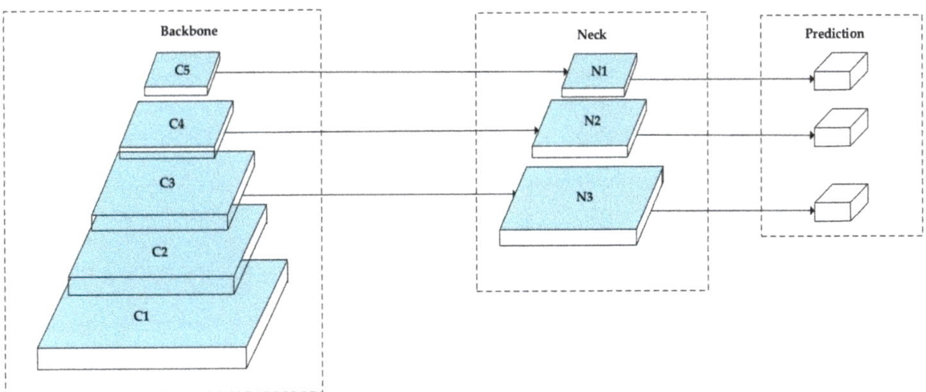

Figure 3. Overviews of multi-scale detection in YOLOv4.

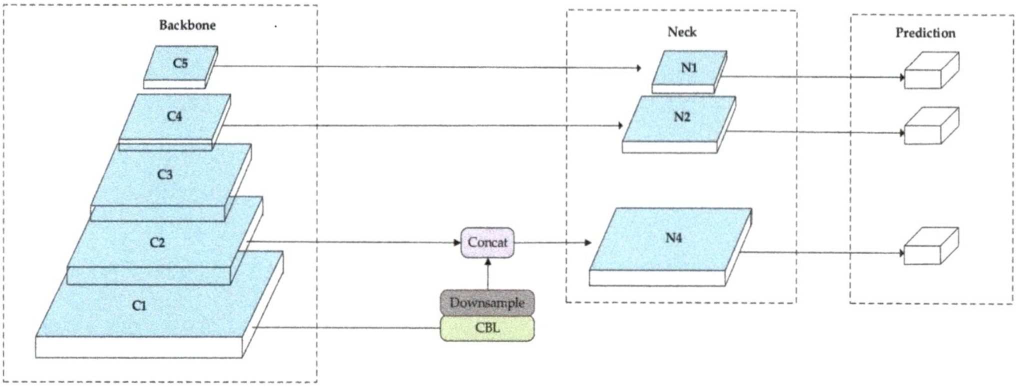

Figure 4. Overview of multi-scale cross-layer detectors and feature fusion networks.

Inspired by Lim et al. [28], we fused higher-level feature maps from the target feature layer to provide context for a given feature map of the target to be detected. For example, in RepDarkNet, given the target features from *R*2, the contextual features come from the *R*1 layer, as shown in Figure 4. However, *R*1 and *R*2 have different dimensions, so the network channels are downscaled by downsample in order to ensure that the size of the feature information is the same on both sides before fusing the features, and the fused feature information is convolved for detection by a multi-scale cross-layer detector.

3.4. Options for Improving Accuracy of Small-Target Detection

3.4.1. GIoU

As a loss of the Bbox regression, GIoU [15] (Generalized Intersection over Union) is always less than or equal to IoU [29] (Intersection over Union), where $0 \leq \text{IoU} \leq 1$ and $-1 \leq \text{GIoU} \leq 1$. When the two shapes coincide exactly, GIoU = IoU = 1. GIoU and IoU can be expressed as:

$$\text{GIoU} = \text{IoU} - \frac{|C/(A \cup B)|}{|C|}, \tag{9}$$

$$\text{IoU} = \frac{|A \cap B|}{|A \cup B|}, \tag{10}$$

where *A* and *B* are arbitrary shapes. In this algorithm, *A* and *B* identify the precision $= \frac{TP}{TP+FP}$ marker box in the image, and the algorithm predicts the anchor selection box. *C* is the minimum closed shape.

Both IoU and GIoU serve as distances between *A* and *B*; if $|A \cap B| = 0$, then IoU(*A*, *B*) = 0, so IoU does not indicate whether the two shapes are nearby or far-away.

3.4.2. Larger Input Size

Larger input size increases the size of the target area and makes it easier to preserve the features of smaller targets. However, increasing the input size takes up more memory during training. In our experiments, we increased the maximum input size only from 608×608 to 640×640 because of experimental constraints. We could also increase the input size while decreasing the batch value, although this would significantly increase the training time.

4. Experiments, Results, and Discussion

This section introduces the dataset that we used, the evaluation criteria, the details of the experiments, and the results. The experimental environment for this work is based on the Ubuntu 16.04 operating system. The hardware platform was an Intel(R) Xeon(R) Silver 4114 CPU @ 2.2 GHz and a Quadro P4000 8 GB * 2 GPU.

In the process of training, the momentum is set to 0.949, initial learning rate is set to 0.0013, and batch is set to 64. There are different numbers of iterations for different datasets for training. In the Dior-vehicle dataset, when the number for training iterations is 3200 and 3600, the learning rate is adjusted to 0.00013 and 0.000013, respectively. We have iterated 4000 times of this algorithm. Finally, in this hardware environment, the training time is about 12 h.

4.1. Dataset

Dior [30] and NWPU VHR-10 [30] were the datasets used to test our method. To show more intuitively the results of the algorithm to detect small targets, we tested the Dior-vehicle [31] dataset using the ablation experiment method.

4.1.1. Dior-Vehicle Dataset

The Dior-vehicle is a separate class of vehicle in Dior; the dataset has mostly small targets, and car targets consuming fewer than 10 pixels are flagged. Here, we follow COCO [32] for object-size classification, which small objects area is less than 32×32 pixels. The Dior-vehicle dataset has 6421 images and over 32,000 targets. In our experiments, we divide the training and test sets in the ratio 6:4. We then test the feasibility of the proposed method via an ablation experiment [4].

4.1.2. Dior Dataset

Dior is a large-scale benchmark dataset for target detection in optical remote sensing images consists of 23,463 images and 190,288 instances. It contains 20 categories. In the experiments reported herein, we use images from "trainval" for training and images from test for testing. The "trainval" is a TXT document. The author of the Dior dataset wrote the name of the image which is used for training into this document.

Besides, the size of objects varies over a wide range, not only in terms of spatial resolution, but also in terms of inter- and intra-class size variation between objects. Next, the images were obtained under different imaging conditions, weather, seasons, and image quality, so there are large differences in the targets therein. Some images may have noise and some targets in motion would be blurred. Finally, it has a high degree of inter-class similarity and intra-class diversity. These characteristics greatly increase the difficulty of detection.

If you want to find out more information about the dataset, please visit http://www.escience.cn/people/gongcheng/DIOR.html, accessed on 7 October 2020.

4.1.3. NWPU VHR-10 Dataset

The NWPU VHR-10 dataset is a research-only public dataset containing 10 classes. NWPU VHR-10 has 650 positive optical images and 150 negative optical images. In our experiments, we divided the 650 positive optical images with annotations at a 5:5 rate to produce a training set and a test set. If you want to find out more information about the dataset, please http://www.escience.cn/people/gongcheng/NWPU-VHR-10.html (accessed on 10 May 2021).

4.2. Evaluation Standards

The F1 score, precision, recall, and IoU are also considered in our measurement in the ablation experiment to test the Dior-vehicle dataset. The formula for IoU is given in Equation (1), and the F1 score, precision, and recall can be expressed by:

$$\text{precision} = \frac{TP}{TP + FP}, \tag{11}$$

$$\text{recall} = \frac{TP}{TP + FN}, \tag{12}$$

$$\text{F1 score} = \frac{2(\text{precision} \times \text{recall})}{\text{precision} + \text{recall}}. \tag{13}$$

TP, FP, and FN are the number of true positives, false positives, and false negatives, respectively. In this paper, we use AP (average precision) and mAP (mean average precision) to evaluate all methods. These can be expressed as follows:

$$AP_i = \int_0^1 \text{precision}_i(r_i)d(r_i), \tag{14}$$

$$mAP = \frac{1}{n}\sum_{i=1}^{n} AP_i. \tag{15}$$

When testing the results, a threshold value for IoU is set, such as 0.5 or 0.75. When the detected box and the actual annotated box exceed this threshold, they are considered true positives; otherwise, they are false positives. When the detected box is not present in the actual annotation, they are false negatives.

4.3. Experimental Results and Analyses

4.3.1. Result of Dior-Vehicle and Analyses

To investigate the impression of the method we used on the detection results, we used an ablation experiment to test the dataset. As shown in Table 1, "a" represents the Rep backbone, "b" the larger input, "c" the GIoU, "d" the multi-scale cross-layer detector, and "e" the feature fusion network. For example, RepDarkNet-A2 uses both "a" and "c". RepDarknet-B3 uses all the methods and is our final algorithm.

Table 1. Improvement options and comparison of results.

	a	b	c	d	e	AP@0.5(%)	AP@0.75(%)
YOLOv4	-	-	-	-	-	60.33	26.35
RepDarkNet-A0	√	-	-	-	-	71.58	34.16
RepDarkNet-A1	√	√	-	-	-	74.51	38.51
RepDarkNet-A2	√	-	√	-	-	74.16	37.10
RepDarkNet-B1	√	-	-	√	-	75.82	33.32
RepDarkNet-B2	√	-	-	√	√	74.87	35.37
RepDarkNet-B3	√	√	√	√	√	75.52	38.40

In the first step, the RepDarkNet-A0, we add only the RepDarkNet backbone on top of the YOLOv4 algorithm. As can be seen in Table 1, the AP is greatly improved (60.33–71.58%). Next, we use larger input and RepDarkNet backbone, and the AP is increased by 2.93%. Thirdly, we use the GIoU and RepDarkNet backbone, and compared to the RepDarkNet-A0, the AP is increased by 2.58%. Then, we use RepDarkNet backbone and the multi-scale cross-layer detector, and compared to the RepDarkNet-A0, the AP is increased by 4.24%. This is the highest promotion method other than RepDarkNet backbone. In step five, we use the RepDarkNet backbone, the multi-scale cross-layer detector and the feature fusion network. Compared to RepDarkNet-B1, AP@50 is 0.95% less, but AP@75 is 2.05% more. Finally, we added all the methods. The proposed method approaches the best performance for the Dior-vehicle dataset in both the AP@50 and AP@75 cases. Figure 5 shows the comparison of loss and mAP between the YOLOv4 and RepDarkNet-B3. It can be seen that RepDarkNet-B3's loss value is lower than YOLOv4 after 1200 iterator times. At the same time, the mAP value is higher than YOLOv4. Besides, Figure 6 shows the variation in AP@50 and AP@75 for each method.

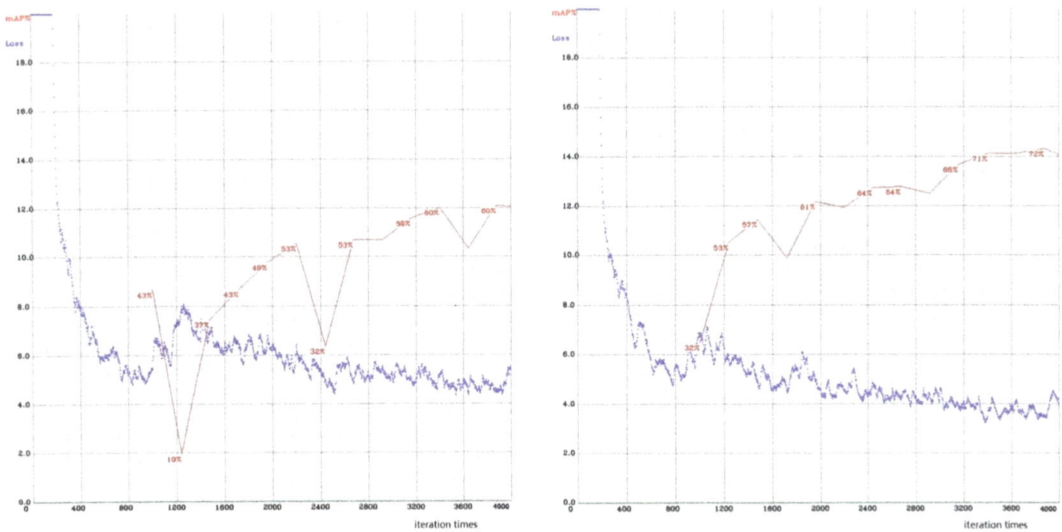

(a) (b)

Figure 5. YOLOv4 (**a**) and RepDarkNet-B3 (**b**) loss comparison.

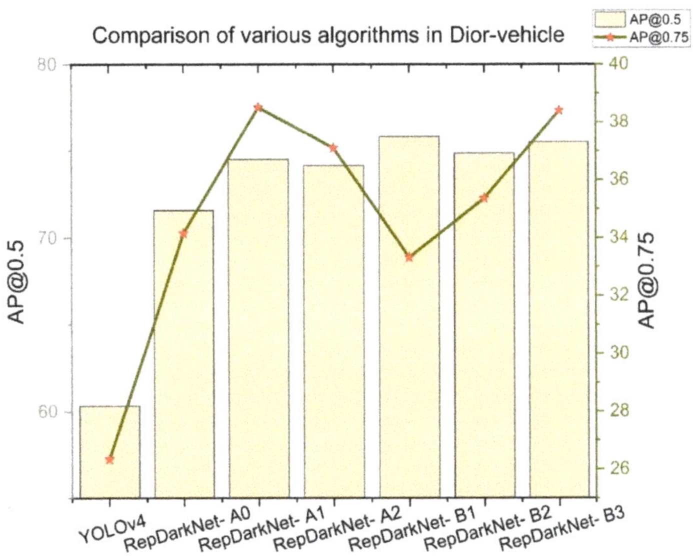

Figure 6. Comparison of various algorithms in Dior-vehicle.

As can be seen in Figure 7(a1,a2,b1,b2), the YOLOv4 algorithm incorrectly detects the ship as a car, whereas RepDarkNet accurately frames the car targets of the images. It is clear from Figure 7(a3,a4,b3,b4) that RepDarkNet detects small targets much better than YOLOv4 does. The results in Tables 1 and 2 demonstrate the clear improvement offered by the proposed method for small targets. Figure 7(a5,a6) show that the YOLOv4 algorithm incorrectly detects road signs as cars, whereas RepDarkNet correctly ignores the road signs. Considering the inconspicuous markings of the misdetection in Figure 7(a5,a6), we have marked it with a yellow circle in order for the reader to notice this information more quickly. This demonstrates that RepDarkNet is more robust than YOLOv4.

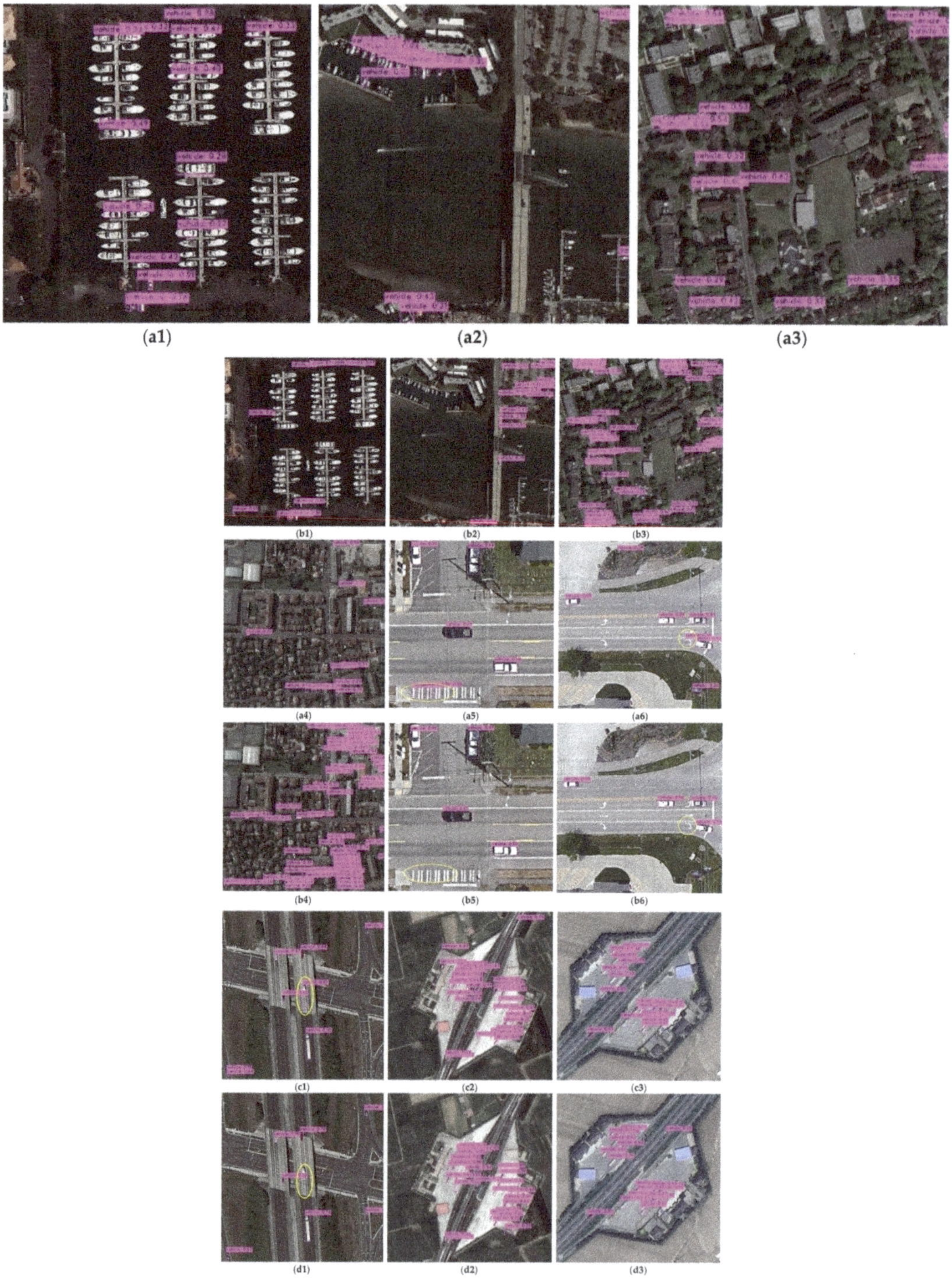

Figure 7. Comparison of detection results between YOLOv4 (**a1–a6,c1–c3**) and RepDarkNet (**b1–b6,d1–d3**).

Table 2. Comparison of recall, precision, F1-score, and IoU by methods.

Method	F1-Score	Recall	Precision	IoU
YOLOv4	58%	59%	57%	42.90%
RepDarkNet-A0	69%	68%	71%	53.89%
RepDarkNet-A1	71%	72%	70%	53.84%
RepDarkNet-A2	72%	72%	71%	54.76%
RepDarkNet-B1	70%	75%	66%	49.38%
RepDarkNet-B2	72%	72%	72%	54.75%
RepDarkNet-B3	72%	70%	77%	59.46%

Furthermore, there are more special categories of vehicles, such as lorries, characteristics of which are different from those of ordinary cars. In the dataset, although lorries are also labeled, the number of lorries in the data is relatively much smaller. As shown in Figure 7(c1,d1), the detection accuracy of our proposed algorithm is higher than that of YOLOv4 for lorries.

Of course, our algorithm also has some shortcomings. By looking at Figure 7(c2,c3,d2,d3), we can see that it is easy to mistake containers as cars, regardless of RepDarkNet or YOLOv4. This is a problem we want to solve in our future research.

Finally, to test the generality of the proposed detector, we directly detect the vehicle class in the HRRSD dataset [16] without any training using Dior's weights. The final detection result of $AP = 50.73\%$ was obtained. Although far from expected, this far exceeds the result of $AP = 23.78\%$ obtained by YOLOv4. Of course, we also hope that in future research, there is a greater breakthrough in the generality performance of the detector.

4.3.2. Results of the Dior Dataset and Analyses

As shown in Table 3, to better display the results, we number each category in the dataset.

Table 3. Categories in the DIOR dataset and their corresponding numbers.

c1	Airplane	c6	Chimney	c11	Ground track field	c16	Storage tank
c2	Airport	c7	Dam	c12	Harbor	c17	Tennis court
c3	Baseball field	c8	Expressway service area	c13	Overpass	c18	Train station
c4	Basketball court	c9	Expressway toll station	c14	Ship	c19	Vehicle
c5	Bridge	c10	Golf court	c15	Stadium	c20	Windmill

Table 4 shows clearly that the proposed algorithm performs the best for the Dior dataset, far outperforming other algorithms for the detection of small targets, such as cars or ships. For relatively large targets, such as bridges or basketball courts, the detection accuracy is also improved slightly relative to other algorithms, which demonstrates the strong applicability of the proposed algorithm. Compared with the YOLOv4 algorithm, most classes in our test are more accurate, which demonstrates convincingly that RepDarkNet is an excellent target-detection algorithm.

Figure 8a–i shows visualization results of the proposed method applied to the Dior dataset. RepDarkNet performs well not only for small targets such as ships, planes, and storage tanks in the dataset but also for other classes such as tennis and basketball courts. The results in Table 4 also show that in addition to dam, expressway toll station, tennis court and train station, the proposed approach produces the best results compared with other algorithms. Only in dam, expressway toll station and tennis court, YOLOv4 is better than RepDarkNet.

Table 4. Comparison of the results of the algorithms on the Dior dataset.

Method	Backbone	c1	c2	c3	c4	c5	c6	c7	c8	c9	c10	c11	c12	c13	c14	c15	c16	c17	c18	c19	c20	mAP
SSD [30]	VGG16	59.5	72.7	72.4	75.7	29.7	65.8	56.6	63.5	53.1	65.3	68.6	49.4	48.1	59.2	61.0	46.6	76.3	55.1	27.4	65.7	58.6
YOLOv3 [30]	Darknet-53	72.2	29.2	74.0	78.6	31.2	69.7	26.9	48.6	54.4	31.1	61.1	44.9	49.7	87.4	70.6	68.7	87.3	29.4	48.3	78.7	57.1
Faster RCNNwith	ResNet-50	54.1	71.4	63.3	81.0	42.6	72.5	57.5	68.7	62.1	73.1	76.5	42.8	56.0	71.8	57.0	53.5	81.2	53.0	43.1	80.9	63.1
FPN [30]	ResNet-101	54.0	74.5	63.3	80.7	44.8	72.5	60.0	75.6	62.3	76.0	76.8	46.4	57.2	71.8	68.3	53.8	81.1	59.5	43.1	81.2	65.1
Mask-RCNNwith	ResNet-50	53.8	72.3	63.2	81.0	38.7	72.6	55.9	71.6	67.0	73.0	75.8	44.2	56.5	71.9	58.6	53.6	81.1	54.0	43.1	81.1	63.5
FPN [30]	ResNet-101	53.9	76.6	63.2	80.9	40.2	72.5	60.4	76.3	62.5	76.0	75.9	46.5	57.4	71.8	68.3	53.7	81.0	62.3	43.0	81.0	65.2
RetinaNet [30]	ResNet-50	53.7	77.3	69.0	81.3	44.1	72.3	62.5	76.2	66.0	77.7	74.2	50.7	59.6	71.2	69.3	44.8	81.3	54.2	45.1	83.4	65.7
	ResNet-101	53.3	77.0	69.3	85.0	44.1	73.2	62.4	78.6	62.8	78.6	76.6	49.9	59.6	71.1	68.4	45.8	81.3	55.2	44.4	85.5	66.1
PANet [30]	ResNet-50	61.9	70.4	71.0	80.4	38.9	72.5	56.6	68.4	60.0	69.0	74.6	41.6	55.8	71.7	72.9	62.3	81.2	54.6	48.2	86.7	63.8
	ResNet-101	60.2	72.0	70.6	80.5	43.6	72.3	61.4	72.1	66.7	72.0	73.4	45.3	56.9	71.7	70.4	62.0	80.9	57.0	47.2	84.5	66.1
CF2PN [33]	VGG16	78.3	78.3	76.5	88.4	37	71	59.9	71.2	51.2	75.6	77.1	56.8	58.7	76.1	70.6	55.5	88.8	50.8	36.9	80.4	67.3
MFPnet [34]	VGG16	76.6	83.4	80.6	82.1	44.3	75.6	68.5	85.9	63.9	77.3	77.2	62.1	58.8	77.2	76.8	60.3	86.4	64.5	41.5	80.2	71.2
CANet [35]	ResNet-101	70.3	82.4	72	87.8	55.7	79.9	67.7	83.5	77.2	77.3	83.6	56.0	63.6	81.0	79.8	70.8	88.2	67.6	51.2	89.6	74.3
YOLOv4	CSPDarkNet	96.0	87.9	94.7	91.9	60.0	90.8	**69.9**	92.1	**87.5**	87.6	83.7	55.6	68.7	94.6	83.8	88.4	**95.7**	44.4	62.1	90.4	81.3
ours	RepDarkNet	**97.7**	**89.5**	**94.9**	**92.5**	**62.2**	**91.4**	68.2	**94.5**	85.8	**87.1**	**91.4**	**62.9**	**72.9**	**95.7**	**92.3**	**89.5**	95.5	56.7	**71.4**	**93.6**	**84.3**

Figure 8. Detection results of RepDarkNet in the Dior dataset (**a**–**i**).

4.3.3. Results of NWPU VHR-10 and Analyses

Table 5 and Figure 9a–i illustrate the performance of RepDarkNet compared with those of other published methods when applied to the NWPU VHR-10 dataset. We first introduce the abbreviations of classes in the Table 5. The categories are Airplane (PL), Ship (SP), Storage tank (ST), Baseball diamond (BD), Tennis court (TC), Basketball court (BC), Ground track field (GT), Harbor (HB), Bridge (BR), and Vehicle (VH).

Table 5. Comparison of results from the NWPU VHR-10 dataset.

Method	*AP* (%) for Each Target Category										*mAP* (%)
	PL	SH	ST	BD	TC	BC	GT	HA	BR	VE	
SSD512 [36]	90.40	60.90	79.80	89.90	82.60	80.60	98.30	73.40	76.70	52.10	78.40
SAPNet [36]	97.80	87.60	67.20	94.80	99.50	99.50	95.90	96.80	68.00	85.10	89.20
StAN-Enh [36]	94.80	79.10	98.20	96.70	89.10	89.60	93.50	91.00	62.70	93.80	88.9
CANet [35]	100.0	81.9	94.6	90.3	90.7	90.6	99.8	89.8	93.9	89.9	92.2
YOLOv4	99.99	81.33	98.80	97.41	97.40	95.52	99.37	82.27	77.30	93.70	92.31
ours	99.96	**93.71**	98.07	**97.47**	**99.56**	99.17	99.59	84.94	74.24	**94.29**	**94.1**

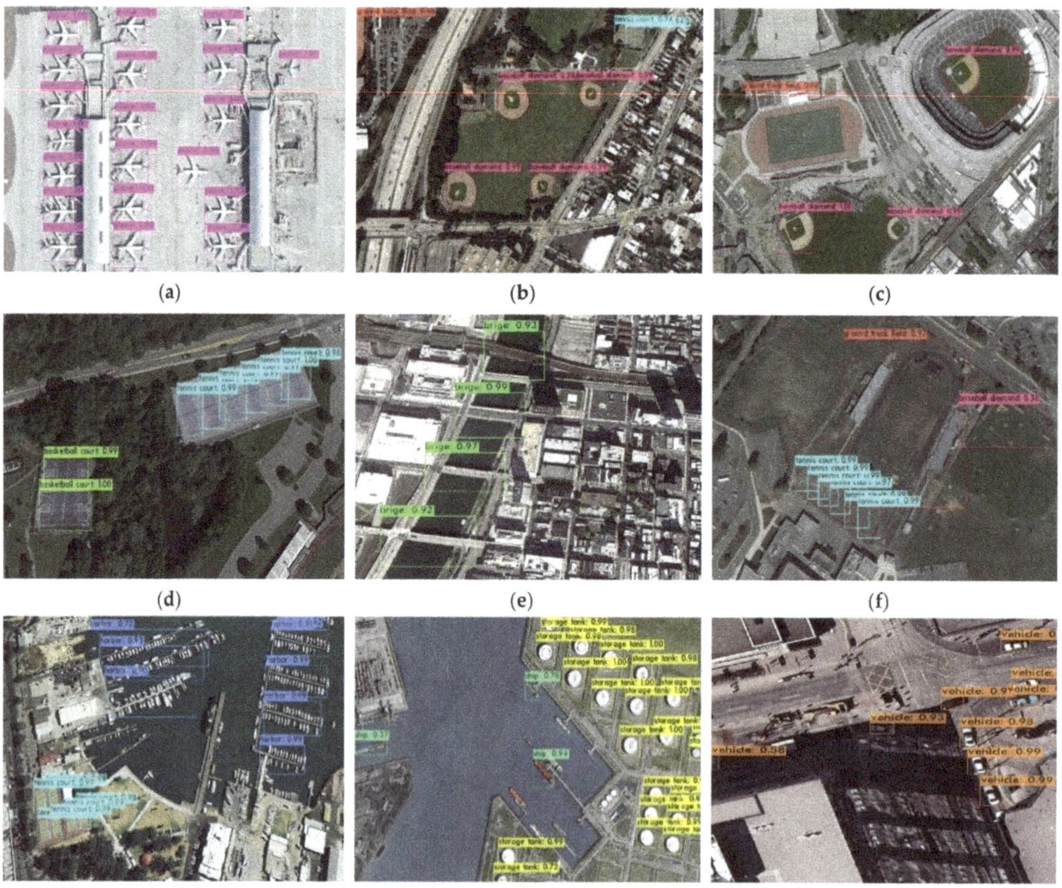

Figure 9. RepDarkNet detection results in the NWPU VHR-10 dataset (**a**–**i**).

The proposed method produces $mAP = 94.1\%$, which is the best performance of all the methods. Note that no small targets appear in the dataset and that the classes SH, BD, TC, and VE which are bolded in Table 5 have the highest accuracy in the experimental results. Overall, mAP is the highest too, which is good evidence of the strong applicability of our method.

5. Conclusions

We propose a new backbone feature-extraction network called "RepDarkNet" that provides improved target detection by considering network throughput and computation time. In addition, for small targets in the vehicle class of optical remote sensing images, we propose a multi-scale cross-layer detector and feature fusion network. Finally, in experiments, RepDarkNet achieves $AP@0.5 = 75.52\%$ and $AP@0.75 = 38.4\%$, both of which are near-optimal, and a series of ablation experiments justify our approach. In extended experiments, RepDarkNet running under Quadro P4000 obtains $mAP = 84.3\%$ when applied to the Dior dataset and $mAP = 94.1\%$ when applied to the NWPU VHR-10 dataset, outperforming other algorithms and demonstrating the broad applicability of the proposed algorithm. To test the algorithm for generalized rows, we obtained $AP = 50.73\%$ performance in HRRSD dataset using the training weights of the Dior-vehicle dataset with no other additional training. Under the same conditions, YOLOv4 obtains $AP = 23.78\%$ performance. Thus, in terms of generality, RepDarkNet is much better than YOLOv4. By optimizing and evaluating the proposed small-target network models under three datasets (Dior-vehicle, Dior, and NWPU VHR-10), these results have practical implications for improving detection techniques based on optical remote sensing images.

Author Contributions: Conceptualization, Liming Zhou and Chang Zheng; Methodology, Liming Zhou and Chang Zheng; Software, Haoxin Yan; Validation, Liming Zhou, Chang Zheng and Xianyu Zuo; Writing—original draft, Liming Zhou and Chang Zheng; Writing—review & editing, Chang Zheng, Yang Liu, Baojun Qiao and Yong Yang. All authors have read and agreed to the published version of the manuscript.

Funding: This work is supported by grants from National Basic Research Program of China (Grant number 2019YFE0126600); the Major Project of Science and Technology of Henan Province (Grant number 201400210300); the Key Scientific and Technological Project of Henan Province (Grant number 212102210496); the Key Research and Promotion Projects of Henan Province (Grant numbers 212102210393; 202102110121, 202102210368, 19210221009) and Kaifeng science and technology development plan (Grant number 2002001); National Natural Science Foundation of China (62176087) and Shenzhen Science and Technology Innovation Commission (SZSTI)-Shenzhen Virtual University Park (SZVUP) Special Fund Project (2021Szvup032).

Institutional Review Board Statement: Not applicable.

Informed Consent Statement: Not applicable.

Data Availability Statement: The following are available online at http://www.escience.cn/people/gongcheng/NWPU-VHR-10.html.NWPU VHR-10 dataset, accessed on 10 May 2021; www.escience.cn/people/gongcheng/DIOR.html, Dior dataset, accessed on 7 October 2020.

Conflicts of Interest: The authors declare no conflict of interest.

References

1. Cheng, G.; Han, J. A survey on object detection in optical remote sensing images. *ISPRS J. Photogramm. Remote Sens.* **2016**, *117*, 11–28. [CrossRef]
2. Girshick, R.; Donahue, J.; Darrell, T.; Malik, J. Rich Feature Hierarchies for Accurate Object Detection and Semantic Segmentation. In Proceedings of the IEEE Conference on Computer Vision and Pattern Recognition, Columbus, OH, USA, 23–28 June 2014.
3. Girshick, R. Fast R-CNN. In Proceedings of the IEEE International Conference on Computer Vision (ICCV), Santiago, Chile, 11–18 December 2015; pp. 1440–1448.

4. Ren, S.; He, K.; Girshick, R.; Sun, J. Faster R-CNN: Towards Real-Time Object Detection with Region Proposal Networks. *IEEE Trans. Pattern Anal. Mach. Intell.* **2017**, *39*, 1137–1149. [CrossRef]
5. Liu, W.; Anguelov, D.; Erhan, D.; Szegedy, C.; Reed, S.; Fu, C.Y.; Berg, A.C. SSD: Single Shot MultiBox Detector. In *European Conference on Computer Vision*; Springer: Cham, Switzerland, 2016.
6. Redmon, J.; Farhadi, A. YOLOv3: An Incremental Improvement. *arXiv* **2018**, arXiv:1804.02767.
7. Bochkovskiy, A.; Wang, C.Y.; Liao, H. YOLOv4: Optimal Speed and Accuracy of Object Detection. *arXiv* **2020**, arXiv:2004.10934.
8. He, K.; Zhang, X.; Ren, S.; Sun, J. Deep Residual Learning for Image Recognition. In Proceedings of the 2016 IEEE Conference on Computer Vision and Pattern Recognition (CVPR), Las Vegas, NV, USA, 27–30 June 2016.
9. Rabbi, J.; Ray, N.; Schubert, M.; Chowdhury, S.; Chao, D. Small-Object Detection in Remote Sensing Images with End-to-End Edge-Enhanced GAN and Object Detector Network. *Remote Sens.* **2020**, *12*, 1432. [CrossRef]
10. Simonyan, K.; Zisserman, A. Very Deep Convolutional Networks for Large-Scale Image Recognition. *arXiv* **2014**, arXiv:1409.1556.
11. Howard, A.G.; Zhu, M.; Chen, B.; Kalenichenko, D.; Wang, W.; Weyand, T.; Andreetto, M.; Adam, H. MobileNets: Efficient Convolutional Neural Networks for Mobile Vision Applications. *arXiv* **2017**, arXiv:1704.04861.
12. Sandler, M.; Howard, A.; Zhu, M.; Zhmoginov, A.; Chen, L.C. MobileNetV2: Inverted Residuals and Linear Bottlenecks. In Proceedings of the 2018 IEEE/CVF Conference on Computer Vision and Pattern Recognition (CVPR), Salt Lake City, UT, USA, 18–23 June 2018.
13. Howard, A.; Sandler, M.; Chu, G.; Chen, L.C.; Chen, B.; Tan, M.; Wang, W.; Zhu, Y.; Pang, R.; Vasudevan, V. Searching for MobileNetV3. *arXiv* **2019**, arXiv:1905.02244.
14. Ding, X.; Zhang, X.; Ma, N.; Han, J.; Ding, G.; Sun, J. RepVGG: Making VGG-style ConvNets Great Again. In Proceedings of the IEEE/CVF Conference on Computer Vision and Pattern Recognition, Nashville, TN, USA, 20–25 June 2021; pp. 13733–13742.
15. Rezatofighi, H.; Tsoi, N.; Gwak, J.; Sadeghian, A.; Reid, I.; Savarese, S. Generalized Intersection Over Union: A Metric and a Loss for Bounding Box Regression. In Proceedings of the IEEE/CVF Conference on Computer Vision and Pattern Recognition (CVPR), Long Beach, CA, USA, 15–20 June 2019.
16. Zhang, Y.; Yuan, Y.; Feng, Y.; Lu, X. Hierarchical and Robust Convolutional Neural Network for Very High-Resolution Remote Sensing Object Detection. *IEEE Trans. Geosci. Remote Sens.* **2019**, *57*, 5535–5548. [CrossRef]
17. Etten, A.V. You Only Look Twice: Rapid Multi-Scale Object Detection in Satellite Imagery. *arXiv* **2018**, arXiv:1805.09512.
18. Ding, J.; Xue, N.; Long, Y.; Xia, G.S.; Lu, Q. Learning RoI Transformer for Detecting Oriented Objects in Aerial Images. *arXiv* **2018**, arXiv:1812.00155.
19. Yang, X.; Yang, J.; Yan, J.; Zhang, Y.; Fu, K. SCRDet: Towards More Robust Detection for Small, Cluttered and Rotated Objects. In Proceedings of the 2019 IEEE/CVF International Conference on Computer Vision (ICCV), Seoul, Korea, 27 October–2 November 2019.
20. Zhou, L.; Wei, H.; Li, H.; Zhao, W.; Zhang, Y. Arbitrary-Oriented Object Detection in Remote Sensing Images Based on Polar Coordinates. *IEEE Access* **2020**, *8*, 223373–223384. [CrossRef]
21. Audebert, N.; Le Saux, B.; Lefèvre, S. Segment-before-Detect: Vehicle Detection and Classification through Semantic Segmentation of Aerial Images. *Remote Sens.* **2017**, *9*, 368. [CrossRef]
22. Zhang, Z.; Liu, Y.; Liu, T.; Lin, Z.; Wang, S. DAGN: A Real-Time UAV Remote Sensing Image Vehicle Detection Framework. *IEEE Geosci. Remote Sens. Lett.* **2020**, *17*, 1884–1888. [CrossRef]
23. Shi, F.; Zhang, T.; Zhang, T. Orientation-Aware Vehicle Detection in Aerial Images via an Anchor-Free Object Detection Approach. *IEEE Trans. Geosci. Remote Sens.* **2020**, *59*, 5221–5233. [CrossRef]
24. Szegedy, C.; Ioffe, S.; Vanhoucke, V.; Alemi, A. Inception-v4, Inception-ResNet and the Impact of Residual Connections on Learning. In Proceedings of the Thirty-First AAAI Conference on Artificial Intelligence, San Francisco, CA, USA, 4–9 February 2017.
25. Huang, G.; Liu, Z.; Laurens, V.; Weinberger, K.Q. Densely Connected Convolutional Networks. In Proceedings of the IEEE Conference on Computer Vision and Pattern Recognition, Honolulu, HI, USA, 21–26 July 2016.
26. Veit, A.; Wilber, M.; Belongie, S. Residual networks behave like ensembles of relatively shallow networks. In Proceedings of the Advances in Neural Information Processing Systems, Barcelona, Spain, 5–10 December 2016.
27. Qing, Y.; Liu, W.; Feng, L.; Gao, W. Improved YOLO Network for Free-Angle Remote Sensing Target Detection. *Remote Sens.* **2021**, *13*, 2171. [CrossRef]
28. Lim, J.S.; Astrid, M.; Yoon, H.J.; Lee, S.I. Small Object Detection using Context and Attention. In Proceedings of the 2021 International Conference on Artificial Intelligence in Information and Communication (ICAIIC), Jeju Island, Korea, 20–23 April 2021.
29. Yu, J.; Jiang, Y.; Wang, Z.; Gao, Z.; Huang, T. UnitBox: An Advanced Object Detection Network. In Proceedings of the 24th ACM International Conference on Multimedia, Amsterdam, The Netherlands, 15–19 October 2016.
30. Li, K.; Wan, G.; Cheng, G.; Meng, L.; Han, J. Object Detection in Optical Remote Sensing Images: A Survey and A New Benchmark. *ISPRS J. Photogramm. Remote Sens.* **2020**, *159*, 296–307. [CrossRef]
31. Cheng, G.; Han, J.; Zhou, P.; Guo, L. Multi-class geospatial object detection and geographic image classification based on collection of part detectors. *ISPRS J. Photogramm. Remote Sens.* **2014**, *98*, 119–132. [CrossRef]

32. Lin, T.-Y.; Maire, M.; Belongie, S.; Hays, J.; Perona, P.; Ramanan, D.; Dollar, P.; Zitnick, C. Microsoft coco: Common objects in context. In Proceedings of the European Conference on Computer Vision, Zurich, Switzerland, 6–12 September 2014; pp. 740–755.
33. Huang, W.; Li, G.; Chen, Q.; Ju, M.; Qu, J. CF2PN: A Cross-Scale Feature Fusion Pyramid Network Based Remote Sensing Target Detection. *Remote Sens.* **2021**, *13*, 847. [CrossRef]
34. Yuan, Z.; Liu, Z.; Zhu, C.; Qi, J.; Zhao, D. Object Detection in Remote Sensing Images via Multi-Feature Pyramid Network with Receptive Field Block. *Remote Sens.* **2021**, *13*, 862. [CrossRef]
35. Li, Y.; Huang, Q.; Pei, X.; Chen, Y.; Jiao, L.; Shang, R. Cross-Layer Attention Network for Small Object Detection in Remote Sensing Imagery. *IEEE J. Sel. Top. Appl. Earth Obs. Remote Sens.* **2021**, *14*, 2148–2161. [CrossRef]
36. Xiong, S.; Tan, Y.; Li, Y.; Wen, C.; Yan, P. Subtask Attention Based Object Detection in Remote Sensing Images. *Remote Sens.* **2021**, *13*, 1925. [CrossRef]

isprs International Journal of *Geo-Information*

MDPI

Article

Cloud and Snow Segmentation in Satellite Images Using an Encoder–Decoder Deep Convolutional Neural Networks

Kai Zheng [1], Jiansheng Li [1,*], Lei Ding [2], Jianfeng Yang [1], Xucheng Zhang [1] and Xun Zhang [1]

1 Institute of Surveying and Mapping, PLA Strategic Support Force Information Engineering University, Zhengzhou 450001, China; kaizheng90@foxmail.com (K.Z.); marsyjf@126.com (J.Y.); banditex1990@126.com (X.Z.); zx121777@163.com (X.Z.)

2 Department of Information Engineering and Computer Science, University of Trento, 38123 Trento, Italy; lei.ding@unitn.it

* Correspondence: ljs2021@vip.henu.edu.cn

Abstract: The segmentation of cloud and snow in satellite images is a key step for subsequent image analysis, interpretation, and other applications. In this paper, a cloud and snow segmentation method based on a deep convolutional neural network (DCNN) with enhanced encoder–decoder architecture—ED-CNN—is proposed. In this method, the atrous spatial pyramid pooling (ASPP) module is used to enhance the encoder, while the decoder is enhanced with the fusion of features from different stages of the encoder, which improves the segmentation accuracy. Comparative experiments show that the proposed method is superior to DeepLabV3+ with Xception and ResNet50. Additionally, a rough-labeled dataset containing 23,520 images and fine-labeled data consisting of 310 images from the TH-1 satellite are created, where we studied the relationship between the quality and quantity of labels and the performance of cloud and snow segmentation. Through experiments on the same network with different datasets, we found that the cloud and snow segmentation performance is related more closely to the quantity of labels rather than their quality. Namely, under the same labeling consumption, using rough-labeled images only performs better than rough-labeled images plus 10% fine-labeled images.

Keywords: satellite image; semantic segmentation; encoder–decoder; CNN; TH-1; cloud and snow detection; label quality

Citation: Zheng, K.; Li, J.; Ding, L.; Yang, J.; Zhang, X.; Zhang, X. Cloud and Snow Segmentation in Satellite Images Using an Encoder–Decoder Deep Convolutional Neural Networks. *ISPRS Int. J. Geo-Inf.* **2021**, *10*, 462. https://doi.org/10.3390/ijgi10070462

Academic Editors: Gloria Bordogna and Cristiano Fugazza

Received: 21 May 2021
Accepted: 1 July 2021
Published: 6 July 2021

1. Introduction

With satellites becoming indispensable infrastructures for the development of the national economy, the acquisition of remote sensing images (RSIs) has become easier. RSIs have been widely used in a variety of fields such as infrastructure, agriculture, forestry, geology, hydrology, transportation, disaster prediction, etc. However, 66.7% of the Earth's surface is covered by clouds [1], which is a major factor restricting the application of optical RSIs. Additionally, due to the similar characteristics (e.g., high reflectivity rate) of cloud and snow in optical bands, some traditional methods such like threshold-based methods cannot distinguish them and often lead to misjudgment. This greatly hinders the automatic processing of RSIs. Furthermore, there is a deeper need to detect cloud and snow such like the construction of the atmospheric reflectance database, which can be used to serve the retrieval of atmospheric aerosols [2]. Therefore, it is of great significance to segment cloud and snow quickly, accurately, and automatically.

A number of image segmentation methods have been proposed since the 1970s, among which the most classic one is the Otsu [3] proposed by Nobuyuki Otsu in 1979. It uses the exhaustive method to determine the threshold that results in the maximum variance between objects in images, thus segmenting images into foreground and background images. Due to the high reflectivity of clouds in optical bands, we can use the Otsu method to segment the cloud and background from the visible images. However, cloud and snow often share similar characteristics in optical bands, making it hard to distinguish them from each other, and it is almost impossible to segment cloud from other objects using Otsu all at once. Although many methods have been proposed to improve the Otsu (such as the multi-Otsu), they suffer from limitations such as large amount of calculations, low robustness, etc. Figure 1 shows the cloud segmentation result of the Otsu, where it produces totally different results on similar images.

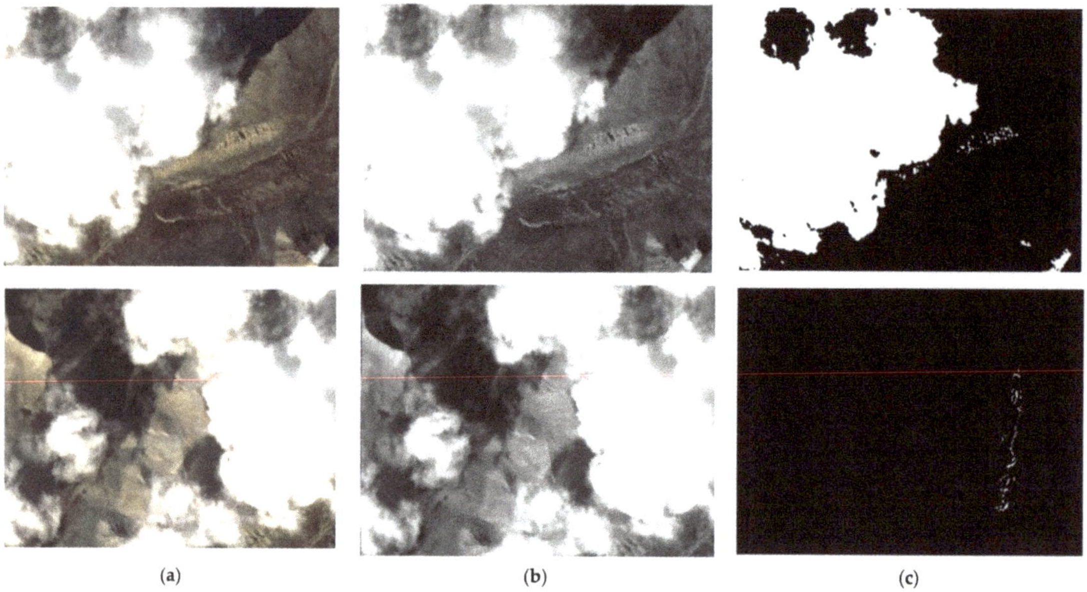

Figure 1. The segmentation examples of the Otsu algorithm. (**a**) Cloud; (**b**) binarization image; (**c**) segmentation result.

The cloud and snow segmentation method based on elevation assistance [4] captures the difference of elevation among cloud, snow, and other objects. The three-dimensional geometric features of cloud are obtained through the dense matching of multiple images, after which the differences between clouds and background objects are exploited by comparing the existing elevation information. Although this kind of methods has high accuracy, it involves many conditional and time-consuming operations such as dense matching and digital elevation model registration.

In recent decades, with the development of pattern recognition and machine learning technologies, researchers have studied intelligent methods for the cloud segmentation and have achieved good results. Amato et al. [5] applied a principal component analysis (PCA) for image cloud detection based on statistical theory. Merchant et al. [6] proposed a cloud detection algorithm based on full probability Bayes theory. Zhao Xiao [7] used fuzzy C-Means clustering to complete sample iteration clustering by minimizing the objective function and used support vector machine (SVM) to perform the classification, which has the advantage of better segmenting results under the condition of empirical knowledge, while human intervention greatly hinders the segmentation efficiency. Additionally, there are sparse perceptual classifiers, automatic codec, and other methods. Generally speaking, the time cost of machine learning based algorithms increases linearly with the number of pixels in the image. Therefore, for large-scale RSIs, the operation time of such an algorithm is often very long and it is hard to satisfy the needs in real-world applications.

Recently, deep learning has made great progress in image classification, image segmentation, object detection, and other vision tasks. In 2015, Long et al. [8] proposed the fully convolution neural network (FCN) and applied it to image semantic segmentation. Unlike the classical CNN, which uses the fully connected layer after the convolution layer to obtain a fixed length feature vector for classification, FCN operates on images of flexible size and performs pixel-level segmentation. For example, Shao et al. [9] proposed a method based on the multiscale feature model MF-CNN, which can detect thin clouds and thick clouds in RSIs, which results in good detection accuracy on the Landsat 8 data. To summarize, the methods based on deep learning are emerging in cloud and snow segmentation in RSIs.

ResNet [10] is a feature extraction backbone proposed by He et al., which is built on the idea of residual learning to solve the problem of gradient disappearance/explosion and network degradation in traditional CNN (when the number of network increases). ResNet module bypasses the input information to the output through an additional connection channel to protect the integrity of the input. The whole network only needs to learn the part of the difference between input and output, which simplifies the difficulty of learning when the network deepens and helps to retain more original semantic information.

DeepLab [11–13] are deep learning semantic segmentation models proposed by Google, which contain encoder–decoder FCN structure and the ASPP module to fuse multiscale features and make better use of image features (compared to the plain FCN). The encoder–decoder structure is reintroduced into DeepLabV3. Combining the Xception backbone [14] and the ASPP module as the encoder, which uses a different expansion rate of perforated convolution to extract features. After feature fusion and $4\times$ upsampling, it fuses with the low-level features extracted by Xception in the decoder and $4\times$ upsampling are used to get the segmentation results. Due to the use of ASPP in DeepLabV3+, it improves the ability of multiscale access to semantic information and achieves state-of-the-art accuracy on multiple datasets.

TH-1 [15], which is the first generation of the transmission stereo mapping satellite in China, carries three types of five camera loads, including a three linear array CCD camera, two-meter resolution panchromatic camera, and ten-meter RGB camera. Our research data consist of 470 tiles of RSIs with RGB bands taken by the TH-1 satellite from 2018 to 2019. A total of 200 tiles were collected on 20 September, 2018, while the others were collected on 5 April 2019. In order to ensure the generalization ability of the network, the selected images cover various underlying surfaces, acquisition seasons and time phases, considering different geographical locations, climatic conditions, and cloud features. The geographic longitude and latitude range is $28^{\circ}35'$ E–$120^{\circ}05'$ E, $5^{\circ}25'$ N–$60^{\circ}05'$ N, respectively, covering different ground surfaces such as deserts, grasslands, cities, and mountains. Additionally, in order to segment clouds and snow simultaneously, 175 scene images with snow are selected.

The threshold-based methods are still used in the actual production. To solve the above problems, we propose a cloud and snow segmentation method based on DCNN with an encoder–decoder structure. Compared with traditional methods, it does not rely on prior knowledge in feature selection and extraction. We combine the advantages of ResNet50 [10] and encoder–decoder structure and improve the decoder to realize the simultaneous segmentation of cloud and snow in RSIs. By improving the network structure, using the exponential activation function (ELU) [16] and focal loss function [17], our objective is to get optimization of cloud edge segmentation and enhancement of the generalization ability of the network. On the other hand, with the DCNN becoming the mainstream of image semantic segmentation, the production of a high-quality pixel level label for semantic segmentation has attracted more and more attention. However, accurate image annotation requires a lot of manpower and the existing datasets are not all fine-labeled. At present, there are more and more DCNN networks, and the accuracy of image semantic segmentation is higher and higher, which, however, based on open-source datasets such as Microsoft COCO [18] and PASCAL-VOC-2012 [19], is mainly studied to innovate the network structure and improve the segmentation accuracy without analysis on the impact of label quality and quantity on the results. So, the other direction of this paper is to explore the influence of different data quality and quantity on the performance of cloud and snow segmentation on RSIs. Our major research contributions are summarized as follows:

First, we propose an end-to-end DCNN framework with encoder–decoder architecture, ED-CNN, which improves the decoder by fusion of features from different encoding stages. The outputs of ASPP, which after Conv 1 × 1 and 4 × upsampling, are concatenated with the enhanced low-level features from the enhanced decoder. Then, the concatenated feature maps are sent to Conv 3 × 3 and 4 × upsampling to recover their original size to segment the image pixels. Second, we present a TH-1 satellite dataset, which contains 23,520 coarse-labeled images with annotations. Additionally, a fine-labeled dataset of 300 images is added to support our experiments. Third, experiments have been conducted based on different datasets, including TH-1 images of a different temporal phase and Google Earth images, which demonstrates that the proposed network is superior to DeepLabV3+ with Xception and the ResNet50 and can be applied to multisource RSIs. Finally, we discuss the effects of labeling quality and quantity in the dataset through extensive experiments with the proposed network. It is demonstrated that the performance of cloud and snow segmentation is positively related mainly to the labeling quantity. Namely, the smaller rough-labeled dataset plus some fine-labeled images, which is 10% of the total images, is equal to the larger rough-labeled dataset with the same total image quantity. Furthermore, under the same labeling consumption, the larger rough-labeled dataset exceeds the smaller rough-labeled dataset plus a few fine-labeled images.

2. Methodology

2.1. Datasets Establishment

First, we produced a BMP image. The data file of TH-1 RSIs was saved from the TIF format with four channels to the "BMP" image with a resolution of 6000 pixels × 6000 pixels. Second, the dataset was divided and the images were cut. The 470 tiles of images were divided into the training set, the validation set, and the testing set according to the numeric ratio of 3:1:1. Since the original image was large in size and took up too much memory, it could not be trained directly in the network. Therefore, we cut the images into patches with 480 pixels × 360 pixels. A total of 23,520 images were generated, including 13,924 training images, 4798 validation images, and 4798 testing images. Third, images were rough-labeled. Labelme [20] was used to roughly mark the cloud area of each image, generate JSON files, and convert the JSON files into label marked images with the same label size in batches. The rough-labeled image and its mask are shown in Figure 1, where the red color represents the cloud, green represents the snow, and black represents the background. Fourth, some images were fine-labeled. In order to verify the influence of the fine annotation image and rough annotation image on the training results (a detailed description is in Section 2.3.2), 310 extra images were randomly selected and then labeled carefully to generate JSON files with the time of 6 times that rough-labeling costs on each image, which were then transformed into labeled images as Step 3. Fine-labeled images had more accurate edge marking (errors less than 5 pixels). The fine-labeled image and its mask are shown in Figure 2, where label 1 represents the cloud, label 2 represents the snow, and label 0 represents the background located in the lower right corner of the image. Fifth, image preprocessing was performed. In Reference [21] it is proved that the network performance can be effectively improved by data augmentation. In order to enhance the generalization ability of the network and prevent overfitting, the dataset was augmented. The augmentation operations include a vertical flip, horizontal flip, contrast change, etc. The original and augmented images are shown in Figures 3 and 4.

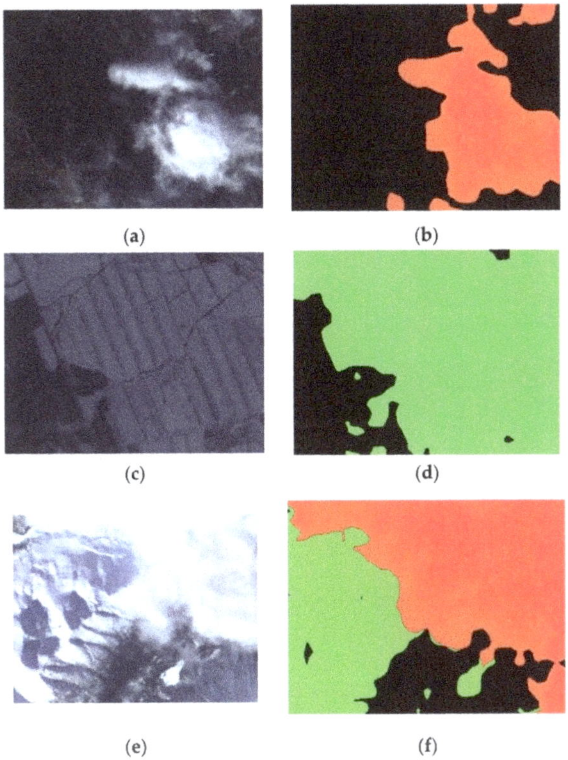

Figure 2. The coarse-labeled images and their masks. (**a**) Cloud; (**b**) mask; (**c**)snow; (**d**) mask; (**e**) cloud and snow; (**f**) mask.

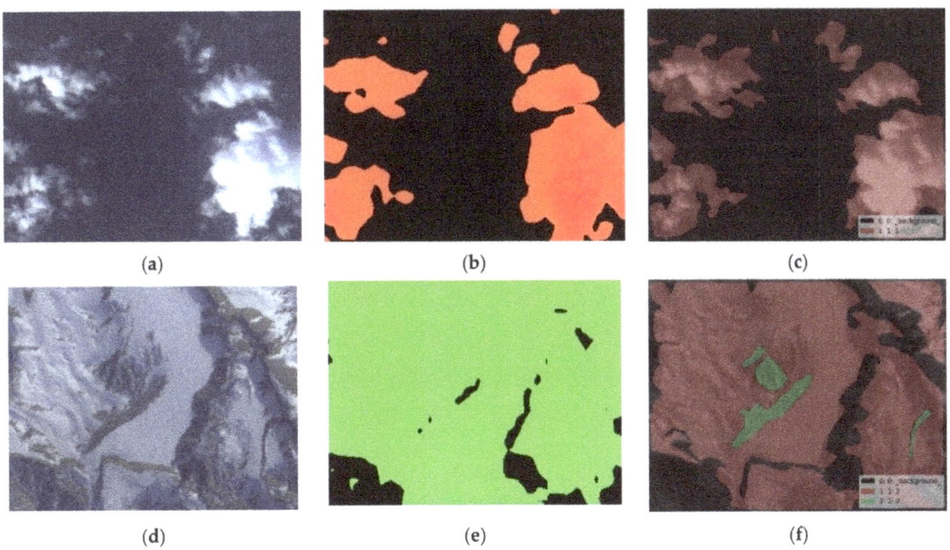

Figure 3. *Cont.*

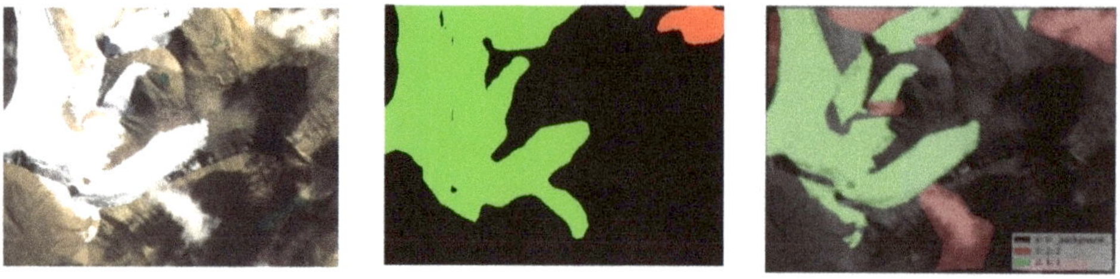

(**g**) (**h**) (**i**)

Figure 3. The fine-labeled images and their masks. (**a**) Cloud, (**b**) rough mask, (**c**) fine mask, (**d**) snow, (**e**) rough mask, (**f**) fine mask, (**g**) cloud and snow, (**h**) rough mask, and (**i**) fine mask.

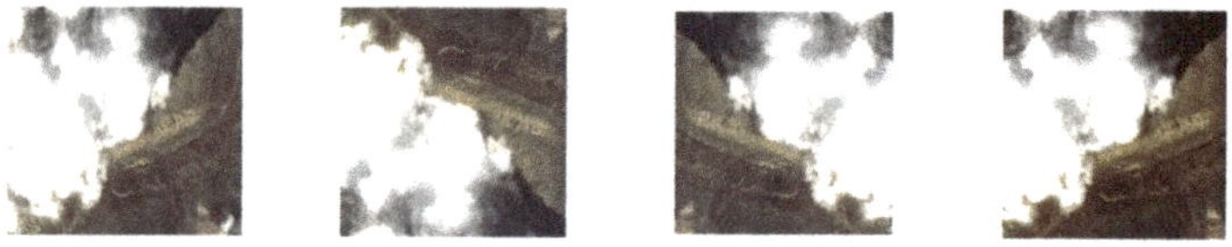

(**a**) (**b**) (**c**) (**d**)

Figure 4. Original and augmented images. (**a**) Original image, (**b**) vertical flip, (**c**) horizontal flip, and (**d**) contrast change.

2.2. *Methods*

2.2.1. Network Backbone

How to extract features from a different kind of RSIs properly and effectively is a key problem. Cloud and snow in RSIs mostly present a planar structure. Their semantic information is simple, whereas their detailed information is rich, which puts forward a high demand for the detail extraction ability. For example, with the parameter amount of 22.8 M [14], Xception has a large number of parameters suitable for segmentation tasks with many kinds of objects, Meanwhile, it requires huge computational resources and is difficult to train, thus it is obviously not fully suitable for the task of cloud and snow segmentation. In this paper, Resnet50 backbone was selected as the encoder to extract features of cloud and snow (as shown in Figure 5). The parameter size of ResNet50 was only 0.85 M [10] and more direct connections were added in the network. Considering its advantages such as less parameters, easy training, and fast convergence, it is more suitable for cloud and snow segmentation compared to Xception.

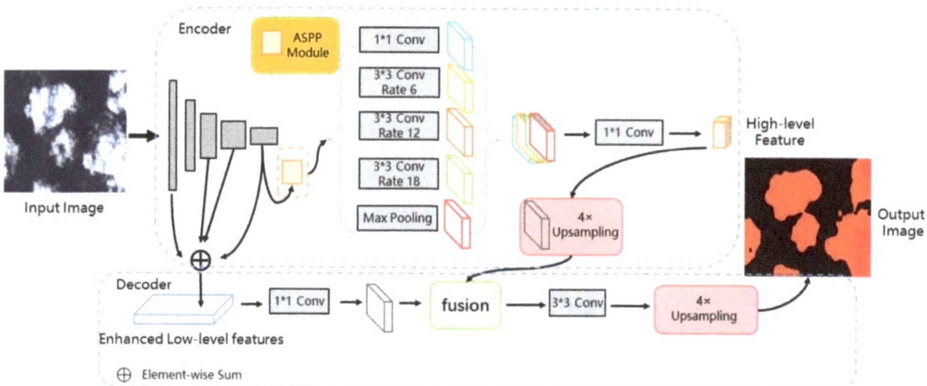

Figure 5. Network architecture of the proposed method.

2.2.2. Enhanced Decoder

In encoder–decoder architectures such like DeepLabV3+, the decoder subnet gradually recovers the spatial information, which is usually not as powerful as the encoder. In this regard, besides replacing the backbone to the ResNet50, we added skip connections in the decoder. To be specific, we selected features in the stages 1, stage 3, stage 4, and stage 5 of the ResNet50 to construct a top–down connection feature map pyramid, which enriched the semantic representation of low-level features to better utilize the spatial information as shown in Figure 5. The low-level feature maps with high resolution and high-level feature maps with rich semantic information were fused, which can quickly construct the decoder with better semantic information from 4 stages instead of a single stage without obvious cost increases.

As Figure 6 shows, first the feature map from stage 5 was 2 $\times$ upsampled after 1 $\times$ 1 convolution, which was added with the output from stage 4, after 1 $\times$ 1 convolution too. The dimension numbers of these stages were all set to 256, ensuring that these feature maps could be added. Second, the added feature map was 2 $\times$ upsampled the second time to be the same size of the output from stage 3, then the process was repeated to get the fused feature map with output from stage 1. Third, the feature map of the enhanced decoder was concatenated with the feature map generated from the ASPP module after the 3 $\times$ 3 convolution, batch-normalization, and ELU operations. Finally, the segmentation map was obtained by 3 $\times$ 3 convolution and 4 $\times$ upsampling.

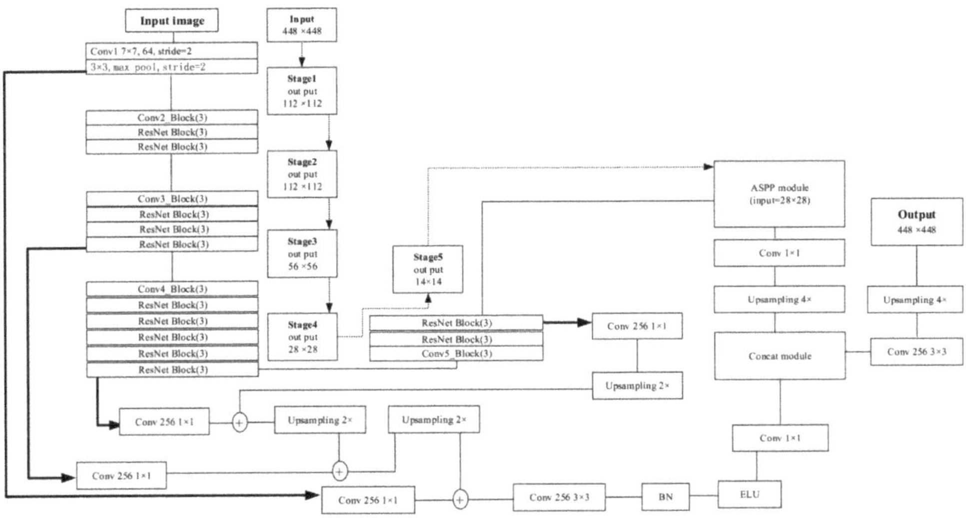

Figure 6. Network workflow of the proposed method (⊕ means element-wise sum).

2.2.3. Loss Function

In practice, there are many kinds of clouds with different shapes. Generally, the proportion of thin clouds and cirrus clouds is less than that of thick clouds. The training dataset in this paper also reflected the characteristics of less data of thin clouds, cirrus clouds, and snow. Cross entropy (CE) loss cannot balance the learning of fewer samples. Its formula is as follows,

$$CE(p_t) = -\log(p_t) \tag{1}$$

Through the combination of different parameters, focal loss [17] can solve the problem of sample imbalance in the semantic segmentation task, which is an improved version of the CE loss by adding a weight. Its formula is as follows:

$$FL(p_t) = -(\lambda - p_t)^{\gamma} \log(p_t) \quad (2)$$

where λ and γ are two hyperparameters and p_t is the prediction probability of the label. $(\lambda - p_t)$ can be regarded as the weight of Equation (1). The paper [17] sets γ = 2 and λ = 1, when the prediction of a certain category is accurate, that is, close to 1, the value is close to 0. The more inaccurate the prediction is, that is, close to 0, the closer to 1 it will be. For the samples that are easy to distinguish, the weight corresponding to the loss will be small, whereas for objects that are difficult to distinguish, their corresponding weight will be larger in order to retain the loss value of difficult samples and reduce the loss value of simple samples. We set γ = 2 and λ = 1.

2.2.4. Activation Function

The exponential linear unit (ELU) was proposed in the paper [16], which can make the mean value of output close to 0, thus accelerating the convergence of the network and effectively overcome problems such as gradient vanishing. If the output of a node is X, the output after passing through the ELU layer is illustrated in Equation (3). We adopted ELU activation after convolution layers.

$$f(x) = \begin{cases} x, x \geq 0 \\ \alpha(e^x - 1), x < 0. \end{cases} \quad (3)$$

2.2.5. Evaluation Metrics

Here we used pixel accuracy (PA) and the mean intersection over union (MIOU) as the evaluation metrics.

1. P.A: the simplest and direct indicator, which only calculates the ratio of the number of correctly classified pixels to the number of all pixels. The calculation is shown in Equation (4).

$$PA = \frac{\sum_{i=0}^{k} p_{ii}}{\sum_{i=0}^{k} \sum_{j=0}^{k} p_{ij}} \quad (4)$$

2. MIOU: it calculates the coincidence ratio between the intersection and union of two sets, that is, the intersection union ratio between real segmentation and algorithm segmentation. This ratio can be redefined as the number of true positive cases (intersections) divided by the total number (including true positive cases, false negative cases, and false positive cases (Union)). MIOU is calculated by class, and then averaged. The calculation is shown in Equation (5).

$$MIoU = \frac{1}{k+1} \sum_{i=0}^{k} \frac{p_{ii}}{\sum_{j=0}^{k} p_{ij} + \sum_{j=0}^{k} p_{ji} - p_{ii}.} \quad (5)$$

2.3. *Experiments*

2.3.1. Experimental Platform

The experimental hardware is the Lenovo workstation with 2.1 GHz CPU frequency of the Intel Xeon processor, and the GPU is NVIDIA Titan XP. We used the Tensorflow + Keras framework to build deep learning models. The network training was carried out on the basis of data augmentation. The initial learning rate was set to $3e^{-4}$ and the learning rate decreasing drop was set to 0.1. Using the ELU activation function and Adam [22] optimizer, the batch size was set to 8 according to the capability of the GPU.

2.3.2. Group Experiment Settings

We set the experiment into two stages numbered from 0 to 4 shown in Table 1. First, the performance of a different network was evaluated on dataset 1, which included 23,520 images. Second, another smaller dataset 2 including 6660 rough-labeled images and 310 fine-labeled images was chosen from dataset 1 to explore the influence of the quantity and quality of image labels on the accuracy of cloud and snow segmentation.

Table 1. Experiment setting.

Group	Group Number	Data	Total	Train	Validation	Test
Stage 1 Dataset 1	0	Rough-labeled	23,520	13,924	4798	4798
Stage 2 Dataset 2	1	Rough-labeled Fine-labeled	5000	3000 0	1000 0	1000 0
	2	Rough-labeled Fine-labeled	4700	2700 0	1000 0	1000 0
	3	Rough-labeled Fine-labeled	5010	2700 310	1000 0	1000 0
	4	Rough-labeled Fine-labeled	6660	4660 0	1000 0	1000 0

3. Results and Analysis

3.1. Training Process and Results

First, we used dataset 1 to train the proposed network, ResNet50, and Xception for 60 epochs, respectively. Figure 7 shows the loss and accuracy comparison between three networks on the training set and the validation set. The red line represents the proposed method, while the yellow and green lines represent ResNet50 and Xception, respectively.

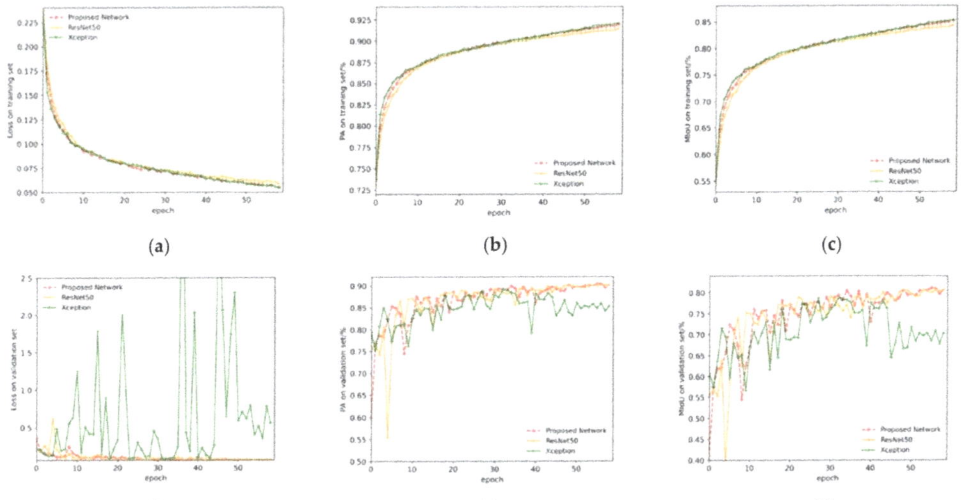

Figure 7. Results of three networks on dataset 1. (**a**) Training loss; (**b**) training PA; (**c**) training MIoU; (**d**) validation loss; (**e**) validation PA; (**f**) validation MIoU.

Second, dataset 2 of 6660 images was randomly chosen from set 1, dividing into 4660 for the training set, 1000 for the validation set, and 1000 for the testing set. Another 310 fine-labeled images were added to the training set. Four group of experiments were performed to explore the influence of data use strategy on the proposed network. Figure 8 shows the results.

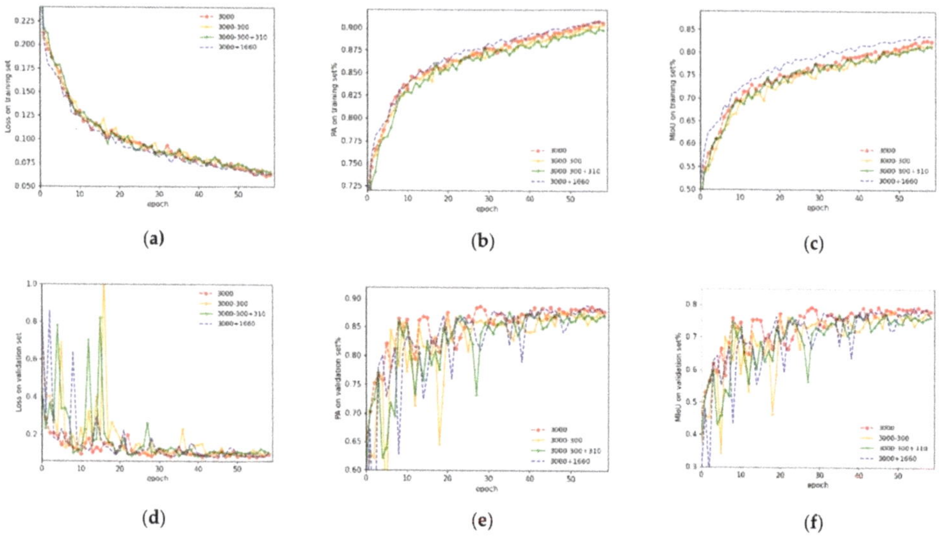

Figure 8. Results of the proposed network on dataset 2. (**a**) Training loss; (**b**) training PA; (**c**) training MIoU; (**d**) validation loss; (**e**) validation PA; (**f**) validation MIoU.

3.2. Comparison and Analysis

First, the testing results of the three methods on dataset 1 are shown in Table 2. In the testing set of 4798 images, the PA of cloud and snow segmentation obtained by the proposed method was 90.3%, while the PA of Xception, ResNet50, and DANet [23] were 87.4%, 89.2%, and 88.9% respectively. The MIoU of the proposed method was 81.1%, exceeding Xception and ResNet50 by 4.2%, 0.6%, and 0.8% respectively. When the proposed enhanced decoder was used to enhance low-level features, the PA and MIoU of the network increased obviously.

Table 2. Results of the three networks on dataset 1.

Method	PA	MIoU
Xception	87.4	76.9
ResNet50	89.2	80.5
DANet	88.9	80.3
ED-CNN	90.3	81.1

Second, four groups of experiments were conducted on dataset 2 to explore the influence of different data quality and quantity on the proposed network. The results are shown in Table 3, in which group 1 and group 2 proved that a 10% reduction in the number of training data will reduce the overall performance. On the contrary, group 2 and group 3 show that a 10% increase of fine-labeled data replacing the reduction had a slight side effect on performance. In our opinion, adding different features of data when the training set is not large enough will lead to the above problem. Additionally, on average, it took six times as long to label a fine-labeled image than making a rough-labeled image. However, group 4 shows that 4660 rough-labeled images resulted in better results compared to the other three groups.

Table 3. Results of a different data use strategy on dataset 2.

Number	Data Use			PA	MIoU
	Rough	Fine	Testing		
1	3000	0	1000	88.5	79.3
2	2700	0		87.4	78.0
3	2700	310		87.2	77.4
4	4660	0		88.8	79.4

Finally, the qualitative analysis of typical images was conducted to compare the results of the proposed method with the Otsu and another two networks (Xception and ResNet50). Figure 9 shows the comparison results of the traditional Otsu method versus the three networks, and the red part is the cloud while the green part is snow. It can be found that when the image illumination conditions were not ideal and the underlying surface contrast was not high, the Otsu created a false alarm in Figure 9b and it could not segment cloud and snow at the same time (as in Figure 9d). The reason is that the Otsu algorithm does not consider the neighborhood information in segmentation and is sensitive to noise. In Figure 9a,c, Xception missed some pixels and got the wrong result in Figure 9b, while getting it totally wrong in Figure 9d. The performance of the proposed network was slightly better than that of the ResNet50 in preserving the spatial details.

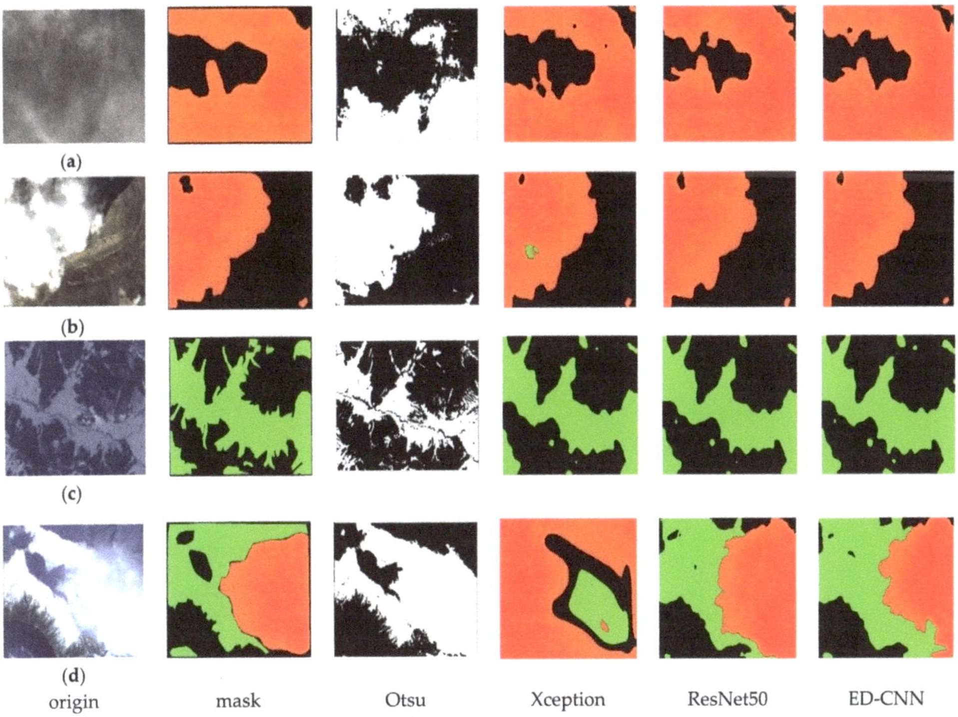

Figure 9. Testing results of the proposed method on the TH-1 image acquired from different temporal phases.

Furthermore, TH-1 images from different time-phases are chosen to verify the generalization performance of the proposed method. The results are shown in Figure 10. It is demonstrated that the proposed method could accurately segment cloud and snow, showing a better generalization ability.

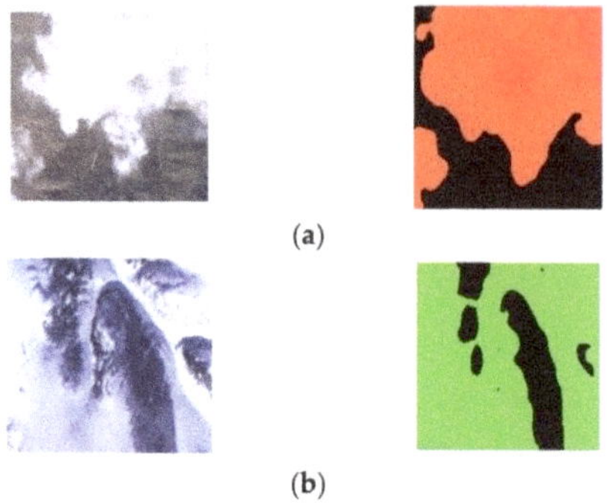

Figure 10. Testing results of the proposed method on the TH-1 image acquired from different temporal phases. (**a**) Cloud; (**b**) snow ground.

4. Conclusions

4.1. The Proposed Method for Cloud and Snow Segmentation

In this paper, we proposed an end-to-end cloud and snow segmentation network for TH-1 RSIs, which combined the advantages of the encoder–decoder architecture and the enhanced decoder. On the one hand, it avoided the shortcomings of traditional cloud detection algorithms (such as they are parameter-dependent, time-consuming, and the scope of application is limited). It achieved the mIoU of 81.1% on 4798 testing images and reduced the segmenting time (on a single 480×360 image) to 49.2 ms, which could basically meet the requirements of image preprocessing. On the other hand, the enhancement of the decoder proved to be a useful way to improve the segmentation performance through exploiting features at different encoder stages, which bridges the gap between different levels of features. Additional experiments show that this method can be used on images acquired from different sensors, as shown in Figure 11.

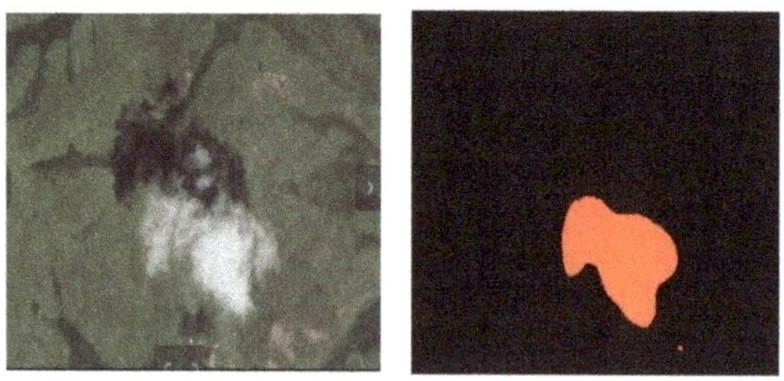

Figure 11. Testing results of the proposed method on a Google Earth image.

There is still room for further improvement in the cloud and snow segmentation method proposed in this paper, such as improving the training accuracy and using additional multisource RSIs for transfer learning. Another research direction is to use multispectral information to solve the problem of distinguishing cloud, fog, and snow in mixed regions.

4.2. Influence of Different Datasets on Segmentation Performance

Given a certain network, we found that its segmentation performance was positively related mainly to the number of training images and labels. Specifically, when the training time was sufficient, more training images led to higher accuracy, whereas a 10% increase of fine-labeled data replacing the original rough-labeled reduction had a slight side effect on performance. Considering fewer categories and a lower complexity of cloud and snow segmentation, our conclusion was that when the same labeling time was considered, we achieved better results by only roughly labeling the data. Instead of spending more manual resources to make fine-labeled masks, roughly labeling more data can lead to the same segmentation accuracy.

There is a margin for further research on the effects of label quality and quantity, such as clarifying the pixel error of coarse-marking labels and exploring the effect of error types and sizes on cloud and snow segmentation results.

Author Contributions: Conceptualization, Kai Zheng and Jiansheng Li; methodology, Kai Zheng; software, Kai Zheng and Lei Ding; validation, Kai Zheng and Jianfeng Yang; formal analysis, Xun Zhang; investigation, Xucheng Zhang; resources, Xucheng Zhang; data curation, Jianfeng Yang; writing—original draft preparation, Kai Zheng; writing—review and editing, Jiansheng Li; visualization, Lei Ding; supervision, Jiansheng Li. All authors have read and agreed to the published version of the manuscript.

Funding: This research received no external funding.

Institutional Review Board Statement: Exclude.

Informed Consent Statement: Exclude.

Data Availability Statement: Not available.

Conflicts of Interest: The authors declare no conflict of interest.

References

1. Zhu, Z.; Woodcock, C.E. Automated cloud, cloud shadow, and snow detection in multitemporal Landsat data: An algorithm designed specifically for monitoring land cover change. *Remote Sens. Environ.* **2014**, *152*, 217–234. [CrossRef]
2. Jian, H. Cloud and Snow Detection Algorithm and Surface Reflectance Database Construction Algorithm of Remote Sensing. Ph.D. Thesis, Henan University, Kaifeng, China, 2017.
3. Xianjun, G.; Youchuan, W.; Shunyi, Z.; Yuan-wei, Y. Real time automatic detection of clouds in aerial photography. *Spectrosc. Spectr. Anal.* **2014**, *34*, 1909–1913. [CrossRef]
4. Shi, T.; Yu, B.; Clothiaux, E.E.; Braverman, A.J. Daytime Arctic Cloud Detection Based on Multi-Angle Satellite Data with Case Studies. *J. Am. Stat. Assoc.* **2008**, *103*, 584–593. [CrossRef]
5. Amato, U.; Antoniadis, A.; Cuomo, V.; Cutillo, L.; Franzese, M.; Murino, L.; Serio, C. Statistical cloud detection from SEVIRI multispectral images. *Remote Sens. Environ.* **2008**, *112*, 750–766. [CrossRef]
6. Merchant, C.J.; Harris, A.R.; Maturi, E.; Maccallum, S. Probabilistic physically based cloud screening of satellite infrared imagery for operational sea surface temperature retrieval. *Q. J. R. Meteorol. Soc.* **2005**, *131*, 2735–2755. [CrossRef]
7. Xiao, Z. Research on Cloud Detection Method of High Resolution Satellite Remote Sensing. Ph.D. Thesis, Harbin Institute of Technology, Harbin, China, 2013.
8. Long, J.; Shelhamer, E.; Darrell, T. Fully convolutional networks for semantic segmentation. *IEEE Trans. Pattern Anal. Mach. Intell.* **2017**, *39*, 640–651.
9. Shao, Z.; Pan, Y.; Diao, C.; Cai, J. Cloud Detection in Remote Sensing Images Based on Multiscale Features-Convolutional Neural Network. *IEEE Trans. Geosci. Remote Sens.* **2019**, *57*, 4062–4076. [CrossRef]
10. He, K.; Zhang, X.; Ren, S.; Sun, J. Deep residual learning for image recognition. In Proceedings of the 2016 IEEE Conference on Computer Vision and Pattern Recognition, Las Vegas, NV, USA, 27–30 June 2016; pp. 770–778.
11. Chen, L.-C.; Papandreou, G.; Kokkinos, I.; Murphy, K.; Yuille, A.L. DeepLab: Semantic Image Segmentation with Deep Convolutional Nets, Atrous Convolution, and Fully Connected CRFs. *IEEE Trans. Pattern Anal. Mach. Intell.* **2018**, *40*, 834–848. [CrossRef] [PubMed]
12. Chen, L.-C.; Papandreou, G.; Schroff, F.; Adam, H. Rethinking atrous convolution for semantic image segmentation. *arXiv* **2017**, arXiv:1706.05587.
13. Chen, L.-C.; Zhu, Y.; Papandreou, G.; Schroff, F.; Adam, H. Encoder-decoder with atrous separable convolution for semantic image segmentation. In Proceedings of the European Conference on Computer Vision (ECCV), 2018—15th European Conference, Munich, Germany, 8–14 September 2018; pp. 801–818.
14. Chollet, F. Xception: Deep learning with depthwise separable convolutions. In Proceedings of the IEEE Conference on Computer Vision and Pattern Recognition, Honolulu, HI, USA, 21–26 July 2017; pp. 1251–1258.
15. Songming, L.; Yan, L.; Jindong, L. TH-1 Transmission Photogrammetry and Remote Sensing Satellite. *Acta Remote Sens. Sin.* **2012**, *16*, 10–16. [CrossRef]
16. Clevert, D.A.; Unterthiner, T.; Hochreiter, S. Fast and accurate deep network learning by exponential linear units (elus). *arXiv* **2015**, arXiv:1511.07289.

17. Lin, T.-Y.; Goyal, P.; Girshick, R.; He, K.; Dollár, P. Focal loss for dense object detection. In Proceedings of the IEEE International Conference on Computer Vision, Venice, Italy, 22–29 October 2017; pp. 2999–3007. [CrossRef]
18. Lin, T.-Y.; Maire, M.; Belongie, S.; Hays, J.; Perona, P.; Ramanan, D.; Dollár, P.; Zitnick, C.L. Microsoft coco: Common objects in context. In Proceedings of the European Conference on Computer Vision, Zurich, Switzerland, 6–12 September 2014; pp. 740–755.
19. Everingham, M.; Eslami, S.M.A.; Van Gool, L.; Williams, C.K.I.; Winn, J.; Zisserman, A. The Pascal Visual Object Classes Challenge: A Retrospective. *Int. J. Comput. Vis.* **2015**, *111*, 98–136. [CrossRef]
20. Russell, B.C.; Torralba, A.; Murphy, K.P.; Freeman, W.T. LabelMe: A Database and Web-Based Tool for Image Annotation. *Int. J. Comput. Vis.* **2008**, *77*, 157–173. [CrossRef]
21. Ronneberger, O.; Fischer, P.; Brox, T. U-Net: Convolutional Networks for Biomedical Image Segmentation. In Proceedings of the International Conference on Medical Image Computing and Computer-Assisted Intervention, Munich, Germany, 5–9 October 2015.
22. Kingma, D.P.; Ba, J. Adam: A method for stochastic optimization. *arXiv* **2014**, arXiv:1412.6980.
23. Fu, J.; Liu, J.; Tian, H.; Li, Y.; Bao, Y.; Fang, Z.; Lu, H. Dual Attention Network for Scene Segmentation. In Proceedings of the 2019 IEEE/CVF Conference on Computer Vision and Pattern Recognition (CVPR), Long Beach, CA, USA, 16–20 June 2019.

International Journal of *Geo-Information*

MDPI

Article

Soft Integration of Geo-Tagged Data Sets in J-CO-QL^{+}

Paolo Fosci and Giuseppe Psaila *

Department of Management, Information and Production Engineering, University of Bergamo, Viale Marconi 5, 24044 Dalmine, Italy
* Correspondence: giuseppe.psaila@unibg.it; Tel.: +39-035-205-2355

Abstract: The possibility offered by the current technology to collect and store data sets regarding public places located on the Earth globe is posing new challenges, as far as the integration of these data sets is concerned. Analysts usually need to perform such an integration from scratch, without performing complex and long preprocessing or data-cleaning tasks, as well as without performing training activities that require tedious and long labeling of data; furthermore, analysts now have to deal with the popular *JSON* format and with data sets stored within *JSON* document stores. This paper demonstrates that a methodology based on soft integration (i.e., data integration performed through soft computing and fuzzy sets) can now be effectively applied from scratch, through the *J-CO* Framework, which is a stand-alone tool devised to process *JSON* data sets stored within *JSON* document stores, possibly by performing soft querying on data sets. Specifically, the paper provides the following contributions: (1) It presents a soft-computing technique for integrating data sets describing public places, without any preliminary pre-processing, cleaning and training, which can be applied from scratch; (2) it presents current capabilities for soft integration of *JSON* data sets, provided by the *J-CO* Framework; (3) it demonstrates the effectiveness of the soft integration technique; (4) it shows how a stand-alone tool able to support soft computing (as the *J-CO* Framework) can be effective and efficient in performing data-integration tasks from scratch.

Keywords: off-line integration of geo-tagged data sets; data sets about public places; soft integration methodology; effective soft integration through a stand-along tool

Citation: Fosci, P.; Psaila, G. Soft Integration of Geo-Tagged Data Sets in J-CO-QL^{+}. *ISPRS Int. J. Geo-Inf.* **2022**, *11*, 484. https://doi.org/10.3390/ijgi11090484

Academic Editors: Wolfgang Kainz and Huayi Wu

Received: 4 June 2022
Accepted: 7 September 2022
Published: 13 September 2022

1. Introduction

Integrating geo-spatial information has become a crucial task in the current world. In fact, in the era of Open Data and Big Data, a plethora of sources can provide both authoritative and non-authoritative data sets concerning places. The situation is further complicated by the fact that social media provide people with tools for describing places in a non-controlled way. For example, *Facebook* provides its users with the functionality to define a "page"; a specific category of the page describes a "public place", such as restaurants, pubs, air dressers, universities, parks and so on; through its API (Application Programming Interface), pages could be queried, on the basis of their category, location, coordinates, and so on. Another interesting service is called *Google Places*: it is a sub-service of *Google Maps*; *Google Places* API can be used to query its corpus to find places of interest, on the basis of category, location, and so on; this corpus is built by *Google Maps* by integrating both authoritative and non-authoritative data, these latter ones given by users through the social interface provided by *Google Maps*.

In the current scenario, it is very easy to collect data sets from multiple sources, such that these data sets provide geo-tagged information about public places. Since current APIs of social media and Open-Data portals provide data as (possibly) geo-tagged *JSON* documents (*JSON* stands for JavaScript Object Notation, see [1]), *JSON* document stores are the natural storage where to save such data sets. Consequently, integrating geo-tagged data sets describing public places asks for suitable tools, which are able to work on *JSON* document stores. This is the reason why at University of Bergamo (Italy), we are devising [2–5] an innovative tool, called *J-CO* Framework, to perform the complex integration and querying of (possibly geo-tagged) *JSON* data sets.

Nevertheless, integrating geo-tagged data sets describing public places is not a novel problem; in general, traditional approaches rely on machine-learning techniques that require a preliminary training phase. In [6], the problem was addressed in a different way, because the context of "online" aggregation was considered; a fuzzy relation was defined, which provides an easy-to-compute metric that is suitable for online integration of data about public places; the experiments demonstrated that the approach is effective and comparable, in terms of effectiveness, with off-line classification techniques. However, the technique presented in [6] was hard-coded within the software prototype; this fact made us able to include some pre-processing steps on strings that were performed on the fly, immediately after place descriptors were acquired. However, the approach seems to be general and could be applied for integrating data sets in an off-line way too, with data sets stored within *JSON* stores. Paradoxically, this apparently small change of context constitutes a significant challenge: in fact, the straightforward solution could be to still hard-code the technique into a software tool, but this approach is not coherent with the world of *JSON* document stores. We think that exploiting a stand-alone tool able to query *JSON* stores is preferable, since it is transparent and comprehensible for analysts; however, a stand-alone tool for processing *JSON* data sets is necessarily less flexible than a programming language. Thus, the challenge is the following: is it possible to identify a stand-alone tool and adapt the integration technique presented in [6] to the case of off-line integration of geo-tagged *JSON* data sets from *JSON* stores?

The current evolution of *J-CO-QL*$^+$, the query language of the *J-CO* Framework, provides constructs for evaluating membership of *JSON* documents to fuzzy sets [7–9]. Thus, the straightforward idea of checking the current capability of *J-CO-QL*$^+$ for integrating *JSON* data sets describing public places has come out: specifically, since *J-CO-QL*$^+$ is able to deal with complex soft querying of *JSON* data sets and the technique presented in [6] for online integration of data sets concerning public places is based on fuzzy relations, we had the intuition of mixing the two approaches. In other words, given two sets of place descriptors represented as *JSON* documents and stored within a *JSON* document store, in this paper, we experiment with the application of the fuzzy technique presented in [6] (to be precise, a slightly evolved version of it) by means of the *J-CO* Framework, in an off-line manner. The goal is to verify that this approach is suitable in an off-line context, without previous training activities and intervention by humans to label data sets for driving the learning phase (typical of classification techniques). Definitely, we want to demonstrate that the availability of a stand-alone tool such as the *J-CO* Framework, which is able to process *JSON* data sets by applying soft computing and fuzzy sets, indeed provides analysts with a powerful tool to address a problem faced by data analysts, in an effective and (possibly) efficient way.

Summarizing, the contribution of the paper is manifold: (1) Presenting a soft-computing technique for integrating data sets describing public places, without any preliminary pre-processing, cleaning and training, which can be applied from scratch; (2) presenting current capabilities for soft integration of *JSON* data sets, as they are provided by *J-CO-QL*$^+$; (3) demonstrating the effectiveness of the soft integration technique in a harder context than that considered in [6]; (4) showing how a stand-alone tool able to support soft computing (as *J-CO-QL*$^+$) can be effective and efficient in performing data-integration tasks from scratch.

The paper is organized as follows. Section 2 presents relevant related work concerned with the paper. Section 3 provides a brief introduction to relevant concepts concerning fuzzy-set theory. Section 4 introduces the main features of the *J-CO* Framework. Section 5 precisely explains the addressed problem and introduces the methodology we follow, which relies on the concept of fuzzy relation. Section 6 presents and discusses the script written by means of *J-CO-QL*$^+$, which practically applies the methodology presented in Section 5; each single instruction is explained, in order to illustrate how it behaves and its contribution within the script. Section 7 reports the results of an experimental evaluation, in which we evaluated effectiveness and, marginally, execution times. Finally, Section 8 draws the conclusions and possible future work.

2. Related Work

This paper embraces two different research lines: soft querying on databases in general and on *JSON* document stores in particular (Section 2.1), as well as the integration of data sets describing public places (Section 2.2).

2.1. Soft Querying on Databases

Providing data users with capabilities for flexibly querying databases is an old challenge. In particular, when selection conditions can rely on vague predicates, queries become "soft", meaning that they are tolerant to thresholds (e.g., given a Boolean predicate `price <= 30` to select cheap products, a product whose price is 30.45 is not selected, while instead it could be of interest) and selected items could be ranked on the basis of their relevance to the selection condition. Fuzzy sets appeared as the formal framework to specify soft selection conditions [10]. Since relational-database technology dominated the panorama of database technology, many works were conducted to propose an extension of SQL (the standard query language for relational databases) towards soft querying based on fuzzy sets. Some popular proposals are *SQLf* [11,12] (for which we can mention an attempt to implement it [13]) and its extension named *SQLf3* (which copes with constructs introduced in *SQL3*), as well as *FQUERY for Access* [14,15] (designed to operate on databases managed by Microsoft Access). Among all these proposals, *SoftSQL* [16–18] provided users with a statement to define non-trivial "linguistic predicates", to be used in the extended SELECT statement to select table rows through linguistic predicates. The interested reader can find various surveys on the topic [19,20]; in particular, the work [21] is a very large handbook that summarizes all research work on this topic.

The advent of *NoSQL* (Not only SQL) databases [22], i.e., databases that do not rely on the classical relational model, has started a novel era in data management. In particular, the popularity obtained by the *JSON* (JavaScript Object Notation) format to represent any kind of complex data is facing the data engineer with the novel (with respect to relational databases) concept of "*JSON* document store", i.e., a database which stores *JSON* documents in a native way. The most famous *JSON* document store is *MongoDB* [23], but many others are available (such as *CouchDB* [24], exploited within the block-chain platform called HyperLedger Fabric [25]). As a result, this novel scenario is revamping the topic of soft querying on databases, this time on *NoSQL* databases in general and on *JSON* document stores in particular.

An extension of MQL, the *MongoDB* query language, is proposed in [26]; in this extension, called "fMQL", "fuzzy labels" can be used to query *JSON* documents, since they are equivalent to linguistic predicates; unfortunately, the work [26] does not provide any indication about how to define fuzzy labels. A further limitation of the proposal is that, for each single *JSON* document, only one membership degree is implicitly evaluated (in contrast, *J-CO-QL*$^+$ allows for dealing with many membership degrees for each single document).

Finally, the work [27] proposes an approach for soft querying *JSON* documents: the corpus of *JSON* documents is preliminarily translated into fuzzy RDF triples [28]; then, the query is translated into *fSPARQL* [29], a fuzzy extension of *SPARQL* [30]. In our opinion, this approach is not suitable for processing *JSON* documents, because it does not work on the original documents, but on an alternative representation of them.

2.2. Integrating Data Sets Describing Public Places

The topic of aggregating information about public places coming from internet sources has been investigated in the last decade. Many different approaches have been followed.

For example, the work [31] adopts the *DAS* technique to integrate data about public places uniquely by exploiting string similarity on names, in particular by comparing the two strings without and with tokenization.

The work [32] compares different string-similarity metrics with various machine learning methods, to solve the problem of toponym matching. The results demonstrate that machine learning methods (in particular, classifiers) perform better than string-similarity metrics. Obviously, they cannot be applied from scratch, without preliminary labeling and training activities. Similarly, the work [33] exploits a neural network to perform "toponym matching", i.e., pairing strings that represent the same location.

The work [34] addresses the problem of "geo-spatial data conflation" (the general name of the problem addressed in this paper) by adopting an entropy-based technique: the key idea is to use phonetic transcriptions, to compensate mistakes in writing names.

Another work that can be considered as related to this paper is [35], in which "semantic aligning" of heterogeneous geo-spatial data sets (GDs) is addressed. Specifically, it proposed an efficient similarity matching technique, which integrates various category systems simultaneously.

Finally, the closest work to this paper is [6]: this work could be considered the natural evolution of it. Specifically, a complex fuzzy relation is defined to perform public-place conflation in an online way. A comparison with a famous classification technique (i.e., "Random-Forest" classifiers) was performed, showing that the fuzzy approach is effective in a comparable way. Here, the definition of the fuzzy relation is improved to cope with not-cleaned names and addresses, as well as it is applied within the context of the *J-CO* Framework for the off-line integration of *JSON* data sets.

3. Basic Notions on Fuzzy Sets

In [36], Zadeh introduced the Fuzzy-Set Theory. It was rapidly clear that it had (and still has) an enormous potentiality to be successfully applied to many areas of computer science, such as decision making, control theory, expert systems, artificial intelligence, natural-language processing, and so on. Here, we report some basic concepts, which constitute the basis to understand the main contribution of this paper.

Definition 1. ***Fuzzy Set.*** *Consider a "universe set" U. A fuzzy set (or type-1 fuzzy set) A in U ($A \subseteq U$) is a mapping $A : U \rightarrow [0,1]$. The value $A(x)$ is referred to as the* membership degree *of the element x to the fuzzy set A. Alternatively, the notation $\mu_A(x) \in [0,1]$ can be used.*

Clearly, given an item $x \in U$, if $A(x) = 0$, this means that x does not belong at all to A; an intermediate value $0 < A(x) < 1$ means that x partially belongs to A (the greater the value, the higher its degree of membership); if $A(x) = 1$, this means that the item x fully belongs to A.

Consequently, a fuzzy set is "empty" if and only if its membership function is identically zero for each $x \in U$.

Furthermore, given two fuzzy sets A in U and B in U, they are "equal" (denoted as $A = B$), if and only if $A(x) = B(x)$ (alternatively, $\mu_A(x) = \mu_B(x)$) for all $x \in U$.

Operators on fuzzy sets can be easily defined, by extending the classical operators on traditional sets.

Definition 2. ***Union, Intersection and Complement.*** *Consider a universe U and two fuzzy sets A in U and B in U.*

The union *of two fuzzy sets A and B, denoted as $S = A \cup B$, generates a novel fuzzy set S whose membership function is $S(x) = max(A(x), B(x))$, for each $x \in U$ (alternatively, $\mu_S(x) = max(\mu_A(x), \mu_B(x))$).*

The Intersection *of two fuzzy sets A and B, denoted as $I = A \cap B$, generates a novel fuzzy set S whose membership function is $I(x) = min(A(x), B(x))$, for each $x \in U$ (alternatively, $\mu_I(x) = min(\mu_A(x), \mu_B(x))$).*

The Complement *of a fuzzy set A, denoted as $C = \overline{A}$, generates a novel fuzzy set C whose membership function is $C(x) = 1 - A(x)$, for each $x \in U$ (alternatively, $\mu_C(x) = 1 - \mu_A(x)$).*

Classical logical operators are mapped onto operators on fuzzy sets: the `OR` operator is mapped onto the union; the `AND` operator is mapped onto the intersection; the `NOT` operator is mapped onto the complement.

Fuzzy sets are useful to represent vague concepts, which characterize many real-life application contexts. For example, if the universe is the set of people, we could think to divide them into "young" and "old". However, is a person whose age is 40 actually young or old? He/she is a little bit young and a little bit old, neither fully young nor fully old.

Various other operators on fuzzy sets can be defined. In the following definition, we introduce the "weighted aggregation" operator.

Definition 3. ***Weighted Aggregation.*** *Given a universe U and two fuzzy sets A in U and B in U, the* weighted aggregation *operator $W = wag_\beta(A, B)$ (with $\beta \in [0,1]$) generates a new fuzzy set W whose membership function is defined as $W(x) = \beta \times A(x) + (1-\beta) \times B(x)$ (alternatively, $\mu_W(x) = \beta \times \mu_A(x) + (1-\beta) \times \mu_B(x)$).*

Example 1. *Through the membership degree, it is possible to denote partial membership of an item $x \in U$ to A; this way, vague linguistic concepts can be modeled. For example, given a public place p, its membership to the $PopularPlaces$ fuzzy set could be partial, denoting a place that is not so popular; thus, the membership degree measures its degree of popularity, for example on the basis of the number of likes obtained on social media.*

Suppose that on the same universe of public places, we conceive the $CheapRestaurants$ fuzzy set, whose membership degree denotes the perception that a public place is cheap (this perception could be induced by analyzing menus published on social media).

We now illustrate how to aggregate the $PopularPlaces$ and the $CheapRestaurants$ fuzzy sets to obtain interesting places.

- *If we are looking for "popular **and** cheap restaurants", we could formulate the search as "$PopularPlaces$ `AND` $CheapRestaurants$" (in terms of fuzzy sets, it is $I = PopularPlaces \cap CheapRestaurants$). Clearly, the lower membership degree determines the actual relevance of a place p.*
- *If we are looking for "popular **or** cheap restaurants", we could formulate the search as "$PopularPlaces$ `OR` $CheapRestaurants$" (in terms of fuzzy sets, it is $S = PopularPlaces \cup CheapRestaurants$). Clearly, the higher membership degree determines the actual relevance of a place p (a place could be not popular, but highly cheap).*
- *If we are looking for "popular **and possibly** cheap restaurants", we could formulate the search as "70% $PopularPlaces$ `AND` 30% $CheapRestaurants$" (in terms of fuzzy sets, it is $W = wag_{0.7}(PopularPlaces, CheapRestaurants)$). Clearly, the final membership degree is dominated by the degree of popularity, but a popular place that is also a cheap restaurant has a higher membership degree than a popular place that is not at all a cheap restaurant.*

The three above-mentioned searches are examples of "soft queries", where selection conditions are expressed in a vague way; the resulting membership degree denotes the "relevance" of an item to the soft query.

Furthermore, notice that when the names given to fuzzy sets linguistically characterize items in a proper way, these names can be used in soft conditions to linguistically express them.

Definition 4. ***Fuzzy Relation.*** *Consider two universes U_1 and U_2. A* fuzzy relation *R on U_1 and U_2, is defined as $R : U_1 \times U_2 \to [0,1]$. $R(x_1, x_2) \in [0,1]$, with $x_1 \in U_1$ and $x_2 \in U_2$, is the membership degree of the relation between x_1 and x_2; the meaning of the relation is linguistically expressed by the name of the relation.*

Through the concept of fuzzy relation, it is possible to model the strength of a relation between two items $x_1 \in U_1$ and $x_2 \in U_2$. Nevertheless, notice that a fuzzy relation is a particular case of fuzzy set in the universe $U = U_1 \times U_2$. Thus, we can reformulate the relation as $R : U \to [0,1]$, where $x = \langle x_1, x_2 \rangle \in U$; consequently, we can write, in an equivalent way, either $R(x_1, x_2)$ or $R(\langle x_1, x_2 \rangle)$.

In this paper, we work on the universe of *JSON* documents. So, given a document $d \in U$, the focus will be on the evaluation of its membership degrees to one or more fuzzy sets.

4. The *J-CO* Framework

The *J-CO* Framework is a suit of software tools able to process *JSON* data sets, in a way that is independent of the data source. In fact, it is able to obtain data sets both from *JSON* document stores (such as *MongoDB*) and from web sources. It is built around the *J-CO-QL*$^+$ language and it contains various tools, as illustrated in Figure 1.

- The *J-CO-QL*$^+$ *Engine* actually processes *J-CO-QL*$^+$ queries. It obtains data to process from *JSON* document stores and from web sources; it is able to store results again into *JSON* document stores.
- *J-CO-DS* is a simplified *JSON* document store [37]: it is designed to provide users with the capability of storing large single *JSON* documents (usually, popular *JSON* document stores are not able to deal with very-large single *JSON* documents). *J-CO-DS* does not provide any internal computational capability, i.e., it does not provide a query language: in fact, it is part of the *J-CO* Framework, in which the component that provides computational capabilities is the *J-CO-QL*$^+$ *Engine*.
- *J-CO-UI* is the user interface of the framework. It provides users with a graphical interface to interactively write *J-CO-QL*$^+$ queries in a step-by-step way; users can also inspect intermediate results.

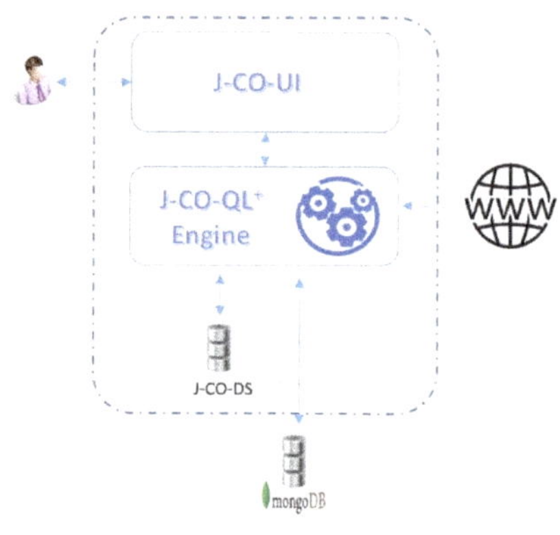

Figure 1. The *J-CO* Framework.

4.1. The Query Language

J-CO-QL$^+$ is the current evolution of the original *J-CO-QL* (see [3–5]): as its predecessor, it is designed to provide high-level and declarative statements, which does not require programming skills to be used; by means of them, it is possible to specify complex procedures (scripts) that are able to retrieve, integrate, transform and save *JSON* data sets. With respect to its predecessor, *J-CO-QL*$^+$ maintains the same approach, but revises syntax and semantics of statements, to improve their usability and effectiveness. Hereafter, we present its data model and its execution model.

4.1.1. Data Model

Here, we present the data model on which *J-CO-QL*$^+$ relies.

- The basic item to process is a *JSON* document. A document is represented within a pair of braces "`{`" and "`}`"; it is a sequence of fields separated by commas.
  A field is a "name: value" pair, where "name" is the field name, while "value" is the value of the field; the name is always enclosed within double quotes (e.g., `"name"`); the value can be a number, a string (enclosed either within double quotes or single quotes), a Boolean value, a nested sub-document (enclosed within a pair of braces "`{`" and "`}`") or an array (enclosed within square brackets "`[`" and "`]`", whose items can be any kind of *JSON* value, separated by commas).
- *J-CO-QL*$^+$ gives a special meaning to root-level fields whose name becomes with "~"; these names are compliant with *JSON* naming rules, but *J-CO-QL*$^+$ considers some of them in a special way, as illustrated hereafter.
  - The root-level `~fuzzysets` field is used to represent membership degrees of a d document to fuzzy sets. It works as a "key-value" map: given a field within `~fuzzysets`, the field name is the name of the fuzzy set to which the membership degree has been evaluated; the value is a real number in the range $[0, 1]$, which denotes the membership degree. This way, given a d document, it is possible to represent its membership to many fuzzy sets.
  - The root-level `~geometry` field represents geometries (also called "geo-tagging") of spatial entities represented as *JSON* documents. In this paper, we do not make use of geometries (the interested reader can refer to [5]).
- A "collection" is an unordered multi-set of heterogeneous documents, i.e., it can contain multiple copies of the same document.

4.1.2. Execution Model

The execution model is the same presented in previous publications [5,7]. Hereafter, we briefly summarize it.

- A "query" $q = (i_1, i_2, \ldots, i_n)$ is a sequence of instructions i_j, with $1 \leq j \leq n$. A query is a "pipe of instructions".
- Each instruction i_j receives an input "query-process state" s_{j-1} and generates a new query-process state s_j.
- A "query-process state" s_j (with $0 \leq j \leq n$) is a tuple $s_j = \langle tc, IR, DBS, FO, JSF \rangle$.
  - tc is called "temporary collection", since it is a collection of *JSON* documents that passes through the pipe of instructions, that contains temporary results of the query process.
  - IR is the "Intermediate-Results databaase", i.e., a database that is exclusive for the query process, to store intermediate results to be used later.
  - DBS is the set of "database descriptors", used to handle connections with external *JSON* document stores.
  - FO is the set of "Fuzzy Operators" defined within the query; they allows for evaluating membership degrees to fuzzy sets (see Section 6.2).
  - JSF is the set of user-defined "JavaScript Functions"; they are defined throughout the query to complete computational capabilities of the query language (see [38]).
- The initial query-process state is $s_0 = \langle tc : \emptyset, IR : \emptyset, DBS : \emptyset, FO : \emptyset, JSF : \emptyset \rangle$. Each instruction possibly modifies one member of the query-process state.

5. Problem and Methodology

In this section, we discuss the premises from which the paper has originated and introduce the problem we addressed as a case study (Section 5.1). Then, we present the methodological framework, which this work relies on (Section 5.2).

5.1. Premises and Problem

In [6], a fuzzy method for the online aggregation of POIs (Points of Interest) is presented. The problem addressed in that paper can be summarized as follows: if a web application has to integrate descriptors of public places (or POIs) caught on the fly from external services, the decision whether two descriptors actually describe the same public place or not must be taken in real time: techniques that require off-line work cannot be used.

In [6], it was proved that the technique can obtain very high levels of accuracy, absolutely comparable with off-line techniques; consequently, here, we argue that the same technique could be effectively adopted to integrate two data sets describing public places, in an off-line way. In particular, the novel support for soft querying [7] provided by *J-CO-QL*$^+$ (the query language of the *J-CO* Framework) has modified the scenario: in fact, the *J-CO* Framework is a stand-alone tool designed for manipulating and querying collections of *JSON* data sets. Consequently, it is straightforward to explore the possibility to exploit it for applying the fuzzy technique presented in [6] for integrating two collections of public-place descriptors coming from two different sources, by adopting a database approach (querying data by means of a query language) instead of hard coding the methodology with a programming language.

Hereafter, we present the problem. Then, Section 5.2 presents an improved formulation of the fuzzy technique presented in [6] that will be applied in *J-CO-QL*$^+$ scripts (discussed in Section 6).

Problem 1. *Consider two collections of descriptors D_1 and D_2. A* descriptor *d (such that either $d \in D_1$ or $d \in D_2$) describes a public place; we assume that d is a tuple whose minimal shape is $d = \langle name, address, lat, lon \rangle$, where $d.name$ is the name of the public place, $d.address$ is the raw address (i.e., as it is provided by the data source, without any pre-processing or cleaning) of the public place, while $d.lat$ and $d.lon$ are, respectively, the latitude and longitude of the public place. Depending on the source, these fields could be missing (either null value or zero-length string).*

Supposing that descriptors in D_1 and D_2 are related to the same municipality, we want to build the collection $SP = \{p_1, p_2, \dots\}$ of pairs of descriptors $p_i : \langle d_{1,h}, d_{2,k} \rangle$ (with $d_{1,h} \in D_1$ and $d_{2,k} \in D_2$) such that it is very likely that $d_{1,h}$ and $d_{2,k}$ actually describe the same public place.

5.2. Fuzzy Relation for Matching Public Places

The key contribution of [6] is a fuzzy relation called $MatchingPlaces$. Given two descriptors d_1 and d_2, it is written as $MatchingPlaces(d_1, d_2)$. Its membership degree denotes the possibility that d_1 and d_2 describe the same place. If we consider the universe $P = D_1 \times D_2$ of pairs $p_i : \langle d_{1,h}, d_{2,k} \rangle$, through the $MatchingPlaces$ fuzzy relation we want to build the fuzzy set PRP in P of Possibly-Relevant Pairs for which the membership degree of p_i is $PRP(p_i) > 0$.

To actually decide whether descriptors in a p_i pair actually describe the same public place, a minimum threshold $\alpha \in [0, 1]$ is used to focus on Relevant Pairs $RP \subseteq PRP$, where $RP(p_i) \geq \alpha$.

However, given a descriptor $d_{1,h} \in D_1$, it could appear several times in RP, because there might be many relevant pairs in which it is involved. For each $d_{1,h} \in D_1$, the subset $\overline{RP}_{1,h} \subseteq RP$ is the set of pairs $p_i \in RP$ such that $p_1.d_1 = d_{1,h}$; if $\overline{RP}_{1,h}$ is not empty, the pair $p_{1,h} \in \overline{RP}_{1,h}$ such that $\overline{RP}_{1,h}(p_{1,h}) \geq \overline{RP}_{1,h}(p_i)$, for all $p_i \in \overline{RP}_{1,h}$, appears in SP (because the two paired descriptors are actually supposed to describe the same place).

In the remainder of this section, we introduce the complete formal framework.

5.2.1. Basic Functions and Relations

The *MatchingPlaces* fuzzy relation is defined by means of some basic functions and fuzzy relations.

Given two pairs of coordinates, i.e., lat_1, lon_1 and lat_2, lon_2 (denoting latitude and longitude of two points on the earth globe), the *Distance* function computes the "Geodesic Distance" [39] between the two points in km; it is denoted as $Distance(lat_1, lon_1, lat_2, lon_2)$. On this basis, it is possible to define the *Close* membership function that, given a distance *dist* (in km) determines whether the distance denotes that two points are close; it is denoted as $Close(dist)$; an example of a typical membership function for this concept (the same exploited in [6]) is depicted in Figure 2: notice that, on the basis of the geodesic distance between the two points, the membership degree is 1 when the distance is between 0 and 50 m; then, it linearly decreases from 50 up to 1000 m. In Section 6.2, we will define a more sophisticated membership function.

The *Close* membership function can be used as the basis for defining the *ClosePlaces* fuzzy relation: given two place descriptors d_1 and d_2, $ClosePlaces(d_1, d_2) = Close(Distance(d_1.lat, d_1.lon, d_2.lat, d_2.lon))$.

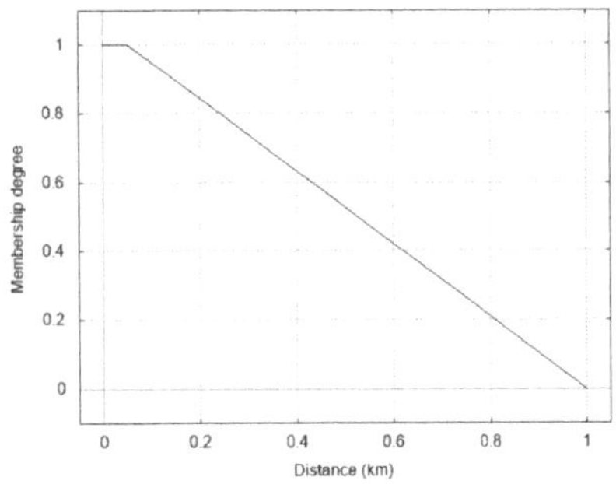

Figure 2. Sample *Close* membership function taken from [6].

Given two strings s_1 and s_2, the *Similar* fuzzy relation is denoted as $Similar(s_1, s_2)$. As a membership function, any string-similarity metric whose value is in the range $[0, 1]$ could be used; in [6], the Jaro-Winkler similarity metric [40–43] was used; here, we still use it, but in a more sophisticated way (see Section 6.2).

Based on the *Similar* relation, which is defined on the universe of strings, it is possible to define two derived relations that are defined on the universe of descriptor pairs $P = D_1 \times D_2$.

The *SimilarAddress* fuzzy relation denotes the extent to which the *address* fields of the two descriptors are similar; it is defined as $SimilarAddress(d_1, d_2) = Similar(d_1.address, d_2.address)$.

The *SimilarName* fuzzy relation denotes the extent to which the *name* fields of the two descriptors are similar; it is defined as $SimilarName(d_1, d_2) = Similar(d_1.name, d_2.name)$.

5.2.2. The *SameLocation* Relation

The *MatchingPlaces* relation is obtained by previously evaluating the *SameLocation* fuzzy relation. It is denoted as $SameLocation(d_1, d_2)$. Its membership function changes, depending on the fact that fields concerning geographical aspects (i.e., address and coordinates) in d_1 and d_2 are missing or not. Hereafter, we provide three different definitions of the *SameLocation* relation, one for each sub-case to deal with.

- **Case A: missing address(es).** If $d_1.address$ is missing, or $d_2.address$ is missing or both, but the two pairs of coordinates are available, only these latter ones can be used to evaluate the *SameLocation* relation.

- **Case B: missing coordinate(s).** When one or more coordinates in d_1 and d_2 are missing, but both addresses $d_1.address$ and $d_2.address$ are available, only these latter ones can be used to evaluate the *SameLocation* relation.
- **Case C: addresses and coordinates are all available.** When in d_1 and d_2 all geographical fields (i.e., address and coordinates) are available, they all contribute to the evaluation of the *SameLocation* relation.

Once the three cases of interest have been identified, it is possible to define the *SameLocation* relation.

Definition 5. ***Case A.*** *Given two descriptors d_1 and d_2, for which either $d_1.address$ or $d_2.address$ or both are missing, while the lat and lon fields are not null in both d_1 and d_2, the SameLocation relation is defined as follows:*

$$SameLocation(d_1, d_2) = ClosePlaces(d_1, d_2)$$

i.e., the membership degree of the SameLocation relation coincides with the membership degree of the ClosePlaces relation.

Definition 6. ***Case B.*** *Given two descriptors d_1 and d_2 for which at least one among $d_1.lat$, $d_1.lon$, $d_2.lat$ and $d_2.lon$ is null, while both $d_1.address$ and $d_2.address$ are available, the SameLocation relation is defined as follows:*

$$SameLocation(d_1, d_2) = SimilarAddress(d_1, d_2)$$

i.e., the membership degree of the SameLocation relation coincides with the membership degree of the SimilarAddress relation.

Definition 7. ***Case C.*** *Given two descriptors d_1 and d_2, for which all the fields $d_1.address$, $d_1.lat$, $d_1.lon$, $d_2.address$, $d_2.lat$ and $d_2.lon$ are available, the SameLocation relation is defined as follows:*

$$SameLocation(d_1, d_2) = wag_{\beta_{geo}}(SimilarAddress(d_1, d_2), ClosePlaces(d_1, d_2))$$

i.e., the membership degree of the SameLocation relation is the weighted aggregation of the Similar Address relation and of the ClosePlaces relation; $\beta_{geo} \in [0, 1]$ is the weight of the first term (the similarity between addresses).

In Section 6.3, we use $\beta_{geo} = 0.55$: this way, the similarity between addresses slightly prevails over closeness; indeed, if two addresses are very similar, their similarity contributes more than coordinates; this way, the effect of erroneous coordinates that give rise to high distances is mitigated.

5.2.3. Global *MatchingPlaces* Relation

At this point, we can define the global *MatchingPlaces* relation.

Definition 8. *Given two descriptors d_1 and d_2, for which both fields $d_1.name$ and $d_2.name$ are available, and for which the SameLocation relation is defined and $SameLocation(d_1, d_2) \geq \alpha_{geo}$ (with $\alpha_{geo} \in [0, 1]$), the MatchingPlaces relation is defined as follows:*

$$MatchingPlaces(d_1, d_2) = wag_{\beta_{name}}(SimilarName(d_1, d_2), SameLocation(d_1, d_2))$$

i.e., the membership degree of the MatchingPlaces relation is obtained by aggregating the membership degrees of the SimilarName relation and of the SameLocation relation, by means of the weighted aggregator with weight β_{name} for similarity of names.

In Section 6.4, we set $\beta_{name} = 0.6$: this way, the similarity between names prevails over the membership degree of the *SameLocation* relation. The rationale is the following: given two similar names, they contribute only for the 60%; the remaining 40% is given by the geographical contribution. However, in order to avoid the two descriptors whose geographical contribution is not significant, the α_{geo} threshold is introduced: if the membership degree of the *SameLocation* fuzzy relation is less than α_{geo}, d_1 and d_2 are no longer considered eligible to be the same place: two places can have very similar names (even identical—imagine two restaurants of the same chain), but if there is the doubt that they are reasonably close, they could be a wrong pair. In Section 6.3, we set this threshold as $\alpha_{geo} = 0.4$.

The membership degree of the *MatchingPlaces* fuzzy relation is used to determine whether a pair actually belongs to the *RP* set of relevant pairs, i.e., $RP(p_i) \geq \alpha$ means $MatchingPlaces\ (p_1.d_1, p_1.d_2) \geq \alpha$. In Section 6.4, we set this threshold as $\alpha = 0.8$, because in our experiments (see Section 7.1), we found this is the threshold that gives the best effectiveness.

6. Presenting the Script

In this section, we provide the technical contribution of the paper. Specifically, we demonstrate how the current version of *J-CO-QL*$^+$ is able to perform the soft integration of two collections containing *JSON* documents that describe public places, obtained from two different data sources.

6.1. Data Set

A *MongoDB* database called `ijgiDb` contains two collections of *JSON* documents: the first one is called `FacebookDescriptors` and its documents are descriptors of pages that present public places mostly located in the area of Manchester (UK); the second collection is called `GoogleDescriptors` and its documents are descriptors of places mostly located in the area of Manchester (UK) as well, obtained from *Google Places*. The `FacebookDescriptors` collection contains 5738 documents, while the `GoogleDesciptors` collection contains 5214 documents. Figure 3a shows a sample document in the `FacebookDescriptors` collection, while Figure 3b reports a sample document in the `GoogleDescriptors` collection. The reader can notice that *Facebook* descriptors clearly distinguish the address (in the `fbStreet` field) from the city name (in the `fbCity` field) from the ZIP code (in the `fbZip` field). In contrast, within a *Google Places* descriptor, the content of the `gAddress` field is less clean, because it contains the city name too. This also demonstrates that we are working on names and addresses as they are provided by *Facebook* and *Google Places*, without any pre-processing or cleaning (in [6], addresses were cleaned from numbers and urban designations, such as "street"). Consequently, here, we are addressing a less favorable situation.

```
{
  "id"            : 266,
  "idLink"        : "1761andLilysBar"
  "fbName"        : "1761 & Lily's Bar",
  "fbCity"        : "Manchester",
  "fbCountry"     : "United Kingdom",
  "fbLatitude"    : 53.4802297,
  "fbLongitude"   : -2.2435781,
  "fbStreet"      : "2 Booth Street",
  "fbZip"         : "M2 4AT",
}
```

(**a**)

```
{
  "gId"           : "ChIJ--wWob6xe0gRBF_8...",
  "gName"         : "Hope Studios"
  "gAddress"      : "52 Newton Street, Ma...",
  "gCity"         : "Manchester",
  "gLatitude"     : 53.4821537,
  "gLongitude"    : -2.2322586,
}
```

(**b**)

Figure 3. Examples of documents representing place descriptors. (**a**) Example of document in the `FacebookDescriptors` collection. (**b**) Example of document in the `GoogleDescriptors` collection.

6.2. Defining Fuzzy Operators

We start presenting the *J-CO-QL*$^+$ script. The first part of the script is reported in Listing 1.

The key concept provided by *J-CO-QL*$^+$ to evaluate membership degrees of *JSON* documents is the concept of "fuzzy operator". Such an operator is called within soft conditions: given some actual parameters (expressions based on document fields), the operator returns a membership degree. This degree will be used to evaluate the overall membership degree of a document to a specific fuzzy set.

Listing 1. *J-CO-QL$^+$* script: fuzzy operators.

```
1. CREATE FUZZY OPERATOR Close
     PARAMETERS
       distance TYPE Float
     PRECONDITION
         distance >= 0
     EVALUATE
         distance
     POLYLINE
       [ (0.00, 1.00), (0.05, 1.00), (0.20, 0.50), (0.60, 0.10), (1.00, 0.00)  ];

2. CREATE FUZZY OPERATOR Similar
     PARAMETERS
       st1 TYPE String,
       st2 TYPE String
     EVALUATE
         JARO_WINKLER_SIMILARITY(st1, st2)
     POLYLINE
       [ (0.00, 0.00), (0.60, 0.40), (0.70, 0.80), (0.80, 1.00), (1.00, 1.00)  ];

3. CREATE FUZZY OPERATOR WeightedAggregationBeta
     PARAMETERS
       f1   TYPE Float,
       f2   TYPE Float,
       beta TYPE Float
     PRECONDITION
       f1   IN_RANGE [0, 1] AND
       f2   IN_RANGE [0, 1] AND
       beta IN_RANGE [0, 1]
     EVALUATE
       f1*beta + f2*(1-beta)
     POLYLINE
       [ (0.00, 0.00), (1.00, 1.00)  ];
```

6.2.1. The `Close` Fuzzy Operator

The instruction on line 1 of the *J-CO-QL$^+$* script in Listing 1 defines the `Close` fuzzy operator: it evaluates the degree of closeness of two places, on the basis of the distance between them. Hereafter, we describe the instruction in details.

- The `PARAMETERS` clause defines the formal parameters of the operator. Specifically, only the `distance` parameter is defined.
- The `PRECONDITION` clause defines a condition on the parameters: if the condition is not satisfied, the evaluation of the fuzzy operator stops and an error signal is raised. Specifically, the precondition says that the distance must be no less than 0.
- The `EVALUATE` clause specifies a mathematical expression on the parameters, whose value is used as x-axis coordinate against the membership function defined by the subsequent `POLYLINE` clause. In the `Close` fuzzy operator, the expression simply takes the value of the `distance` parameter.
- The `POLYLINE` clause specifies the membership function actually used to compute the membership value. The function is defined as a polyline, by a sequence of pairs (x_i, y_i), where x_i can be any real value, while $y_i \in [0,1]$; given two consecutive points (x_i, y_i) and (x_{i+1}, y_{i+1}), it must be $x_i < x_{i+1}$. Each pair of consecutive points defines a segment. Given an x value, if it is between x_1 and x_n (in the case of n points), the corresponding y value is considered as a membership degree; if $x < x_1$, the membership degree is y_1; if $x > x_n$, the membership degree is y_n.

Figure 4a reports the polyline defined for the `Close` fuzzy operator. Notice that it is not the same defined in [6] (reported in Figure 2): in fact, we opted for a function that immediately penalizes distances that are between 50 m and 600 m, because two places in the same neighborhood are not perceived as very close when their distance becomes greater than 100 m.

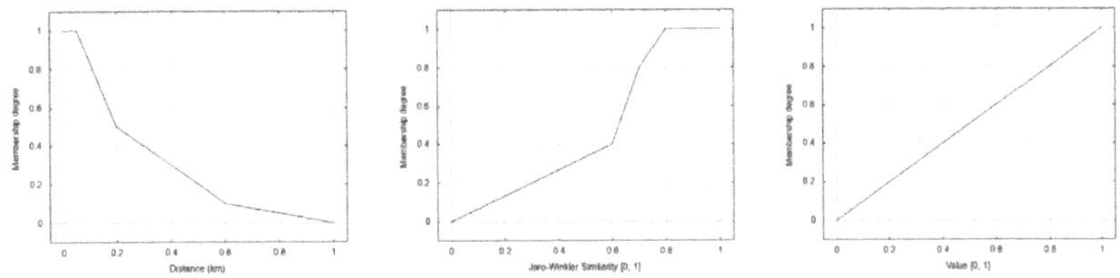

(a) (b) (c)

Figure 4. Membership functions for the fuzzy operators in Listing 1. (**a**) `Close`; (**b**) `Similar`; (**c**) `WeightedAggregationBeta`.

6.2.2. The `Similar` Fuzzy Operator

The instruction on line 2 of the *J-CO-QL*$^+$ script in Listing 1 creates the `Similar` fuzzy operator. Its goal is to evaluate a membership degree on the basis of the similarity degree of two strings. The operator is described in detail hereafter.

- The operator receives two parameters, called `st1` and `st2`; they are the two strings to compare.
- No precondition is specified: in the case of empty or null strings, the operator returns 0 as membership degree, because the similarity degree is 0.
- The `EVALUATE` clause calls the built-in (i.e., provided by *J-CO-QL*$^+$) function named `JARO_WINKLER_SIMILARITY`, to obtain the similarity degree of the two strings. The similarity degree is a value in the range $[0, 1]$. In the case of null or zero-length strings, the returned similarity degree is 0.
- The `POLYLINE` clause defines the membership function depicted in Figure 4b. Notice that it penalizes similarity degrees that are less than 0.7, while membership degrees that are greater than 0.8 are rewarded: this is due to the sometimes bizarre behavior of the Jaro-Winkler similarity, that returns high similarity degrees even when strings only shares some characters, but are not actually similar; furthermore, for strings such as "`The Gray Horse`" and "`GrayHorse`", the similarity degree is around 0.7, although they clearly have to be considered very similar. With this shape, we try to compensate the behavior of the Jaro-Winkler similarity, so as to deal with raw addresses and names (i.e., not cleaned from articles, numbers, punctuation, and so on).

6.2.3. The `WeightedAggregationBeta` Fuzzy Operator

The instruction on line 3 of the *J-CO-QL*$^+$ script in Listing 1 defines the third fuzzy operator. This is called `WeightedAggregationBeta` and its goal is to perform the "weighted aggregation" wag_β (see Definition 3). In fact, *J-CO-QL*$^+$ does not provide such an operator in its language; through the `WeightedAggregationBeta` fuzzy operator, we show how to introduce novel fuzzy concepts. The fuzzy operator is described in detail hereafter.

- The operator receives three parameters: `f1` and `f2` are the two values in the range $[0, 1]$ to aggregate, while `beta` is the aggregation weight (in the range $[0, 1]$ too) of `f1` with respect to `f2`.
- The `PRECONDITION` clause ensures that the actual values of the three parameters are in the range $[0, 1]$ (notice the `IN_RANGE` predicate).
- The `EVALUATE` clause actually performs the weighted aggregation.
- The `POLYLINE` clause defines a very simple membership function, which is reported in Figure 4c: it is a straight segment from the point $(0, 0)$ to the point $(1, 1)$; this way, the value computed by the `EVALUATE` clause is returned, as it is, as membership degree.

6.3. Retrieving and Pairing Descriptors

Once the three fuzzy operators are defined, it is time to start working on the data set. This is conducted by the second part of the *J-CO-QL*$^+$ script, which is reported in Listing 2.

Listing 2. *J-CO-QL*$^+$ script: retrieving and joining collections.

```
4. USE DB ijgiDb
     ON SERVER MongoDB 'http://127.0.0.1:27017';

5. JOIN OF COLLECTIONS
     FacebookDescriptors@ijgiDb AS f, GoogleDescriptors@ijgiDb AS g
    CASE
     // Case A: missing address(es)
     WHERE ( FIELD  .f.fbStreet IS NULL  OR
             FIELD  .g.gAddress IS NULL  OR
                    .f.fbStreet = ""     OR
                    .g.gAddress = ""   )     AND
           FIELD .f.fbLatitude  IS NOT NULL AND
           FIELD .f.fbLongitude IS NOT NULL AND
           FIELD .g.gLatitude   IS NOT NULL AND
           FIELD .g.gLongitude  IS NOT NULL
       GENERATE
         CHECK FOR
           FUZZY SET ClosePlaces    USING
                      Close ( GEODESIC_DISTANCE(  .f.fbLatitude, .f.fbLongitude,
                                                  .g.gLatitude,   .g.gLongitude ) ),
           FUZZY SET SameLocation  USING ClosePlaces
         ALPHACUT 0.4 ON SameLocation

     // Case B: missing coordinate(s)
     WHERE  FIELD  .f.fbStreet      IS NOT NULL AND
            FIELD  .g.gAddress      IS NOT NULL AND
                   .f.fbStreet      != ""       AND
                   .g.gAddress      != ""       AND
          ( FIELD .f.fbLatitude  IS NULL  OR
            FIELD .f.fbLongitude IS NULL  OR
            FIELD .g.gLatitude   IS NULL  OR
            FIELD .g.gLongitude  IS NULL  )
        GENERATE
          CHECK FOR
            FUZZY SET SimilarAddress  USING Similar (.f.fbStreet, .g.gAddress),
            FUZZY SET SameLocation    USING SimilarAddress
          ALPHACUT 0.4 ON SameLocation

     // Case C: addresses and coordinates all available
     WHERE  FIELD .f.fbStreet     IS NOT NULL AND
            FIELD .g.gAddress     IS NOT NULL AND
                  .f.fbStreet     != ""       AND
                  .g.gAddress     != ""       AND
            FIELD .f.fbLatitude  IS NOT NULL AND
            FIELD .f.fbLongitude IS NOT NULL AND
            FIELD .g.gLatitude   IS NOT NULL AND
            FIELD .g.gLongitude  IS NOT NULL
       GENERATE
         CHECK FOR
           FUZZY SET SimilarAddress  USING Similar ( .f.fbStreet, .g.gAddress ),
           FUZZY SET ClosePlaces     USING
                     Close ( GEODESIC_DISTANCE(  .f.fbLatitude, .f.fbLongitude,
                                                 .g.gLatitude,   .g.gLongitude ) ),
           FUZZY SET SameLocation    USING
                       WeightedAggregationBeta(  MEMBERSHIP_OF(SimilarAddress),
                                                 MEMBERSHIP_OF(ClosePlaces), 0.55 )
         ALPHACUT 0.4 ON SameLocation;
```

The instruction on line 4 connects the query process to the database. After this instruction, it will be possible to access the `ijgiDb` database to retrieve and store collections.

The `JOIN OF COLLECTIONS` instruction on line 5 retrieves the two source collections (called `FacebookDescriptors` and `GoogleDescriptors`) and creates all possible pairs of documents contained in the two collections. Then, the subsequent `CASE` clause evaluates a pool of conditions on these pairs to possibly evaluate fuzzy sets on the actually-interesting pairs and discards the others. The instruction is explained in detail hereafter.

- The instruction retrieves the `FacebookDescriptors` collection from the `ijgiDb` database and aliases it as `f`; similarly, it retrieves the `GoogleDescriptors` collection from the same database and aliases it as `g`.
  For each f document from the `f` collection and for each g document from the `g` collection, a new d document is created. This document contains two fields: the first one is called `f` and its value is the source f document; the second one is called `g` and its value is the source g document. The d document is further processed by the subsequent `CASE` clause.

Figure 5 reports an example of d document, which is obtained by joining the two sample documents reported in Figure 3; notice the names of the root-level fields.

```
{
  "f" : {
    "id"           : 266,
    "idLink"       : "1761andLilysBar"
    "fbName"       : "1761 & Lily's Bar",
    "fbCity"       : "Manchester",
    "fbCountry"    : "United Kingdom",
    "fbLatitude"   : 53.4802297,
    "fbLongitude"  : -2.2435781,
    "fbStreet"     : "2 Booth Street",
    "fbZip"        : "M2 4AT",
  },
  "g"  : {
    "gId"          : "ChIJ--wWob6xe0gRBF_8…",
    "gName"        : "Hope Studios"
    "gAddress"     : "52 Newton Street, Ma…",
    "gCity"        : "Manchester",
    "gLatitude"    : 53.4821537,
    "gLongitude"   : -2.2322586,
  }
}
```

Figure 5. Example of document generated by the `JOIN OF COLLECTIONS` instruction on line 5 of the *J-CO-QL*$^+$ script, before the `CASE clause`.

- The `CASE` clause evaluates a pool of selection conditions expressed within a `WHERE` clause; if a d document is selected by a condition, it is processed according to the subsequent sub-clauses. Many `WHERE` branches are possible: a d document is processed by the branch associated with the first `WHERE` condition that it satisfies; if no condition is satisfied, d is discarded (it will not appear in the output temporary collection).
  Specifically, the `CASE` clause in the instruction on line 5 in Listing 2 contains three `WHERE` branches: each of them deals with one of the three situations considered for defining the $SameLocation$ relation by Definitions 5–7. Hereafter, we separately discuss the behavior of the three branches.
  - The first `WHERE` branch deals with the case A of the $SameLocation$ fuzzy relation, defined in Definition 5. The condition is true if either the value for the `fbStreet` field is missing or the value for the `gAddress` field is missing or both are missing, and all coordinates are available. If a d document meets the condition, the `GENERATE` block further processes d through the `CHECK FOR` clause, whose goal is to evaluate the membership degrees of d to fuzzy sets.
    Specifically, two `FUZZY SET` branches are present: the former evaluates the `ClosePlaces` fuzzy set, the latter evaluates the `SameLocation` fuzzy set.
    The membership degree to the `ClosePlaces` fuzzy set is obtained by the associated `USING` clause: this is a "soft condition", in which fuzzy operators (such as those defined in Section 6.2) and fuzzy-set names can be composed by the usual (fuzzy) logical operators `AND`, `OR` and `NOT`; the resulting membership degree is the membership degree to the evaluated fuzzy set. If this is the first membership degree evaluated for d, then d does not have the special `~fuzzysets` field: in this case, the field is added and within it only one single field is present, having the same name of the evaluated fuzzy set, whose value is the computed membership degree. In contrast, if the `~fuzzysets` field is already present, it is extended with one extra internal field, describing the membership degree to the new evaluated fuzzy set.
    Specifically, the first branch evaluates the membership degree to the `ClosePlaces` fuzzy set, by means of the `Close` fuzzy operator (see Listing 1), which is called passing the geodesic distance computed by the `GEODESIC_DISTANCE` built-in function.
    The second `FUZZY SET` branch evaluates the membership degree to the `SameLocation` fuzzy set, by assuming that it coincides with the `ClosePlaces` fuzzy set (see Definition 5). Finally, the `ALPHACUT` clause discards the d document from the output temporary collection if its membership degree to the `SameLocation` fuzzy set is less than 0.4; remember that this is the α_{geo} threshold mentioned within Definition 8. Figure 6a reports a sample document generated

by the first `WHERE` branch; notice the presence of the `~fuzzysets` field and its inner fields.

- The second `WHERE` branch deals with case *B* of the *SameLocation* relation (see Definition 6), i.e., at least one coordinate is null but both addresses are present. In this case (see Definition 6), the membership degree to the `SimilarAddress` fuzzy set is evaluated by means of the `Similar` fuzzy operator, which evaluates the fuzzy similarity relation between two strings (in this case, the two addresses). Then, as defined by Definition 6, the second `FUZZY SET` branch tells that the `SameLocation` fuzzy set coincides with the `SimilarAddress` fuzzy set. Again, the `ALPHACUT` clause puts the *d* document into the output temporary collection if the membership degree to the `SameLocation` fuzzy set is no less than 0.4 (the α_{geo} threshold in Definition 8). Figure 6b shows a sample document generated by the second `WHERE` branch.
- The third `WHERE` branch deals with case C of the *SameLocation* fuzzy relation (see Definition 7), i.e., both all addresses and all coordinates are available. Consequently, the membership degrees to three different fuzzy sets are evaluated: the first one is the `SimilarAddress` fuzzy set, by means of the `Similar` fuzzy operator applied to addresses; the second one is the `ClosePlaces` fuzzy set, by means of the `Close` fuzzy operator applied to the geodesic distance between the two points.
The third `FUZZY SET` branch evaluates the membership degree to the fuzzy set named `SameLocation`: according to Definition 7, it is obtained by calling the `WeightedAggregationBeta` fuzzy operator, whose goal is to perform the weighted aggregation: it receives the two values (in the range $[0, 1]$) to aggregate and the β weight.
The `USING` soft condition calls the `WeightedAggregationBeta` fuzzy operator, passing the membership values to the `SimilarAddress` fuzzy set and to the `ClosePlaces` fuzzy set, which are obtained by means of the `MEMBERSHIP_OF` built-in function (that extracts the membership degree from within the `~fuzzysets` field). The third parameter is the constant value 0.55: thisis the β_{geo} weight presented and discussed in Definition 7. The `ALPHACUT` clause discards the evaluated document if its membership degree to the `SameLocation` fuzzy set is less than 0.4 (the α_{geo} threshold mentioned in Definition 8).
Figure 6c reports a sample document generated by the third branch; notice that the `~fuzzysets` field has three inner fields.


```
{
 "f" : {
   "id"     : 233,
   "idLink" :"pages/Mustaphs/16487095…",
   "fbName"      : "Mustaph's",
   "fbStreet"    : null,
   "fbZip"       : "SK3 9",
   "fbCity"      : "Stockport",
   "fbCountry"   : "United Kingdom",
   "fbLatitude"  : 53.40252971,
   "fbLongitude" : -2.163522497
 },
 "g" : {
   "gId"     : "ChIJ-ZU5iW-ze0gR7YewCI-…",
   "gName"      : "Mustaphs",
   "gAddress"  : "11 Castle Street, St…",
   "gCity"      : "Stockport",
   "gLatitude" : 53.4024049,
   "gLongitude": -2.1636548,
 },
 "~fuzzysets" : {
    "ClosePlaces"    : 1.0,
    "SameLocation"   : 1.0
 }
}
```


(a)


```
{
 "f" : {
   "id"     : 528,
   "idLink" : "bettereastmanchester",
   "fbName" : "East Manchester Leisure…",
   "fbStreet"    : "189 Grey Mare Lane",
   "fbZip"       : "M11 3ND",
   "fbCity"      : "Manchester ",
   "fbCountry"   : "",
   "fbLatitude"  : null,
   "fbLongitude" : null
 },
 "g" : {
   "gId"     : "ChIJAdYbKW2xe0gRCVJcmIJ…",
   "gName"  : "East Manchester Leisure…",
   "gAddress"  : "189 Grey Mare Lane,…",
   "gCity"      : "Manchester",
   "gLatitude" : 53.4775529,
   "gLongitude": -2.1955198
 },
 "~fuzzysets" : {
   "SameLocation"    : 1.0,
   "SimilarAddress" : 1.0
 }
}
```


(b)


```
{
 "f" : {
   "id"            : 4193,
   "idLink"        : "FoodCycle-Manchest…",
   "fbName"   : "FoodCycle Manchester",
   "fbStreet": "The Roby, 307 Dicken…",
   "fbCity"        : "Manchester",
   "fbCountry"     : "United Kingdom",
   "fbZip"         : "M13 0NG",
   "fbLatitude"  : 53.4554234,
   "fbLongitude" : -2.205004
 },
 "g"  : {
   "gId" : "ChIJ3T0-vdWze0gRBPlxJfafcnY",
   "gName"    : "FoodCycle Manchester",
   "gAddress": "307 Dickenson Road, M…",
   "gCity"    : "Manchester",
   "gLatitude"   : 53.4552679,
   "gLongitude"  : -2.205912
 },
 "~fuzzysets"  : {
   "ClosePlaces"     : 0.958144669398663,
   "SameLocation"    : 0.81180003253166,
   "SimilarAddress" : 0.692063511458657
 }
}
```


(c)

Figure 6. Examples of documents generated by the `JOIN OF COLLECTIONS` instruction on line 5. (**a**) Example for Case A. (**b**) Example for Case B. (**c**) Example for Case C.

The temporary collection produced by the instruction on line 5 of Listing 2 contains heterogeneous documents, as far as the structure of the `~fuzzysets` field is concerned, but all have the inner `SameLocation` field, denoting the membership degree to the `SameLocation` fuzzy set; it will be used in the next instruction, to evaluate the membership degree to the `MatchingPlaces` fuzzy set.

Furthermore, notice that `SameLocation`, `ClosePlaces` and `SimilarAddresses` are called "fuzzy sets", while they were defined in Section 5 as "fuzzy relations": this is not a mistake, but the consequence of the fact that *JSON* documents represent pairs of descriptors; consequently, fuzzy relation on pairs are translated into fuzzy sets on *JSON* documents.

6.4. Relevant Pairs

All documents contained in the temporary collection produced by the instruction on line 5 (Listing 2) have the membership degree to the `SameLocation` fuzzy set no less than $\alpha_{geo} = 0.4$, as required by Definition 8. The `FILTER` instruction on line 6 in Listing 3 actually evaluates the membership degree to the `MatchingPlaces` fuzzy set, which corresponds to the *MatchingPlces* relation defined in Definition 8. The `FILTER` instruction on line 6 is described in detail hereafter.

- The `FILTER` statement takes the temporary collection as input and generates a new temporary collection by applying a `CASE` clause. The behavior of this clause is the same as in the `JOIN OF COLLECTIONS` statement.
- On line 6, only one `WHERE` branch is present: if a document does not meet the selection condition, it is discarded from the output temporary collection.
  Specifically, the selection condition selects those documents having both the names in the two paired descriptors, so as to evaluate the membership degree to the `SimilarName` fuzzy set.
- The first `FUZZY SET` branch in the `CHECK FOR` clause evaluates the membership degree to the `SimilarName` fuzzy set; again, in the `USING` soft condition, the `Similar` fuzzy operator (see Listing 1) is called, this time passing names (instead of addresses).
- The second `FUZZY SET` branch can finally evaluate the membership degree to the `MatchingPlaces` fuzzy set, corresponding to the *MatchingPlaces* fuzzy relation defined by Definition 8. Remember that the fuzzy relations named *SameLocation* and *SimilarName* are aggregated by means of the weighted aggregation operator. In Listing 1, we defined the `WeightedAggregationBeta` fuzzy operator, which here is used to aggregate the `SimilarName` fuzzy set and the `SameLocation` fuzzy set; the `SimilarName` fuzzy set weights for the 60% of the final membership degree (this is the β_{name} weight mentioned in Definition 8), so that similarity between names moderately prevails over geographical similarity (whose goal is to cofirm that two places having similar or identical names are actually the same place). The resulting membership degree becomes the membership degree to the `MatchingPlaces` fuzzy set.
  The three sample documents reported in Figure 6 become as reported in Figure 7; notice the presence of the `MatchingPlaces` inner field within the `~fuzzysets` field.

Listing 3. *J-CO-QL*$^+$ script: matching places.

```
6. FILTER
    CASE
      WHERE  WITH .f.fbName, .g.gName  AND
             KNOWN FUZZY SETS SameLocation
        GENERATE
          CHECK FOR
            FUZZY SET SimilarName       USING Similar( .f.fbName, .g.gName ),
            FUZZY SET MatchingPlaces  USING
              WeightedAggregationBeta  ( MEMBERSHIP_OF  (SimilarName),
                                         MEMBERSHIP_OF  (SameLocation),   0.60)
          ALPHACUT 0.8 ON MatchingPlaces
          BUILD{
            .f    : .f,
            .g    : .g,
            .rank: MEMBERSHIP_OF(MatchingPlaces)   }
          DEFUZZIFY;

7. SAVE AS RelevantPairs@ijgiDb;
```

```
{
 "f" : {
  "id"      : 233,
  "idLink" :"pages/Mustaphs/16487095…",
  "fbName"       : "Mustaph's",
  "fbStreet"     : null,
  "fbZip"        : "SK3 9",
  "fbCity"       : "Stockport",
  "fbCountry"    : "United Kingdom",
  "fbLatitude"  : 53.40252971,
  "fbLongitude" : -2.163522497
 },
 "g" : {
  "gId"    : "ChIJ-ZU5iW-ze0gR7YewCI-…",
  "gName"  : "Mustaphs",
  "gAddress"  : "11 Castle Street, St…",
  "gCity"     : "Stockport",
  "gLatitude" : 53.4024049,
  "gLongitude": -2.1636548,
 },
 "~fuzzysets"  : {
   "ClosePlaces"     : 1.0,
   "MatchingPlaces" : 1.0,
   "SameLocation"   : 1.0,
   "SimilarName"    : 1.0
 }
}
```

```
{
 "f" : {
  "id"      : 528,
  "idLink" : "bettereastmanchester",
  "fbName" : "East Manchester Leisure…",
  "fbStreet"    : "189 Grey Mare Lane",
  "fbZip"       : "M11 3ND",
  "fbCity"      : "Manchester ",
  "fbCountry"   : "",
  "fbLatitude"  : null,
  "fbLongitude" : null
 },
 "g" : {
  "gId"     : "ChIJAdYbKW2xe0gRCVJcmIJ…",
  "gName"  : "East Manchester Leisure…",
  "gAddress"   : "189 Grey Mare Lane,…",
  "gCity"      : "Manchester",
  "gLatitude" : 53.4775529,
  "gLongitude": -2.1955198
 },
 "~fuzzysets" : {
  "MatchingPlaces" : 1.0,
  "SameLocation"   : 1.0,
  "SimilarAddress" : 1.0,
  "SimilarName"    : 1.0
 }
}
```

```
{
 "f" : {
  "id"            : 4193,
  "idLink"        : "FoodCycle-Manchest…",
  "fbName"   : "FoodCycle Manchester",
  "fbStreet": "The Roby, 307 Dicken…",
  "fbCity"       : "Manchester",
  "fbCountry"    : "United Kingdom",
  "fbZip"        : "M13 0NG",
  "fbLatitude"  : 53.4554234,
  "fbLongitude" : -2.205004
 },
 "g"  : {
  "gId" : "ChIJ3T0-vdWze0gRBPlxJfafcnY",
  "gName"    : "FoodCycle Manchester",
  "gAddress": "307 Dickenson Road, M…",
  "gCity"   : "Manchester",
  "gLatitude"   : 53.4552679,
  "gLongitude"  : -2.205912
 },
 "~fuzzysets"  : {
  "ClosePlaces"     : 0.958144669398663,
  "MatchingPlaces" : 0.924720013012664,
  "SameLocation"   : 0.81180003253166,
  "SimilarAddress" : 0.692063511458657,
  "SimilarName"    : 1.0
 }
}
```

(a) (b) (c)

Figure 7. Examples of documents transformed by the `FILTER` instruction on line 6 before the `BUILD` section. (**a**) Example for Case A. (**b**) Example for Case B. (**c**) Example for Case C.

- At this point, only relevant pairs must be kept, i.e., those pairs whose membership degree to the `MatchingPlaces` fuzzy set is no less than $\alpha = 0.8$. The `ALPHACUT` clause does that.
- The final `BUILD` section (which is optional, this is why it was not present in the `JOIN OF COLLECTIONS` instruction on line 5 in Listing 2) restructures all survived documents. Specifically, a novel `rank` field is added, whose value is the membership degree to the `MatchingPlaces` fuzzy set. This field is necessary, because the subsequent `DEFUZZIFY` option discards the `~fuzzysets` field (as a consequence, documents are "defuzzified"). Figure 8 reports the final state of the three sample documents reported in Figure 7. Notice the presence of the `rank` field, whose value is the membership degree of the `MatchingPlaces` fuzzy set.

```
{
 "f"     : {
  "id"      : 233,
  "idLink" :"pages/Mustaphs/16487095…",
  "fbName"       : "Mustaph's",
  "fbStreet"     : null,
  "fbZip"        : "SK3 9",
  "fbCity"       : "Stockport",
  "fbCountry"    : "United Kingdom",
  "fbLatitude"  : 53.40252971,
  "fbLongitude" : -2.163522497
 },
 "g"     : {
  "gId"    : "ChIJ-ZU5iW-ze0gR7YewCI-…",
  "gName"  : "Mustaphs",
  "gAddress"  : "11 Castle Street, St…",
  "gCity"     : "Stockport",
  "gLatitude" : 53.4024049,
  "gLongitude": -2.1636548,
 },
 "rank" : 1.0,
}
```

```
{
 "f"     : {
  "id"      : 528,
  "idLink" : "bettereastmanchester",
  "fbName" : "East Manchester Leisure…",
  "fbStreet"    : "189 Grey Mare Lane",
  "fbZip"       : "M11 3ND",
  "fbCity"      : "Manchester ",
  "fbCountry"   : "",
  "fbLatitude"  : null,
  "fbLongitude" : null
 },
 "g"     : {
  "gId"     : "ChIJAdYbKW2xe0gRCVJcmIJ…",
  "gName"  : "East Manchester Leisure…",
  "gAddress"   : "189 Grey Mare Lane,…",
  "gCity"      : "Manchester",
  "gLatitude" : 53.4775529,
  "gLongitude": -2.1955198
 },
 "rank" : 1.0,
}
```

```
{
 "f"     : {
  "id"            : 4193,
  "idLink"        : "FoodCycle-Manchest…",
  "fbName"   : "FoodCycle Manchester",
  "fbStreet": "The Roby, 307 Dicken…",
  "fbCity"       : "Manchester",
  "fbCountry"    : "United Kingdom",
  "fbZip"        : "M13 0NG",
  "fbLatitude"  : 53.4554234,
  "fbLongitude" : -2.205004
 },
 "g"     : {
  "gId" : "ChIJ3T0-vdWze0gRBPlxJfafcnY",
  "gName"    : "FoodCycle Manchester",
  "gAddress": "307 Dickenson Road, M…",
  "gCity"   : "Manchester",
  "gLatitude"   : 53.4552679,
  "gLongitude"  : -2.205912
 },
 "rank" : 0.924720013012664
}
```

(a) (b) (c)

Figure 8. Examples of documents generated by the `FILTER` instruction on line 6. (**a**) Example for Case A. (**b**) Example for Case B. (**c**) Example for Case C.

The instruction on line 7 in Listing 3 saves the temporary collection into the `ijgiDb` database, with name `RelevantPairs`. Its documents contain the most promising pairs of descriptors (remember the *RP* set mentioned in Section 5.2), but it could happen that, e.g., the same *Google Places* descriptor is associated with more than one *Facebook* descriptor. Clearly, it is the case to choose the pair having the highest rank (i.e., building the final *SP* set mentioned in Section 5.2). This is discussed in Section 6.5.

6.5. Choosing the Best Pairs

The last part of the *J-CO-QL*$^+$ script is reported in Listing 4. It actually chooses the best pairs that involves each single *Google Places* descriptor obtained by line 6 in Listing 3. Indeed, the original *J-CO-QL* language (from which *J-CO-QL*$^+$ derives) was designed to cope with this kind of task too (see [3,4]). Hereafter, we briefly describe this last part of the script.

- The `GET COLLECTION` instruction on line 8 again obtains the `RelevantPairs` collection from the database, again making it the temporary collection.
- The `GROUP` instruction on line 9 groups documents in the temporary collection, on the basis of the `gId` field, which is the identifier of *Google Places* descriptors. For each group, a novel document is generated into the output collection, such that it has the `gId` field and an array called `gGroup`, in which all grouped documents are reported. This array is sorted in reverse order of value of the `rank` field within grouped documents. Figure 9a reports an example of the grouped document.
- The `EXPAND` instruction on line 10 unnests again all grouped documents. For each output document, the `gPair` field is added to the global ones (apart from the expanded array); this new field contains two inner fields: the `item` field contains the unnested document; the `position` field denotes the position occupied by the unnested item in the `gGroup` array.
  As a result, the temporary collection generated by line 10 contains as many documents in the `RankedPairs` collection, but now they are tagged with the relative order for *Google Places* descriptors on the basis of the `rank` field. Figure 9b reports an example of an unnested document.
- The `FILTER` instruction on line 11 actually selects only documents that previously occupied the first position in their group (based on the reverse order of `rank`, they are the ones with the highest rank). The `BUILD` section builds the same structure again, as in the `RelevantPairs` collection. Figure 9c reports an example of resulting document.
- Finally (on line 12) the last temporary collection is saved into the `ijgiDb` database with name `SamePlaces`, which is the desired output of the process.


```
{
  "g" : {
    "gId" : "ChIJH7MYbLKze0gRVIeX3GG1HFQ"
  },
  "gGroup" : [
    {
      "rank" : 1.0,
      "f" : {
        "id"           : 1773,
        "idLink"       : "TheFootparlour",
        "fbName"       : "The Footparlour",
        "fbCity"       : "Manchester",
        "fbCountry"    : "United Kingdom",
        "fbStreet"     : "448 Burnage Lane",
        "fbZip"        : "M19 1LH",
        "fbLatitude"   : 53.42517,
        "fbLongitude"  : -2.20296
      },
      "g" : {
        "gId":"ChIJH7MYbLKze0gRVIeX3GG1HFQ",
        "gName"        : "The Foot Parlour",
        "gAddress"     : "448 Burnage Lan…",
        "gCity"        : "Manchester",
        "gLatitude"    : 53.4252831,
        "gLongitude"   : -2.2031884
      }
    },
    {
      "rank" : 0.810648074150466,
      "f" : {
        "id"           : 1996,
        "idLink"       : "burnageM19",
        "fbName"       : "The Food Box",
        "fbCity"       : "Manchester",
        "fbCountry"    : "United Kingdom",
        "fbStreet"     : "227 Burnage Lane",
        "fbZip"        : "M19 1FN",
        "fbLatitude"   : 53.43231411,
        "fbLongitude"  : -2.200215987
      },
      "g" : {
        "gId":"ChIJH7MYbLKze0gRVIeX3GG1HFQ",
        "gName"        : "The Foot Parlour",
        "gAddress"     : "448 Burnage Lan…",
        "gCity"        : "Manchester",
        "gLatitude"    : 53.4252831,
        "gLongitude"   : -2.2031884
      }
    }
  ]
}
```

(a)

```
{
  "g"     : {
    "gId" : "ChIJH7MYbLKze0gRVIeX3GG1HFQ"
  },
  "gPair" : {
    "position" : 1,
    "item"     : {
      "rank" : 1.0,
      "f"    : {
        "id"           : 1773,
        "idLink"       : "TheFootparlour",
        "fbName"       : "The Footparlour",
        "fbCity"       : "Manchester",
        "fbCountry"    : "United Kingdom",
        "fbStreet"     : "448 Burnage Lane",
        "fbZip"        : "M19 1LH",
        "fbLatitude"   : 53.42517,
        "fbLongitude"  : -2.20296
      },
      "g"    : {
        "gId": "ChIJH7MYbLKze0gRVIeX3GG1HFQ",
        "gName"        : "The Foot Parlour",
        "gAddress"     : "448 Burnage Lane…",
        "gCity"        : "Manchester",
        "gLatitude"    : 53.4252831,
        "gLongitude"   : -2.2031884
      }
    }
  }
}
```

(b)

```
{
  "rank" : 1.0,
  "f"    : {
  "id"             : 1773,
    "idLink"       : "TheFootparlour",
    "fbName"       : "The Footparlour",
    "fbCity"       : "Manchester",
    "fbCountry"    : "United Kingdom",
    "fbStreet"     : "448 Burnage Lane",
    "fbZip"        : "M19 1LH",
    "fbLatitude"   : 53.42517,
    "fbLongitude"  : -2.20296
  },
  "g"    : {
    "gId": "ChIJH7MYbLKze0gRVIeX3GG1HFQ",
    "gName"        : "The Foot Parlour",
    "gAddress"     : "448 Burnage Lane, Ma…",
    "gCity"        : "Manchester",
    "gLatitude"    : 53.4252831,
    "gLongitude"   : -2.2031884
  }
}
```


(c)

Figure 9. Examples of documents during selection of `BestPairs` in Listing 4. (**a**) Example of document after `GROUP` instruction on line 9. (**b**) Example of document during `EXPAND` instruction on line 10 before the `BUILD` clause. (**c**) Example of document after `EXPAND` instruction on line 10.

Listing 4. *J-CO-QL*$^+$ script: Selecting the best pairs.

```
8. GET COLLECTION RelevantPairs@ijgiDb;

9. GROUP
    PARTITION WITH .g.gId
      BY    .g.gId
      INTO .gGroup
        ORDER BY .rank TYPE NUMERIC DESC;

10. EXPAND
      UNPACK WITH .gGroup
        ARRAY  .gGroup
        TO     .gPair;

11. FILTER
      CASE WHERE WITH .gPair AND
                 .gPair.position = 1
        GENERATE
          BUILD {
            .f     : .gPair.item.f,
            .g     : .gPair.item.g,
            .rank  : .gPair.item.rank };

12. SAVE AS SamePlaces@ijgiDb;
```

7. Experimental Evaluation

In this section, we report a brief evaluation of the results that can be obtained by the *J-CO-QL*$^+$ script. We exploited the same data set adopted in [6], related with the city of Manchester (UK). Remember, from Section 6.1, that the `FacebookDescriptors` collection contains 5738 descriptors, while the `GoogleDescriptors` collection contains 5214 descriptors. Both collections contain descriptors about a variety of different public places, such as restaurants, pubs, hairdressers, universities, parks and so on.

7.1. Effectiveness

In order to evaluate the effectiveness of the method, we performed a sensitivity analysis by varying the value of the α threshold from 0.5 to 0.99.

We used the same test set again, which was used in the work [6]: it contained a total of 400 pairs, selected among the 5738 $\times$ 5214 total pairs, evaluated by a human as *Good* or *Bad*. We randomly selected 300 pairs from the starting 400 pairs, and from each pair we extracted the related 300 *Google Places* descriptors and 300 *Facebook* descriptors; among all possible pairs, 103 pairs were labeled as *Good* pairs (and obviously the remaining 197 were labeled as *Bad*).

Then, we run the script on these two reduced collections of descriptors. Table 1 reports the results of our experiments. Specifically, the first colum reports the single values for the alpha-cut α; the second and third columns report the number of relevant pairs saved by line 7 of the *J-CO-QL*$^+$ script (Listing 3) into the `RelevantPairs` collection and the number of pairs generated by line 12 of the script (Listing 4) and saved into the `SamePlaces` collection, respectively. The, columns from 4 to 7 reports the number TP of true positive pairs, the number TN of negative pairs, the number FP of false positive pairs and the number FN of false negative pairs, respectively. Finally, the last three columns reports "Precision" (defined as $TP/(TP+FP)$), "Recall" (defined as $TP/(TP+FN)$) and "Accuracy" (defined as $(TP+TN)/(TP+TN+FP+FN)$), respectively. These three latter values are depicted in Figure 10: the x-axis reports the values of the alpha-cut α parameter; precision is depicted by the blue line, recall is depicted by the red line and accuracy is depicted by the black line.

Analyzing Table 1 and Figure 10, it is possible to see that the best combination of values for precision, recall and accuracy were obtained for $\alpha = 0.8$: precision is 0.962, recall is 0.981 and accuracy is 0.980. Indeed, this value for α appears to be the best compromise between the need to keep as many pairs as possible and the fact that those pairs actually describe the same place, even though names, addresses and coordinates are different. The reader can further notice that higher values of α give rise to a precision of 1 with poor recall, while lower values for α give rise to a recall of 1 with poor precision. To conclude this analysis, notice that with $\alpha = 0.85$, the accuracy is the same as the one obtained for $\alpha = 0.8$; it could be considered as a valid alternative choice, with better precision but low recall.

Table 1. Sensitivity analysis.

Alpha Cut	Relevant Pairs	Same Places	TP	TN	FP	FN	Precision	Recall	Accuracy
0.50	253	163	103	137	60	0	0.632	1.000	0.800
0.55	178	139	103	161	36	0	0.741	1.000	0.880
0.60	140	124	103	176	21	0	0.831	1.000	0.930
0.65	119	112	103	188	9	0	0.920	1.000	0.970
0.70	110	108	101	190	7	2	0.935	0.981	0.970
0.75	109	107	101	191	6	2	0.944	0.981	0.973
0.80	106	105	101	193	4	2	0.962	0.981	0.980
0.85	98	97	97	197	0	6	1.000	0.942	0.980
0.90	90	90	90	197	0	13	1.000	0.874	0.957
0.95	80	80	80	197	0	23	1.000	0.777	0.923
0.99	71	71	71	197	0	32	1.000	0.689	0.893

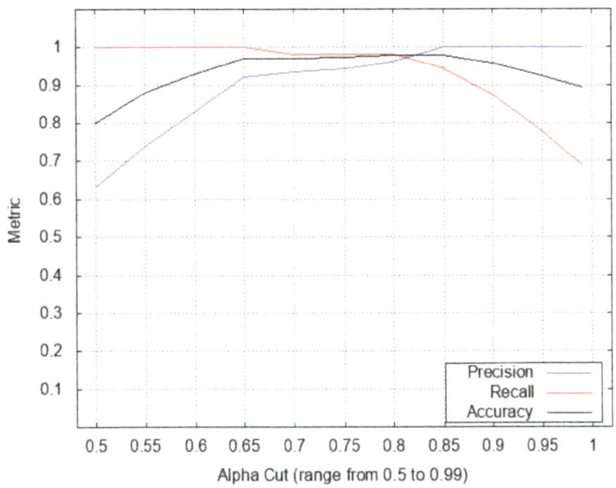

Figure 10. Sensitivity analysis of precision, recall and accuracy.

Thus, we can state that the novel formulation for the *MatchingPklaces* relation and the complex membership functions adopted for the `Similar` fuzzy operator and for the `Close` fuzzy operator are effective, provided that $\alpha = 0.8$.

We can consider the results reported in [6] as a baseline for further evaluating the effectiveness of the novel formulation of the technique.

Remember that the version presented in [6] (for online aggregation) performed pre-processing tasks on names and addresses, so as to clean them from urban designations and numbers. In contrast, the present version does not. The main reason is that such a kind of pre-processing and cleaning is not easy to do within *J-CO-QL*$^+$ scripts; however, the flexibility of the `CREATE FUZZY OPERATOR` statement as far as the possibility to define complex shapes for membership functions seems to be effective. Consequently, the proper baseline to consider is the best result presented in [6]. There, a comparison with a machine-learning technique, namely "Random Forest" classification, was performed, by applying it on the same data set. Results are reported in Table 2: for the three considered techniques, precision, recall and F1-score (defined as $2 \times (precision \times recall)/(precision + recall)$) are reported. Notice that in [6], the proposed technique was as effective as random-forest classifiers; the current version outperform them, even though names and addresses are neither pre-processed nor cleaned. Observe that the old version of the fuzzy technique and the Random-Forest technique, applied on the data set describing public places in Manchester (UK), obtain the same identical effectiveness; this is why [6] states that the two techniques are comparable.

Table 2. Comparison with experimental results presented in [6].

Technique	Precision	Recall	F1-Score
Current version	0.962	0.981	0.971
Best of [6]	0.931	0.931	0.931
Random Forest (from [6])	0.931	0.931	0.931

Consequently, we can say that the current version improves the old one and is suitable to be executed as a *J-CO-QL*$^+$ script. Furthermore, it still maintains the advantage provided by the old version in comparison with classification techniques, i.e., it can be applied from scratch, without knowing the data set; in contrast, classification techniques ask for a training step on previously labeled training sets, which is a time-consuming and critical activity.

7.2. About Execution Times

Before concluding this work, we report some considerations about execution times.

Usually, this aspect is not considered in the literature about integration of geographical data sets: authors were focused on the effectiveness of the proposed techniques, but did not consider efficiency. However, in our opinion, this is not a negligible aspect for the practical use of integration techniques, in particular with large data sets to integrate.

In this paper, the goal is neither to provide the most efficient technique, nor to evaluate execution times of a plethora of techniques proposed in the literature. Here, the goal is to observe "what to expect" while running the *J-CO-QL*$^+$ script on a real data set, such as the one we used for our experiments.

We decided to consider a working environment that could be a common situation: analysts are equipped with stand-alone PCs, on which they perform their daily activities. On these PCs, they might have a running instance of *MongoDB*, which stores their data sets about geographical places, as well as an installation of the *J-CO* Framework; indeed, not necessarily analysts are equipped with super-computers. Consequently, we run experiments on a laptop PC powered by an Intel quad-Core i7-8550-U processor, running at 1.80 GHz, equipped with 16 GB RAM and 250 GB Solid-State Drive and running the Java Virtual Machine version 1.8.0_251 (the *J-CO-QL*$^+$ *Engine* is written in the Java programming language).

Table 3 reports the execution times of the script discussed in this paper. We used the full data set reporting descriptors of public places located in Manchester (UK). Remember that it contains 5738 *Facebook* descriptors and 5214 *Google* descriptors.

Table 3 reports the execution times observed for each single instruction in the script; the right-most column reports the cumulative execution time after each instruction. Clearly, the overall execution time is dominated by the `JOIN OF COLLECTIONS` instruction, which actually builds $5738 \times 5214 = 29,917,932$ pairs of descriptors; it takes 216.61 s. Looking at the other instructions they contribute only less than 4 s, so that the overall execution time is 220.804 s, i.e., about 3.6 min.

In terms of user perception, waiting for about 3 min is acceptable in this context, in which near-real time performances are not expected. Furthermore, we want to remark that once the two data sets to integrate are available, the *J-CO-QL*$^+$ script can be applied from scratch, and a few minutes later the integrated data set is obtained. This is an incredible advantage if compared with the adoption of classification techniques, because there is no need to build training sets labeled by humans; this activity can take from several hours to several days (depending on the size of the training set) and is prone to errors and misunderstanding, as well as its effectiveness depends on the way the training sets are built before labeling.

Table 3. Execution times of the *J-CO-QL*$^{+}$ script.

N.	Instruction	Instruction Time (s)	Incremental Time (s)
1	`CREATE FUZZY OPERATOR`	0.000	0.000
2	`CREATE FUZZY OPERATOR`	0.000	0.000
3	`CREATE FUZZY OPERATOR`	0.000	0.000
4	`USE DB`	0.003	0.003
5	`JOIN OF COLLECTIONS`	216.661	216.664
6	`FILTER`	1.262	217.926
7	`SAVE AS`	0.529	218.455
8	`GET COLLECTION`	0.106	0.106
9	`GROUP`	0.035	0.141
10	`EXPAND`	1.910	2.051
11	`FILTER`	0.091	2.142
12	`SAVE AS`	0.207	2.349
Total Time (s)			220.804

8. Conclusions and Future Work

To conclude, it is time to summarize and discuss the contribution of the paper, as well as to sketch future developments.

8.1. Conclusions

This paper addresses the problem of soft integrating data sets describing public places, when these data sets are represented as *JSON* documents and are stored within a *JSON* document store. Specifically, fuzzy-set theory provides the formal framework for the integration methodology presented in the paper: the *MatchingPlaces* fuzzy relation is the core of the proposed methodology. Then, the soft integration method is applied in a practical way by means of the *J-CO* Framework: a script (or query) written in *J-CO-QL*$^{+}$ (the query language of the *J-CO* Framework) is written, which implements the soft integration method, by exploiting its fuzzy capabilities. This is the main contribution of the paper: showing that a novel stand-alone tool (the *J-CO* Framework), suitable for performing the soft integration of geo-tagged *JSON* data sets stored within *JSON* document stores, is now available for analysts and spatial-data engineers.

Hereafter, we would like to perform some considerations.

- The effectiveness that the script obtains is very interesting: with the value of the α threshold set to $\alpha = 0.8$, it is possible to obtain the best balance between precision and recall, which are slightly less than 100%.
- Execution times are good too: less than 4 min to perform the soft integration is absolutely acceptable (the reader can notice that writing the full script from scratch takes longer).
- The complex membership functions that it is possible to specify in fuzzy operators (see Section 6.2) were exploited to deal with bizarre behavior of the Jaro-Winkler string-similarity metric. In fact, it returns high similarity values for strings that appears to be very different (in the sense that they do not denote similar names or similar addresses). However, we experienced also the opposite behavior, i.e., two strings that were actually very similar obtained not-so-high similarity degree. The complex membership function that we defined for the `Similar` fuzzy operator allowed us to compensate this behavior.

8.2. Future Work

As future works, many activities are planned along the development of the *J-CO* Framework and its application to data integration problems.

- First of all, we are going to complete the extension of all *J-CO-QL*$^+$ statements with support for fuzzy concepts. In particular, we are going to address the problem of defining "soft aggregators" that can be applied on arrays of *JSON* documents; we will adopt the same approach followed for defining fuzzy operators.
- Various types of fuzzy sets have been proposed in the literature (for example, Intuitionistic Fuzzy Sets [44,45] and Type-2 Fuzzy Sets [46–48]. In our perspective evolution, we plan to extend *J-CO-QL*$^+$ to support multiple types of fuzzy sets simultaneously.
- Many challenges concerning integration and processing of geo-tagged data sets are arising. For example, the *GeoJSON* format [49] represents a geographical information layer as a unique, giant, *JSON* document. We conceived the idea of defining a domain-specific language for querying features within *GeoJSON* documents [50], which is translated into *J-CO-QL*$^+$ scripts. We plan to further explore this idea, by identifying other application domains and defining novel domain-specific languages to translate into *J-CO-QL*$^+$. Indeed, the idea of devising the *J-CO* Framework came out while working on an international project [51,52], in which Big Data concerning mobility had to be collected and processed.

The *J-CO* Framework is available on a public GitHub repository (https://github.com/JcoProjectTeam/JcoProjectPage, accessed on 1 September 2022).

Author Contributions: Conceptualization and methodology, Giuseppe Psaila; software, Paolo Fosci; writing—original draft preparation, Giuseppe Psaila; writing—review and editing, Paolo Fosci and Giuseppe Psaila; experiments, Paolo Fosci. All authors have read and agreed to the published version of the manuscript.

Funding: This research received no external funding.

Data Availability Statement: All collection data sets used for the experiments, reported in this paper, can be freely download from the *J-CO* Framework GitHub repository at: https://github.com/JcoProjectTeam/JcoProjectPage/tree/main/papers/dataset/ijgi2022.

Conflicts of Interest: The authors declare no conflict of interest.

References

1. Bray, T. The Javascript Object Notation (JSON) Data Interchange Format. 2014. Available online: https://www.rfc-editor.org/rfc/rfc7159.txt (accessed on 1 September 2022).
2. Bordogna, G.; Capelli, S.; Psaila, G. A big geo data query framework to correlate open data with social network geotagged posts. In Proceedings of the Annual International Conference on Geographic Information Science, Wageningen, The Netherlands, 10–11 May 2017; pp. 185–203.
3. Bordogna, G.; Ciriello, D.E.; Psaila, G. A flexible framework to cross-analyze heterogeneous multi-source geo-referenced information: The J-CO-QL proposal and its implementation. In Proceedings of the International Conference on Web Intelligence, Leipzig, Germany, 23–26 June 2017; pp. 499–508.
4. Bordogna, G.; Capelli, S.; Ciriello, D.E.; Psaila, G. A cross-analysis framework for multi-source volunteered, crowdsourced, and authoritative geographic information: The case study of volunteered personal traces analysis against transport network data. *Geo-Spat. Inf. Sci.* **2018**, *21*, 257–271. [CrossRef]
5. Psaila, G.; Fosci, P. J-CO: A Platform-Independent Framework for Managing Geo-Referenced JSON Data Sets. *Electronics* **2021**, *10*, 621. [CrossRef]
6. Psaila, G.; Toccu, M. A Fuzzy Technique for online Aggregation of POIs from Social Media: Definition and Comparison with Off-Line Random-Forest Classifiers. *Information* **2019**, *10*, 388. [CrossRef]
7. Fosci, P.; Psaila, G. Towards flexible retrieval, integration and analysis of json data sets through fuzzy sets: A case study. *Information* **2021**, *12*, 258. [CrossRef]
8. Fosci, P.; Psaila, G. J-CO, a Framework for Fuzzy Querying Collections of JSON Documents. In Proceedings of the International Conference on Flexible Query Answering Systems, Bratislava, Slovakia, 19–24 September 2021; Springer: Cham, Switzerland; pp. 142–153.
9. Psaila, G.; Marrara, S. A First Step Towards a Fuzzy Framework for Analyzing Collections of JSON Documents. In Proceedings of the IADIS AC 2019, Cagliari, Italy, 7–9 November 2019; pp. 19–28.
10. Blair, D.C. Information Retrieval, 2nd ed. C.J. Van Rijsbergen. London: Butterworths; 1979: 208 pp. Price: $32.50. *J. Am. Soc. Inf. Sci.* **1979**, *30*, 374–375. [CrossRef]
11. Bosc, P.; Pivert, O. SQLf: A relational database language for fuzzy querying. *IEEE Trans. Fuzzy Syst.* **1995**, *3*, 4895977. [CrossRef]

12. Bosc, P.; Pivert, O. SQLf query functionality on top of a regular relational database management system. In *Knowledge Management in Fuzzy Databases*; Springer: Berlin/Heidelberg, Germany, 2000; pp. 171–190.
13. Galindo, J.; Medina, J.M.; Pons, O.; Cubero, J.C. A server for fuzzy SQL queries. In Proceedings of the International Conference on Flexible Query Answering Systems, Roskilde, Denmark, 13–15 May 1998; pp. 164–174.
14. Zadrozny, S.; Kacprzyk, J. Fquery for access: Towards human consistent querying user interface. In Proceedings of the 1996 ACM Symposium on Applied Computing, Philadelphia, PA, USA, 17–19 February 1996; pp. 532–536.
15. Kacprzyk, J.; Zadrożny, S. FQUERY for Access: Fuzzy querying for a Windows-based DBMS. In *Fuzziness in Database Management Systems*; Springer: Berlin/Heidelberg, Germany, 1995; pp. 415–433.
16. Bordogna, G.; Psaila, G. Modeling soft conditions with unequal importance in fuzzy databases based on the vector p-norm. In Proceedings of the IPMU COnference, Malaga, Spain, 22–27 June 2008.
17. Bordogna, G.; Psaila, G. Customizable flexible querying in classical relational databases. In *Handbook of Research on Fuzzy Information Processing in Databases*; IGI Global: Hershey, PA, USA, 2008; pp. 191–217.
18. Bordogna, G.; Psaila, G. Soft Aggregation in Flexible Databases Querying based on the Vector p-norm. *Int. J. Uncertain. Fuzziness Knowl.-Based Syst.* **2009**, *17*, 25–40. [CrossRef]
19. Kacprzyk, J.; Zadrozny, S. SQLf and FQUERY for Access. In Proceedings of the Joint 9th IFSA World Congress and 20th NAFIPS International Conference (Cat. No. 01TH8569), Vancouver, BC, Canada, 25–28 July 2001; Volume 4, pp. 2464–2469.
20. Urrutia, A.; Tineo, L.; Gonzalez, C. FSQL and SQLf: Towards a standard in fuzzy databases. In *Handbook of Research on Fuzzy Information Processing in Databases*; IGI Global: Hershey, PA, USA, 2008; pp. 270–298.
21. Galindo, J. *Handbook of Research on Fuzzy Information Processing in Databases*; IGI Global: Hershey, PA, USA, 2008.
22. Han, J.; Haihong, E.; Le, G.; Du, J. Survey on NoSQL database. In Proceedings of the 2011 6th International Conference on Pervasive Computing and Applications, Port Elizabeth, South Africa, 26–28 October 2011; pp. 363–366.
23. Chodorow, K. *MongoDB: The Definitive Guide: Powerful and Scalable Data Storage*; O'Reilly Media, Inc.: Sebastopol, CA, USA, 2013.
24. Anderson, J.C.; Lehnardt, J.; Slater, N. *CouchDB: The Definitive Guide: Time to Relax*; O'Reilly Media, Inc.: Sebastopol, CA, USA, 2010.
25. Garcia Bringas, P.; Pastor, I.; Psaila, G. Can BlockChain technology provide information systems with trusted database? The case of HyperLedger Fabric. In Proceedings of the International Conference on Flexible Query Answering Systems, Amantea, Italy, 2–5 July 2019; Springer: Cham, Switzerland; pp. 265–277.
26. Abir, B.K.; Amel, G.T. Towards fuzzy querying of NoSQL document-oriented databases. In Proceedings of the DBKDA 2015: The Seventh International Conference on Advances in Databases, Knowledge, and Data Applications, Rome, Italy, 24–29 May 2015; p. 163.
27. Almendros-Jimenez, J.M.; Becerra-Teron, A.; Moreno, G. Fuzzy queries of social networks with FSA-SPARQL. *Expert Syst. Appl.* **2018**, *113*, 128–146. [CrossRef]
28. Manola, F.; Miller, E.; McBride, B. RDF Primer. W3C Recommendation (2004). Available online: http://www.w3.org/TR/rdf-primer (accessed on 1 September 2022).
29. Cheng, J.; Ma, Z.M.; Yan, L. f-SPARQL: A flexible extension of SPARQL. In Proceedings of the International Conference on Database and Expert Systems Applications, Bilbao, Spain, 30 August–3 September 2010; Springer: Berlin/Heidelberg, Germany; pp. 487–494.
30. Pérez, J.; Arenas, M.; Gutierrez, C. Semantics and complexity of SPARQL. *ACM Trans. Database Syst. (TODS)* **2009**, *34*, 16. [CrossRef]
31. Kilinc, D. An Accurate Toponym-Matching Measure Based On Approximate String Matching. *J. Inf. Sci.* **2016**, *42*, 138–149. [CrossRef]
32. Santos, R.; Murrieta-Flores, P.; Martins, B. Learning to combine multiple string similarity metrics for effective toponym matching. *Int. J. Digit. Earth* **2018**, *11*, 913–938. [CrossRef]
33. Rui, S.; Patricia, M.F.; Pavel, C.; Bruno, M. Toponym matching through deep neural networks. *Int. J. Geogr. Inf.* **2018**, *32*, 324–348.
34. Li, L.; Xing, X.; Xia, H.; Huang, X. Entropy-Weighted Instance Matching between Different Sourcing Points of Interest. *Entropy* **2016**, *18*, 45. [CrossRef]
35. Yu, L.; Qiu, P.; Liu, X.; Lu, F.; Wan, B. A Holistic Approach to Aligning Geospatial Data with Multidimensional Similarity Measuring. *Int. J. Digit. Earth* **2018**, *11*, 845–862. [CrossRef]
36. Zadeh, L.A. The concept of a linguistic variable and its application to approximate reasoning—I. *Inf. Sci.* **1975**, *8*, 199–249. [CrossRef]
37. Psaila, G.; Fosci, P. Toward an Anayist-Oriented Polystore Framework for Processing JSON Geo-Data. In Proceedings of the International Conferences on WWW/Internet, ICWI 2018 and Applied Computing 2018, Budapest, Hungary, 21–23 October 2018; IADIS (International Association for Development of the Information Society): Budapest, Hungary, 2018; pp. 213–222.
38. Fosci, P.; Psaila, G. Powering Soft Querying in J-CO-QL with JavaScript Functions. In Proceedings of the International Workshop on Soft Computing Models in Industrial and Environmental Applications, Bilbao, Spain, 22–24 September 2021; Springer: Cham, Switzerland; pp. 207–221.
39. Solomon, J.; Rustamov, R.; Guibas, L.; Butscher, A. Earth mover's distances on discrete surfaces. *ACM Trans. Graph. (ToG)* **2014**, *33*, 67. [CrossRef]
40. Jaro, M.A. *UNIMATCH, a Record Linkage System: Users Manual*; Bureau of the Census: Washington, DC, USA, 1980.

41. Jaro, M.A. Advances in record-linkage methodology as applied to matching the 1985 census of Tampa, Florida. *J. Am. Stat. Assoc.* **1989**, *84*, 414–420. [CrossRef]
42. Winkler, W.E. String Comparator Metrics and Enhanced Decision Rules in the Fellegi-Sunter Model of Record Linkage. In *Proceedings of the Section on Survey Research Methods*; American Statistical Association: Boston, MA, USA, 1990; pp. 354–359.
43. Winkler, W.E. *The State of Record Linkage and Current Research Problems*; Statistical Research Division, U.S. Bureau of the Census: Washington, DC, USA, 1999.
44. Atanassov, K. Intuitionistic fuzzy sets. *Fuzzy Sets Syst.* **1986**, *20*, 187–196. [CrossRef]
45. De, S.K.; Biswas, R.; Roy, A.R. An application of intuitionistic fuzzy sets in medical diagnosis. *Fuzzy Sets Syst.* **2001**, *117*, 209–213. [CrossRef]
46. Karnik, N.N.; Mendel, J.M. Operations on type-2 fuzzy sets. *Fuzzy Sets Syst.* **2001**, *122*, 327–348. [CrossRef]
47. Mendel, J.M. Type-2 fuzzy sets and systems: An overview. *IEEE Comput. Intell. Mag.* **2007**, *2*, 20–29. [CrossRef]
48. Mendel, J.M.; John, R.B. Type-2 fuzzy sets made simple. *IEEE Trans. Fuzzy Syst.* **2002**, *10*, 117–127. [CrossRef]
49. Butler, H.; Daly, M.; Doyle, A.; Gillies, S.; Hagen, S.; Schaub, T. *The GeoJSON Format*; Internet Engineering Task Force (IETF): Fremont, CA, USA, 2016.
50. Fosci, P.; Marrara, S.; Psaila, G. Soft Querying GeoJSON Documents within the J-CO Framework. In Proceedings of the 16th International Conference on Web Information Systems and Technologies (WEBIST 2020), online, 3–5 November 2020; SciTePress—Science and Technology Publications, Lda.: Setubal, Portugal, 2020; pp. 253–265.
51. Burini, F.; Cortesi, N.; Gotti, K.; Psaila, G. The Urban Nexus Approach for Analyzing Mobility in the Smart City: Towards the Identification of City Users Networking. *Mob. Inf. Syst.* **2018**, *2018*, 6294872. [CrossRef]
52. Bordogna, G.; Cuzzocrea, A.; Frigerio, L.; Psaila, G.; Toccu, M. An interoperable open data framework for discovering popular tours based on geo-tagged tweets. *Int. J. Intell. Inf. Database Syst.* **2017**, *10*, 246–268. [CrossRef]

International Journal of *Geo-Information*

Article

Modeling and Querying Fuzzy SOLAP-Based Framework

Sinan Keskin [1,*], **Adnan Yazıcı** [1,2]

[1] Department of Computer Engineering, Middle East Technical University, Ankara 06800, Turkey; yazici@metu.edu.tr
[2] Department of Computer Science, SEDS, Nazarbayev University, Nur-Sultan 010000, Kazakhstan
* Correspondence: keskin.sinan@metu.edu.tr

Abstract: Nowadays, with the rise of sensor technology, the amount of spatial and temporal data is increasing day by day. Modeling data in a structured way and performing effective and efficient complex queries has become more essential than ever. Online analytical processing (OLAP), developed for this purpose, provides appropriate data structures and supports querying multidimensional numeric and alphanumeric data. However, uncertainty and fuzziness are inherent in the data in many complex database applications, especially in spatiotemporal database applications. Therefore, there is always a need to support flexible queries and analyses on uncertain and fuzzy data, due to the nature of the data in these complex spatiotemporal applications. FSOLAP is a new framework based on fuzzy logic technologies and spatial online analytical processing (SOLAP). In this study, we use crisp measures as input for this framework, apply fuzzy operations to obtain the membership functions and fuzzy classes, and then generate fuzzy association rules. Therefore, FSOLAP does not need to use predefined sets of fuzzy inputs. This paper presents the method used to model the FSOLAP and manage various types of complex and fuzzy spatiotemporal queries using the FSOLAP framework. In this context, we describe how to handle non-spatial and fuzzy spatial queries, as well as spatiotemporal fuzzy query types. Additionally, while FSOLAP primarily includes historical data and associated queries and analyses, we also describe how to handle predictive fuzzy spatiotemporal queries, which typically require an inference mechanism.

Keywords: OLAP; fuzzy SOLAP-based framework; fuzzy spatiotemporal queries; fuzzy spatiotemporal predictive query; fuzzy query visualization

Citation: Keskin, S.; Yazıcı, A. Modeling and Querying Fuzzy SOLAP-Based Framework. *ISPRS Int. J. Geo-Inf.* **2022**, *11*, 191. https://doi.org/10.3390/ijgi11030191

Academic Editors: Wolfgang Kainz, Gloria Bordogna and Cristiano Fugazza

Received: 31 January 2022
Accepted: 9 March 2022
Published: 11 March 2022

1. Introduction

Recently, the amount and variety of data used for analytical purposes have greatly increased. In order to improve the data to be analyzed, it is necessary to use expertise and a suitable application for the processing and interpretation of these data. For this purpose, various methods and applications have been developed to analyze large amounts of data. One of the most common developed applications is online analytical processing (OLAP) [1]. OLAP enables data analysis and query processes to help in decision-making about the data source. It is a computational method that allows users to quickly and selectively extract and query data for analysis from different perspectives. OLAP has emerged because classic databases cannot be used in decision-making and require expertise in data access. While traditional databases are concerned with the retention of data and the efficient management of online transactions, OLAP is concerned with the efficient analytics of online data.

In addition, conventional data mining techniques are insufficient in the area of spatiotemporal database applications because they often require intensive computations and involve complex differential equations and computational algorithms [2]. However, we need to perform effective and efficient querying with a colossal amount of spatiotemporal data. One of the widely used geospatial data mining tools is spatial online analytical processing (SOLAP), which enables the exploration of data cubes to extract new information effectively and efficiently [3]. SOLAP can also be defined as a platform supporting fast and easy spatiotemporal querying. It allows data mining following a multidimensional approach comprised of levels of aggregation.

Researchers working with OLAP mainly use numerical and statistical models [4–6], which generally use precise values as input and output. Furthermore, SOLAP provides querying and analysis of numeric and alphanumeric multidimensional data. However, there is a need to support flexible queries on uncertain and fuzzy data, due to the nature of complex applications such as meteorological and spatiotemporal applications. Uncertainty and fuzziness are inherent features of most meteorological applications [7]. That is, spatial and temporal information and various relationships in these applications frequently involve uncertainty and fuzziness. For example, in describing a rainy region, the region's boundary is a fuzzy concept. Likewise, in estimating a weather event, the need to determine its position at a particular time, or its time of occurrence at a specific location, gives rise to fuzzy estimations.

The most common reasons for various types of uncertainties in spatiotemporal applications are:

- Some spatial information is inherently imprecise or fuzzy. The locations of events, spatial relationships, and various geometric and topological properties usually involve multiple forms of uncertainty [7].
- Most natural phenomena have fuzzy boundaries due to the transitional nature of variation in their aspects (e.g., high humidity and low temperature cause precipitation at a certain altitude) [8–10].
- Obtaining precise data is tedious and unnecessary most of the time, and we may only be able to give a range of values within which the exact numbers would fall. For instance, we may need the number of "cloudy" or "partly cloudy" days for some areas within a certain period. In this request, the user specifies the cloudiness criteria in linguistic terms instead of giving numeric degrees of cloudiness (e.g., 2/8 or 7/8) [11].

The use of OLAP is mainly related to querying and analyzing historical data, but we also need to make predictions based on spatiotemporal data. In this study, we describe how to handle predictive fuzzy spatiotemporal queries that require an inference mechanism. We also show that various complex queries, including predictive fuzzy spatiotemporal queries, are effectively and efficiently handled using our fuzzy spatial OLAP framework. We do this with the support of the association rules and fuzzy inference system (FIS) components of the FSOLAP framework. In other words, the FIS component included in the FSOLAP framework supports fuzzy predictive query types.

Spatial–temporal database applications naturally contain hierarchical data structures. Spatial data include hierarchical breakdowns such as country–region–city, while temporal data have hierarchical relationships at levels such as year–month–day. SOLAP was developed to provide effective and efficient analysis and querying of hierarchical data. Spatial and temporal information and various associations in spatial–temporal applications frequently involve uncertainty and fuzziness, which are inherent features of most of these applications [7] (e.g., in describing a rainy region). In addition, since spatial–temporal applications are complex, they are challenging to analyze with conventional logic approaches. Fuzzy logic can be used for situations in which conventional logic technologies are ineffective, such as applications [2,12–21] and systems [22,23] that mathematical models cannot precisely describe, those with significant uncertainties or contradictory conditions, and linguistically controlled applications or systems. The concepts of SOLAP and fuzzy logic can be combined to benefit from both to provide an effective and efficient platform for spatiotemporal applications. The aim of this study is to propose a new framework, FSOLAP, to take advantage of both SOLAP and fuzzy logic to provide analytics and querying of imprecise spatiotemporal data and to extend the framework with inference ability.

Our study aims to find spatiotemporal patterns in data which have spatiotemporal characteristics, in order to perform data analytics and querying. Researchers [24,25] typically use synthetic or semi-synthetic data to demonstrate the performance of their compound models in data science applications. The use of synthetic data makes it impossible to represent the true efficiency and accuracy of the model. Validation of the FSOLAP framework on a big database under the fuzzy spatial–temporal data model is vital. However, it is not easy to find real data to study. In our study, thanks to the Turkish State Meteorological Service, we were able to use a real meteorological dataset containing spatiotemporal features and measurement attributes as a case study to test our framework and models. It was shown that a fuzzy approach is suitable for handling spatiotemporal data. Therefore, we present our approach for dealing with different types of fuzzy spatiotemporal queries using FSOLAP. In this context, the FSOLAP framework is modeled, and the methods for supporting fuzzy non-spatial, fuzzy spatial, and fuzzy spatiotemporal query types using FSOLAP are explained. In general, the FSOLAP framework includes SOLAP, a fuzzy module, a fuzzy knowledge base (FKB), and a fuzzy inference system (FIS), as explained in Section 2.2. This framework allows us to make efficient and flexible fuzzy queries and analyses on spatiotemporal data.

The main contribution of this study is the development of FSOLAP as a new fuzzy SOLAP-based framework, allowing effective and efficient analysis and querying of spatiotemporal data. FSOLAP supports the fuzzy spatiotemporal predictive query, which is a new query type that has not been proposed before, as well as the complex type of fuzzy spatial queries present in the literature.

More specifically, the contributions of this study are as follows. We propose a fuzzy SOLAP-based complex system (FSOLAP) for analytics on fuzzy spatiotemporal data and for predictive analysis of various spatiotemporal events, including support for various querying capabilities, visualization of data, and analysis. The SOLAP server and its multidimensional expression (MDX) query processor is modified to support various flexible and complex queries. An optimal number of fuzzification clusters is calculated and integrated into the FSOLAP framework as an automated process. Moreover, fuzzy sets are generated automatically and used to create fuzzy association rules. The appropriate minsup and minconf values related to fuzzy association rule generation are also determined. In addition, an analysis of the performance of the framework is undertaken using a real meteorological dataset. Average CPU usage, memory usage, and query execution time for running each query type included in the FSOLAP framework are measured. A pruning method based on confidence measures that removes complex rules in the generated fuzzy association rule set to speed up the inference performance is also applied. Additionally, fuzzy association rule weighting for rule-based pruning is performed on the generated rules. Thus, we derive accurate inferences from the fuzzy association rules.

The organization of this paper is as follows. Background information, related works, proposed architecture, and supported query types are given in Section 2. The execution of queries and experimental results are explained in Section 3. In Section 4, the results of the study are discussed and compared with those of previous studies. Finally, in Section 5, the conclusions and future work are presented.

2. Materials and Methods

Here, we first introduce the related work in Section 2.1 and then explain the FSOLAP framework and its components in Section 2.2. FSOLAP query management is presented and the structure of the modules explained in Section 2.3. Brief information about the dataset used to confirm the performance of the framework is given in Section 2.4. Finally, we present the supported complex and fuzzy queries in Section 2.5.

2.1. Background and Related Works

The increase in spatial data and human limitations in analyzing spatial data in detail make querying spatial databases crucial for spatiotemporal applications. In recent years, many studies [2,5,26] have addressed the issue of performing data mining tasks on data warehouses. Some of them [26,27] are explicitly interested in mining patterns and association rules in data cubes. For instance, Imieli'nski et al. [27] state that OLAP is closely intertwined with association rules and shares with association rules the goal of finding patterns in the data. Data mining techniques such as association rule mining can be used together with OLAP to extract knowledge from data cubes. Spatial data mining can be performed in a spatial data cube as well as in a spatial database. For this purpose, J. Han constructed GeoMiner [4], a spatial OLAP and data mining system prototype. Another proposed study [26] considers a framework for mining association rules from data cubes according to a sum-based aggregate measure, which is more general than frequencies provided by the count measure. The mining process is guided by a meta-rule, is context-driven by analysis objectives, and exploits aggregate measures to revisit the definitions of support and confidence. These studies profit from the hierarchical aspect of cube dimensions to mine association rules at different levels of granularity, such as spatial and temporal hierarchies.

Supporting spatial queries is one of the key features in spatial database management systems, due to the broad range of applications. Providing these types of queries involves introducing spatial components such as fuzzy topological relations into relational and object-relational databases. Fuzzy topological relations between fuzzy regions are explained in [28] and shown in Figure 1b. The formal definitions of the fuzzy topological relations can be explained as follows.

Let A be a set of attributes under consideration and let a region be a fuzzy subset defined in two-dimensional space R^2 over A. We can define the membership function of the region as $\mu : X \times Y \times A \rightarrow [0,1]$, where X and Y are the sets of coordinates defining the region. Each point (x,y) within the region is assigned a membership value for an attribute $a \in A$. We show a fuzzy region in Figure 1a, which has a core, an indeterminate boundary, an exterior, and $\alpha - cut$ levels.

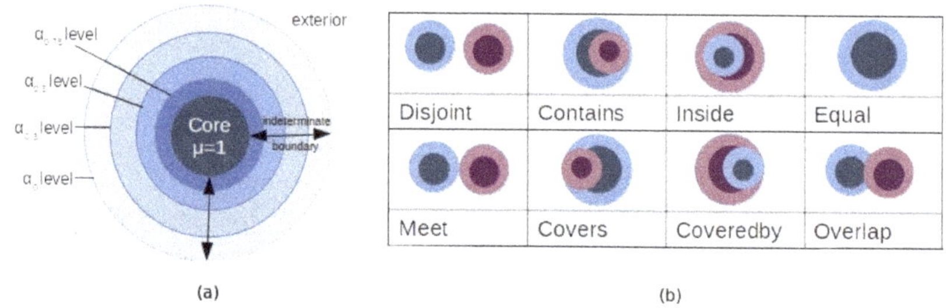

Figure 1. (**a**) Visualization of a simple fuzzy region. (**b**) Examples of topological relations between fuzzy regions.

The concept of the $\alpha - cut$ level region is used to approximate the indeterminate boundaries of a fuzzy region and is defined as follows:

$$R_\alpha = \{(x,y,a)|\mu_R(x,y,a) \geq \alpha\}(0 < \alpha < 1) \tag{1}$$

The degree of the fuzzy relation is measured by aggregating the $\alpha - cut$ levels of fuzzy regions. The basic probability assignment $m(R_{\alpha i})$, which can be interpreted as the probability that $R_{\alpha i}$ is the true representative of R, is defined as in [29,30]:

$$m(R_{\alpha i}) = \alpha_i - \alpha_{i+1}, 1 \leq i \leq n, n \in N, 1 = \alpha_1 > \alpha_2 > \ldots > \alpha_n > \alpha_{n+1} = 0 \tag{2}$$

Assuming that $\tau(R,S)$ is the value representing the topological relation between two fuzzy regions R and S, and $\tau(R_{\alpha i}, R_{\alpha j})$ is the value representing the topological relation between two $\alpha - cut$ level regions $R_{\alpha i}$ and $S_{\alpha j}$, the general relation between two fuzzy regions can be determined by

$$\tau(R,S) = \sum_{i=1}^{n}\sum_{j=1}^{m} m(R_{\alpha i})m(S_{\alpha j})\tau(R_{\alpha i}, S_{\alpha j}) \tag{3}$$

For example, the overlap relation between two fuzzy regions can be approximated by using the formula above as follows:

$$\tau(R,S) = \sum_{i=1}^{n} \sum_{j=1}^{m} m(R_{\alpha i}) m(S_{\alpha j}) \tau_{overlap}(R_{\alpha i}, S_{\alpha j}) \quad (4)$$

Since spatial OLAP querying deals with some concepts expressed in verbal language, fuzziness is frequently involved in spatial OLAP. Hence, the ability to query spatial data under fuzziness is an essential characteristic of any spatial database. The studies in [25,31] discuss the directional and topological relationships in fuzzy concepts. Some earlier works [24,32] provide a basis for fuzzy querying capabilities based on a binary model to support queries of this nature. Another study [33] considers unary operators for querying fuzzy multidimensional databases. The study discusses the properties of unary operators on fuzzy cubes and investigates the combination of several queries to explore the possibility of the definition of an algebra to manipulate fuzzy cubes. All these studies mainly focus on modeling basic fuzzy object types and operations, leaving aside the processing of more advanced queries.

In existing fuzzy OLAP studies [12–15], OLAP mining and fuzzy data mining are combined to take advantage of the fact that fuzzy set theory treats numeric values more naturally, increases understanding, and extracts more generalizable rules. Fuzzy OLAP is performed on fuzzy multidimensional databases. The multidimensional data model of data warehouses is extended to manage the imperfect and imprecise data (e.g., cold days) of the real world. These studies typically focus on finding knowledge about fuzzy spatial data, but more complex queries (e.g., select cold regions) are not considered.

In studies [16,17] on fuzzy spatial querying, neither SOLAP nor MDX query supports are used, but an extension to the standard Structured Query Language (SQL) is used to support spatial and temporal data. The authors combine and extend techniques developed in spatial and fuzzy data mining to deal with the uncertainty of typical spatial data, though they were not concerned about the performance side of the queries. In another study [18], fuzzy logic is integrated into spatial databases to help with decision support and OLAP query processes. In this study, the design of the fuzzy spatial data warehouse methodology is presented, but the effectiveness and efficiency are not discussed.

In addition, there are studies [19,20] on the nearest-neighbor and range types of queries in the field of fuzzy spatial queries. These studies consider range and nearest-neighbor queries in the context of fuzzy objects with indeterminate boundaries. They show that processing these types of queries in spatial OLAP is essential, but the query types are too limited. Support for complex spatial query types is still required.

Special structures have been developed for efficient and effective queries on fuzzy spatiotemporal data [21,34]. In these studies, novel indexes such as R*-tree [35] and X-tree [36] were used for efficient and effective queries, but there were no queries showing the benefits of spatial OLAP.

2.2. FSOLAP Framework

The FSOLAP framework provides for fuzzy spatial–temporal data analytics and flexible and complex querying. The framework includes a multilayered system architecture that consists of four layers. The layers are data sources, structured data, logic, and presentation layers (from the bottom to the top). The system architecture of FSOLAP is represented in Figure 2.

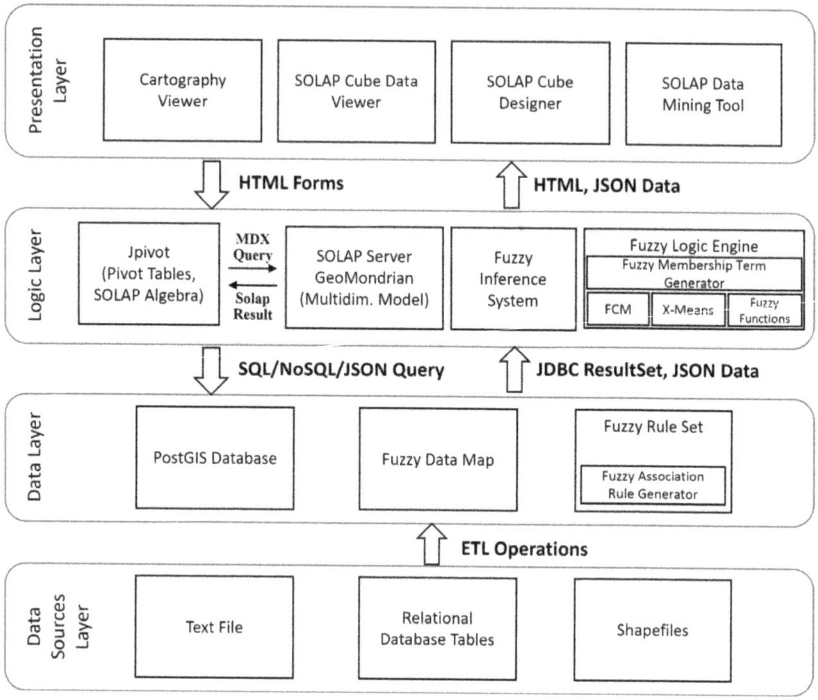

Figure 2. Multilayer framework architecture of FSOLAP.

At the bottom of the system, there are text files, database tables, and shape files. These structures contain the pure data which may be gathered from a web service or collected from a website. Data are migrated to the structured data layer via extract, transform, and load (ETL) operations from this layer. ETL operations are mainly related to reading files, preprocessing data, cleaning data, and validating data operations.

The data layer includes semi-structured or structured data such as a relational database, fuzzified data, and a fuzzy rule set. ETL output data, the fuzzification phase, and fuzzy association rule generation are handled in this layer. The upper layer is called the logic layer, and it requests data from the data layer using SQL or JavaScript Object Notation (JSON) requests. The data layer returns the requested data via SQL tuples, Java Database Connectivity (JDBC) result sets, or JSON responses. The data layer also provides fuzzy querying on PostGIS database data supported by the fuzzy logic module.

The logic layer contains systems that provide spatial, non-spatial, temporal, and fuzzy data mining tools, and a set of fuzzy functions used for fuzzification/defuzzification. It also includes data analytics and visualization platforms that help in visual pattern detection. The reporting tools that provide standard reports on the data are integrated into this layer. The SOLAP server is another central part of this layer that supports SOLAP data cube operations and multidimensional expression (MDX) querying. We integrated a fuzzy inference system and a fuzzy logic module for spatial data mining tasks. The fuzzy logic module was assembled to support fuzzy operations such as membership calculation, fuzzy clustering, and fuzzy class identification.

The presentation layer is shown at the top of our proposed architecture in Figure 2. This layer provides a categorized and simplified system structure. We can demonstrate the data on a map with a cartography viewer. We can also design a new SOLAP cube with hierarchies and measurements using the SOLAP data cube designer. In addition, the SOLAP cube data viewer allows querying of the data using user-friendly query interfaces for data selection. The data selection corresponds to the process of obtaining a subcube from the SOLAP cube via an MDX query. The definition of a subcube is as follows.

Let $D_s \subseteq D$ be a non-empty set of p dimensions $\{D_1, D_2, \ldots, D_p\}$ from data cube $C(p \leq d)$. The p-tuple $\{\Theta_1, \Theta_2, \ldots, \Theta_p\}$ defines a subcube on C according to D_s iff $\forall_i \in \{1, \ldots, p\}, \Theta_i \neq \emptyset$, and there exists a unique j such that $\Theta_i \subseteq A_{ij}$, which can be visualized as shown in Figure 3.

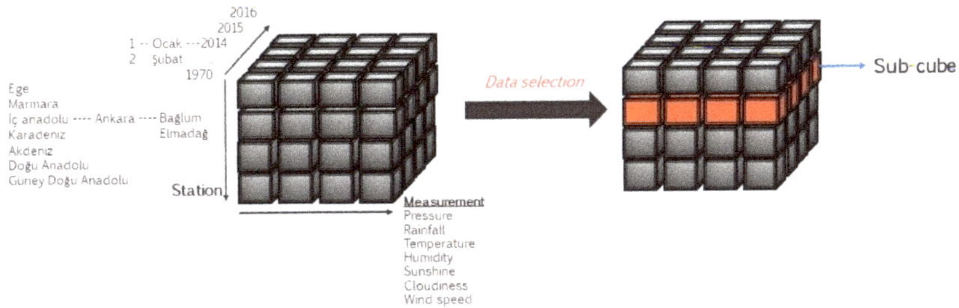

Figure 3. Subcube from SOLAP data selection.

Data selection does not always involve running a simple MDX query; it includes complex fuzzy queries based on the requirements of the data analytics. In data analytics, a hierarchical query is also necessary for certain situations. In this case, it is essential to use structures that support hierarchical querying. SOLAP enables querying and analysis of multidimensional numeric and alphanumeric data. However, there is still a need to support flexible queries on uncertain and fuzzy data due to the nature of complex applications such as meteorological and other spatiotemporal applications. The framework supports data analytics with the management of fuzzy spatiotemporal queries. FSOLAP can handle a variety of complex queries, including fuzzy spatiotemporal queries, which are dealt with effectively and efficiently using our FSOLAP framework.

2.3. *FSOLAP Query Management*

This section describes the architecture and query types that support fuzzy spatiotemporal queries on spatial OLAP-based structures. In the FSOLAP framework, we typically achieve query management through two main structures, as shown in Figure 4. One of these is the data layer, where we prepare, format, and query data. The other is the query module, which contains the frontend presented to the user for querying and query management components.

2.3.1. Data Layer

The raw data are structured after ETL operations and inserted into the PostgreSQL database at the data layer. SOLAP cube metadata are constructed by using the data in the database via the SOLAP cube designer. Then, for each attribute in SOLAP, the appropriate number of clusters is specified using X-means clustering [37].

X-means clustering is a variation of K-means clustering that refines cluster assignments by repeatedly attempting subdivision and keeping the best resulting splits, until some criterion is reached [37]. Algorithm 1, for X-means clustering, consists mainly of two operations repeated until completion.

Algorithm 1 Algorithm of X-means Clustering

Input: given sets of data to be clustered: $d_1, \ldots, d_n$
Output: $K \leftarrow$ number of clusters
1: Improve-Params $\leftarrow$ run conventional K-means to convergence
2: Improve-Structure $\leftarrow$ find out if and where new centroids should appear
3: **if** $K > K_{max}$ **then**
4: stop and report best scoring model found during the search
5: **else if** $K <= K_{max}$ **then**
6: Go to 1
7: **end if**
8: **return** K

The objective function of K-means is as follows:

$$J = \sum_{j=1}^{k} \sum_{i=1}^{n} || x_i^j - c_j ||^2 \quad (5)$$

where $|| x_i^j - c_j ||^2$ is a chosen distance measure between a data point x_i^j and the cluster centre c_j, which is an indicator of the distance of the n data points from their respective cluster centres.

The determined number of clusters is used as input when fuzzifying each attribute with the fuzzy c-means (FCM) clustering algorithm [38,39].

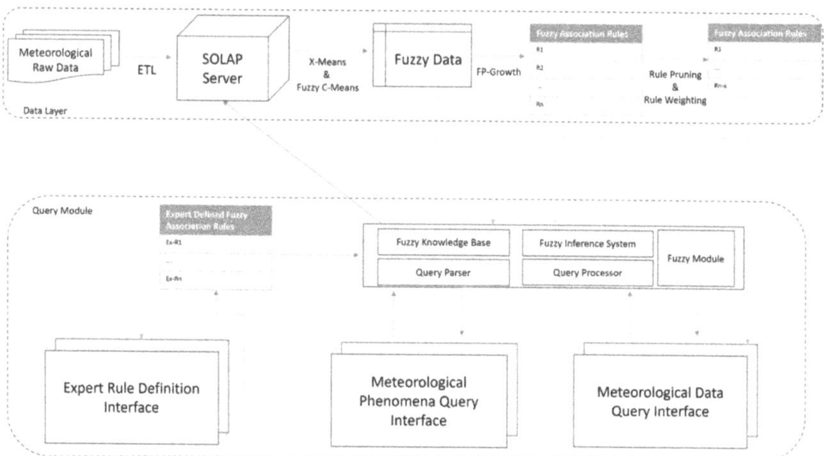

Figure 4. FSOLAP query management.

FCM is based on minimization of the following objective function:

$$J_m = \sum_{i=1}^{N} \sum_{j=1}^{C} u_{ij} || x_i - c_j ||^2, 1 \leq m < \infty \quad (6)$$

where m is any real number greater than 1, u_{ij} is the degree of membership of x_i in the cluster j, x_i is the ith value of d-dimensional measured data, c_j is the d-dimension center of the cluster, and $|| * ||$ is any norm expressing the similarity between any measured data point and the center [39]. Fuzzy partitioning is carried out through an iterative optimization of the objective function shown above, updating the membership u_{ij} and the cluster centers u_j by:

$$u_{ij} = \left(\frac{1}{\sum_{k=1}^{C} u_{ij} (\frac{||x_i - c_j||}{||x_i - c_k||})^{\frac{2}{m-1}})} \right) \quad (7)$$

$$c_j = \frac{\sum_{i=1}^{N} u_{ij}^m \cdot x_i}{\sum_{i=1}^{N} u_{ij}^m} \quad (8)$$

This iteration will stop when $max_{ij} = |u_{ij}^{(k+1)} - u_{ij}^{(k)}| < \delta$, where δ is a termination criterion between 0 and 1, whereas k represents the iteration steps. This procedure converges to a local minimum or a saddle point of J_m [39]. The algorithm is composed of the following steps:

1. Initialize $U = [u_{ij}]$ matrix, $U^{(0)}$
2. At k-step: calculate the center vectors $C^{(k)} = [c_j]$ with $U^{(k)}$

$$c_j = \frac{\sum_{i=1}^{N} u_{ij}^m \cdot x_i}{\sum_{i=1}^{N} u_{ij}^m} \quad (9)$$

3. Update $U^{(k)}, U^{(k+1)}$

$$u_{ij} = \left(\frac{1}{\sum_{k=1}^{C} u_{ij} (\frac{||x_i - c_j||}{||x_i - c_k||})^{\frac{2}{m-1}})} \right) \quad (10)$$

4. If $|| U^{(k+1)} - U^{(k)} || < \delta$ then STOP, otherwise return to step 2.

After determining the fuzzy clusters and membership functions, fuzzy association rules are generated on the fuzzified attributes with the FP-growth algorithm [40]. Association finds rules about items that appear together in an event such as a purchase transaction.

The problem of association rule mining is defined as follows. Let $I = \{i_1, i_2, \cdots, i_n\}$ be a set of n binary attributes called items. Let $D = \{t_1, t_2, \cdots, t_m\}$ be a set of transactions called the database. Each transaction in D has a unique transaction ID and contains a subset of the items in I. A rule is defined as an implication of the form $X \Rightarrow Y$, where $X, Y \subseteq I$. A rule is defined only between a set and a single item, $X \Rightarrow i_j$ for $i_j \in I$. Every rule is composed of two different sets of items, also known as itemsets, X and Y, where:

- X is called the antecedent or left-hand side (LHS);
- Y is called the consequent or right-hand side (RHS).

A heuristic approach is applied to generate a proper number of association rules. First, a different number of rules is generated by parametrically changing the minsup and minconf values for the FP-growth algorithm. After running FP-growth, the generated ruleset is tested with test data for making inferences. Then, the accuracy values of the inferences produced with the test data are calculated. Finally, the proper number of fuzzy association rules is obtained when no change in the accuracy is calculated according to the number of rules. However, this ruleset may contain duplicative rules. We need to reduce the number of rules with confidence-based rule pruning to prevent duplication. We used a rule-based pruning algorithm [41] that removes the unnecessarily complex rules, as shown in Algorithm 2.

Algorithm 2 Algorithm of Fuzzy Association Rule Pruning Based on Confidence

Input: given the sets of several length rules: $S_1, \ldots, S_L$
$L \leftarrow$ max length(R_L), $l = 1, \ldots, M$
J is an empty set
Output: RB: pruned fuzzy association rule base with reduced number of rules
for $i = L, \ldots, 2$ **do**
  for all $R \epsilon S_i$ **do**
    for all $R' \epsilon S_{i-1}$ **do**
      if size($R' \cap R$) = i **then**
        $J \leftarrow J \cup$ index of R'
      end if
    end for
    if $\max(FC(R_J)) > FC(R)) - \varepsilon$ **then**
      delete R from the rule base RB
    end if
  end for
end for
return RB

The pruning method compares the most comprehensive rules with shorter ones. A general rule which contains more minor rules is removed from the rule base when the maximal confidence of a fuzzy association rule (FC) value of the more minor rules is higher than the FC value of the broad rule minus ε, the *correction factor* (initially set to 2 percent). This rule pruning method offers shorter rules in the rule base. Although the pruned rule base contains fewer rules, the new classifier has the same classification accuracy as the unpruned rule base.

The fact that pruned rules have different weights during inference is a factor that affects accuracy. Results produced by association rules that make inferences for the same attribute in proportion to their weights should be considered. For this reason, a weighting process for the rules in the association rule set was performed. This study uses an interest measure called Rule Power Factor (RPF) [42] to give weight to each fuzzy association rule and to mine the fuzzy association rule between them. The equation of the RPF is as follows:

$$RFP(X \rightarrow Y) = support(X \cup Y) * confidence(X \cup Y) \tag{11}$$

where support and confidence are defined as follows:

$$support(X \rightarrow Y) = \frac{number\ of\ tuples\ containing\ both\ X\ and\ Y}{total\ number\ of\ tuples} \tag{12}$$

$$confidence(X \rightarrow Y) = \frac{number\ of\ tuples\ containing\ both\ X\ and\ Y}{number\ of\ tuples\ containing\ X} \tag{13}$$

2.3.2. Query Module

The query module (QM) is the component which handles query operations. Basically, it includes a fuzzy module (FM), a fuzzy knowledge base (FKB), a fuzzy inference system (FIS), a query parser (QPr), a query processor (QPc), and a query interface (QIn), as shown in Figure 4. User queries are entered into the system via the query interface. The QIn component receives user queries and sends these queries to the QPr. After the query is evaluated, the query results are displayed to the user.

There are two user interfaces for querying meteorological phenomena and meteorological data. Before querying meteorological phenomena, it is necessary to determine the association rules of related phenomena. For this purpose, the rules regarding the meteorological phenomenon can be defined with the expert rule definition interface shown in Figure 5.

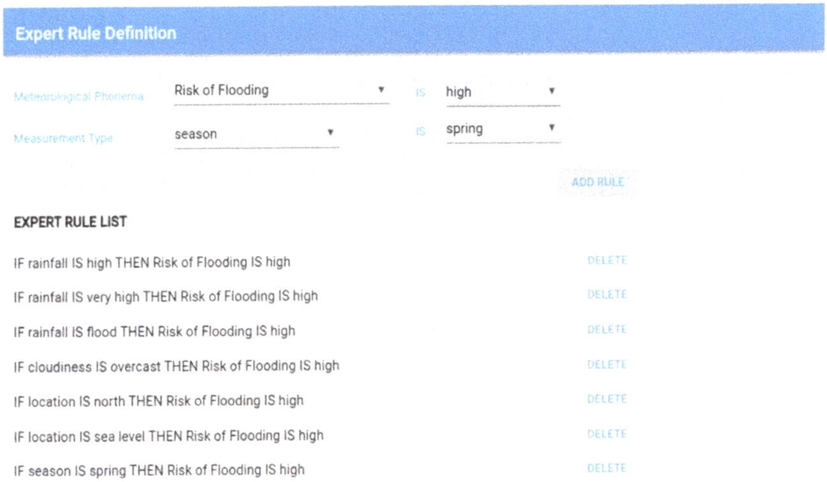

Right Hand Side/antecedent	Left Hand Side/consequent
IF rainfall IS high	THEN Risk of Flooding IS high
IF rainfall IS very high	THEN Risk of Flooding IS high
IF rainfall IS flood	THEN Risk of Flooding IS high
IF cloudiness IS overcast	THEN Risk of Flooding IS high
IF location IS north	THEN Risk of Flooding IS high
IF location IS sea level	THEN Risk of Flooding IS high
IF season IS spring	THEN Risk of Flooding IS high

Figure 5. Expert rule definition UI.

In this interface, after the type and fuzzy class of a phenomenon are determined, the fuzzy association rule is produced by selecting the meteorological attribute and fuzzy class that are the antecedents of the relevant event. These fuzzy association rules are stored in the FKB and then used in the meteorological phenomenon inquiry interface, as shown in Figure 6.

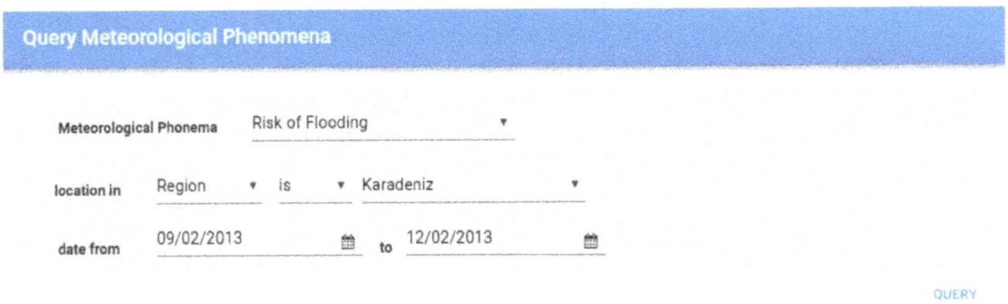

Figure 6. Meteorological phenomena query UI.

In addition, meteorological data can be queried by selecting the attribute and the spatial and temporal criteria using the interface, as shown in Figure 7. The query results are represented in a list, and the spatial information is shown on a map.

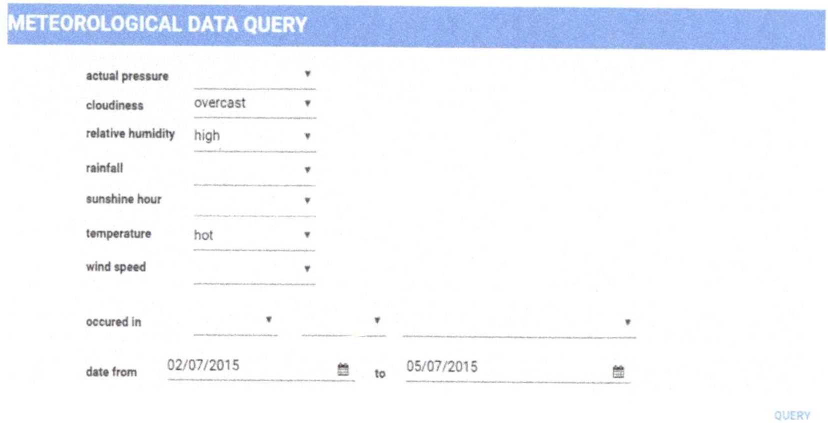

Figure 7. Meteorological data query UI.

In the meteorological phenomenon inquiry process, the association rules of the relevant event are selected from the FKB. In the antecedent part of these rules, fuzzy attributes and classes are determined and used as query criteria. The user can insert the spatial and temporal conditions into the requirements of the MDX query. Query results are fetched after executing the built MDX query on the SOLAP server. Again, query results are displayed in a list, and spatial information is shown on a map. Figure 8 shows how the selected criteria are used in the interface when building the MDX query.

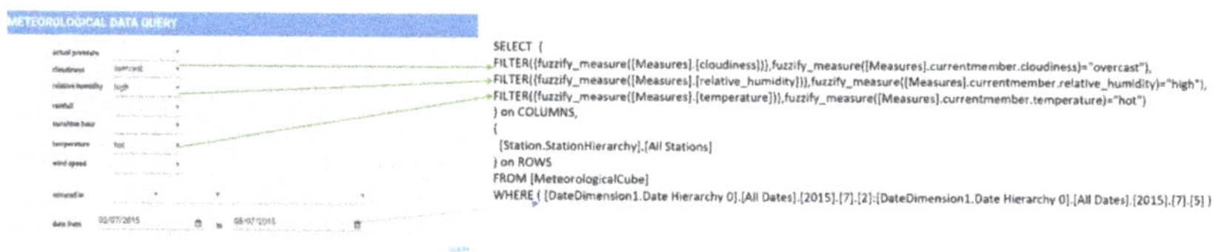

Figure 8. Sample MDX of meteorological data query.

The QPr component parses and interprets the user query and determines which elements will process the query. The QPc module works as a subcomponent responsible for running the query on the related systems and collecting and displaying the results. In other words, the QPc component plays a coordinating role in query processing. QPc performs the communication and interactions between the SOLAP, the FIS, and the fuzzy module. It receives user queries, analyzes them, sends requests to the SOLAP and/or to the FKB/FM, retrieves the results, and sends them to the query interface.

The fuzzy module is the component that provides crisp-to-fuzzy or fuzzy-to-crisp transformations using fuzzification and defuzzification operations. In this module, using the FCM algorithm, fuzzy clustering is performed to generate membership classes and determine membership values. FCM needs the number of clusters as a parameter. Therefore, we used X-means clustering to determine the appropriate number of clusters and to cross-check the cluster with elbow [43] and silhouette [44] methods. In addition, the definitions of uncertain types, similarity relations, and membership functions are stored in the fuzzy data map.

The fuzzy knowledge base (FKB) produces and stores fuzzy association rules. After fuzzifying the meteorological data on SOLAP, the fuzzy association rules are generated with the FP-growth algorithm and stored in the FKB. The resulting extensive list of rules is pruned using a confidence-measure-based pruning method [41] for performance improvement. The rules in the FKB are used in the case of inference as input for the FIS.

The FIS is utilized to support prediction-type queries. While querying, the fuzzy association rule required for each criterion is requested from the FKB and sent to the FIS. In addition, the FM provides the fuzzy membership classes and membership values required for the values in the query as input to the FIS. This interface works as follows. $A' = F(x_0)$, where x_0 is a crisp value defined in the input universe $\cup$, A_0 is a fuzzy set defined in the same universe, and F is a fuzzifier operator. The FIS is based on the application of the generalized modus ponens, an extension of the classical modus ponens proposed by Zadeh, where:

$$\frac{(\text{If } X \text{ is } A \text{ then } Y \text{ is } B) \cap (X \text{ is } A')}{(Y \text{ is } B')} \tag{14}$$

where X and Y are linguistic variables, A and B are fuzzy sets, and B' is the output fuzzy set inferred. To achieve this, the system firstly obtains the degree of matching of each rule by applying a conjunctive operator, and then infers the output fuzzy sets by means of a fuzzy implication operator. The FIS produces the same number of output fuzzy sets as the number of rules collected in the FKB.

The SOLAP server acts as a database server for objects and provides an application that stores measurement results, including spatiotemporal hierarchies, and supports MDX query types. We used the GeoMondrian SOLAP server [45] in our system. After the ETL process, the meteorological data are inserted onto the spatial OLAP server. These data are stored on the spatial OLAP server as spatial, temporal, and measurement-value hierarchies. The spatial hierarchy has region, city, and station breakdowns. Spatial hierarchy can be achieved with a foreign key, as in classical relational databases, or with a minimum bounded rectangle (MBR) structure supporting the spatial structure. The temporal hierarchy is organized according to year, month, and day divisions. Furthermore, each measurement result is available in a hierarchical structure in SOLAP.

We extended the MDX query and modified the GeoMondrian SOLAP server to support fuzzy queries. In general, the user asks for the fuzzy spatial or non-spatial objects that meet the conditions of the predefined rules within a specified time interval, when querying. The rules can be evaluated by examining the topological relations between fuzzy regions and fuzzy objects. To support this, the fuzzify_measure and fuzzify_geo methods are implemented in the MDX query processor of the SOLAP server. The fuzzify_measure method uses the hierarchy for the non-spatial attributes, while the fuzzify_geo method uses the hierarchy for the spatial attributes. The spatial hierarchy is used while detecting the fuzzy relationships such as around, inside, covers, etc., of two different spatial data items that are related to each other, using the fuzzify_geo method. To develop these methods, the geomondrian.jar Java library [45], which is used by the GeoMondrian SOLAP server for querying, was edited. We modified the MondrianServerImpl.java, Query.java, and Parser.java classes in this Java library by adding fuzzify_measure and fuzzify_geo methods. The MondrianServerImpl.java class contains keywords such as Filter, Member, Where, etc., which are used in the query. The fuzzify_measure and fuzzify_geo methods are inserted as keywords to this class. The Query.java class parses the MDX query with the help of the Parser.java class, then determines the query parts and parameters. While parsing the MDX query in the Parser.java class, fuzzy methods are identified using the keywords defined in the MondrianServerImpl.java class. The fuzzy module is integrated with its API while implementing these methods. The parameters of the methods are fuzzified in the fuzzy module via the API. The query results are fetched by processing the fuzzified parameter, and the fuzzy criterion is entered into the query with the relevant operator. While the query processor creates an MDX query, it fuzzifies the parameters that are associated with fuzzy methods and transforms them into a standard MDX query. In the query process, attributes are fuzzified via the fuzzy module and made suitable for the MDX query structure. Similarly, geometric features are fuzzified during queries and handled using the spatial functions provided with PostGIS.

The algorithm for implementing queries is given in Algorithm 3, and some sample queries are defined in Section 2.5.

Algorithm 3 The generic query evaluation algorithm

```
Input: The user query with set of column members CLN and predicates PR
Output: Set of retrieved/predicted objects RSL
  Initialization:
  FT_p ← {}        //fuzzy membership terms
  FAR ← {}         //fuzzy association rules
  SP_t ← {}        //spatial terms
  NSP_t ← {}       //non-spatial terms, measurement
  D_s ← {}         //SOLAP data cube query result holder
  S_O ← {}         //satisfying-objects
Retrieve and Parse (query)
if query includes prediction predicate(PR) then
  Send query to FKB with (CLN, PR)
  Transfer to FIS with (CLN, PR)
  FAR ← Retrieve fuzzy association rules from FKB with (CLN, PR)
  FT_p ← Retrieve fuzzy memberships from FM with (CLN, PR)
  SP_t ← Defuzzify spatial predicates with (CLN)
  NSP_t ← Defuzzify non-spatial predicates with (PR)
  D_s ← Query spatial temporal data from SOLAP with (SP_t, NSP_t)
  S_O ← Make prediction with (FAR, FT_p, D_s)
  return S_O
else
  if query is spatial then
    SP_t ← Defuzzify spatial predicates with (CLN)
    NSP_t ← Defuzzify non-spatial predicates with (PR)
    D_s ← Query spatial temporal data from SOLAP with (SP_t, NSP_t)
    S_O ← Fuzzify satisfying objects with (D_s)
    return S_O
  else
    NSP_t ← Defuzzify non-spatial predicates with (PR)
    D_s ← Query spatial temporal data from SOLAP with (NSP_t)
    S_O ← Fuzzify satisfying objects with (D_s)
    return S_O
  end if
end if
```

2.4. Data Sets

In this study, we utilized a spatiotemporal database including real meteorological measurements that have been observed and collected in Turkey over many years. The spatial extent of Turkey is 36° N to 42° N in latitude and from 26° E to 45° E in longitude. The meteorological data measurement interval of the study was 1970 to 2017. There are seven geographical regions in Turkey. These geographical regions are separated according to their climate, location, flora and fauna, human habitat, agricultural diversities, transportation, topography, etc. The names of the regions are: Mediterranean, Black Sea, Marmara, Aegean, Central Anatolia, Eastern Anatolian, and Southeastern Anatolia. There are meteorological measurement data in our meteorological database from 1161 meteorological observation stations. These stations were selected from different geographical regions. Sample data from different meteorological stations are given in Table 1.

Table 1. Meteorological station samples from station database table.

Station No	Station Name	City	Town	Latitude (°)	Longitude (°)	Altitude (m)
17038	Trabzon	Trabzon	Ortahisar	40.99	39.78	39
17040	Rize	Rize	Merkez	41.04	40.50	3
17050	Edirne	Edirne	Merkez	41.67	26.55	51
17064	İstanbul	İstanbul	Kartal	40.91	29.15	18

Tables in the Meteorological Database

In this study, we used database tables containing ten types of meteorological measurements for our various queries. The types of meteorological measurements were: daily vapor pressure, daily hours of sunshine, daily maximum speed and direction of the wind, daily average actual pressure, daily average cloudiness, daily average relative humidity, daily average speed of the wind, daily average temperature, daily total rainfall—manual and daily total rainfall—omgi. The database table names of the measurement types and the details of each measurement are described in Table 2.

Table 2. Database tables and descriptions.

Table Name	Description	Units
station	Station code, names, city, and coordinates	latitude, longitude, and altitude
vapor-pressure	Daily vapor pressure	hectopascal (1 hPa = 100 Pa)
sunshine-hour	Daily hours of sunshine	hours
speed-direction-wind	Daily max speed and direction of the wind	meter/second and direction
average-pressure	Daily average actual pressure	hectopascal (1 hPa = 100 Pa)
cloudiness	Daily average cloudiness	8 octa
average-humidity	Daily average relative humidity	percentage
average-speed-wind	Daily average speed of the wind	meter per second
average-temperature	Daily average temperature	celsius
total-rainfall-manual	Daily total rainfall—manual	kg per meter square
total-rainfall-omgi	Daily total rainfall—omgi	kg per meter square

These tables contain daily measurements from 1 January 1970 to 1 January 2017. Each table record consists of a station number, measurement type, measurement date, and measurement value. Sample data for the daily average speed of the wind are given in Table 3.

Table 3. Sample data for daily average wind speed table.

Station No	Station Name	Year	Month	Day	The Daily Average Speed of Wind (m/s)
8541	HASSA	1977	1	1	1.3
8541	HASSA	1977	1	2	1.1
8541	HASSA	1977	1	3	3.1
8541	HASSA	1977	1	4	3.4

2.5. *Supported Query Types*

After illustrating the architecture of the proposed environment for fuzzy spatiotemporal querying, we apply the following procedures to handle the various query types employing the given components.

2.5.1. Fuzzy Non-Spatial Query

This query type asks for fuzzy data not dealing with spatial attributes. The QM, the FM, and the SOLAP server components are working in the execution step and the query flow is given in Figure 9:

1. The QM retrieves the user query, parses it, and sends it to the FM.
2. The QM asks the SOLAP server for data using the query. The objects retrieved by the QM are sent to the FM component to fuzzify the result.
3. Fuzzified query results are displayed in the QM component.

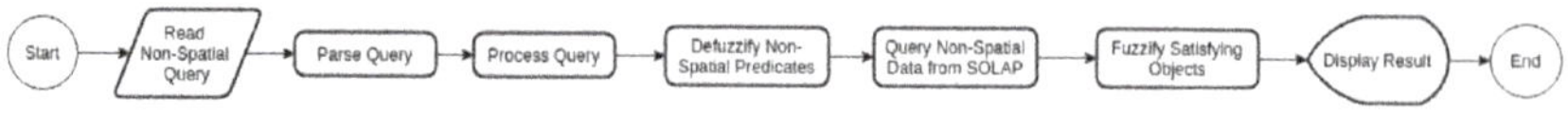

Figure 9. Fuzzy non-spatial query flow.

Query 1: *Find all the cities at risk of flooding.*

The query is expressed in MDX, which is an OLAP query language which provides a specialized syntax for querying and manipulating the multidimensional data stored in OLAP cubes [46]. While it is possible to translate some of these queries into traditional SQL, this would frequently require the synthesis of clumsy SQL expressions, even for elementary MDX expressions. Furthermore, many OLAP vendors have used MDX, and it has become the standard for OLAP systems. While it is not an open standard, it is embraced by a wide range of OLAP vendors. Therefore, we extended MDX with fuzzy operators and wrote the query specified above in MDX form, using the query parameters shown in Figure 10.

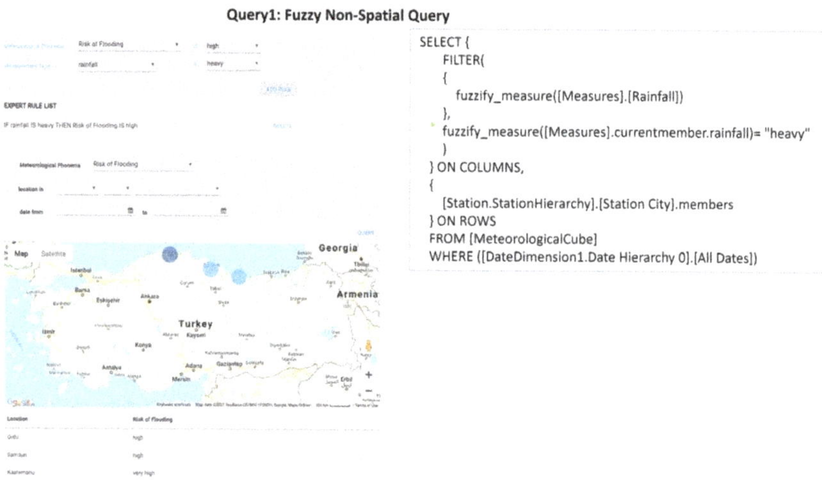

Figure 10. Fuzzy non-spatial query.

To query the database, we first need to defuzzfy the fuzzy expression part of the query. The query processor requests the FM to defuzzify the fuzzy expression in the query. The fuzzy term is defuzzified according to the fuzzy membership function, as shown in Figure 11. The *heavy* class in the query has a triangular-shaped membership function defined by the triple (7.5, 8.5, 9.5) that overlaps the membership function of the *overmuch* class in the range [7.5, 8.5]. In this case, the *heavy* class includes measurements between 8.0 and 9.5. The query processor of the GeoMondrian rearranges the MDX query with the crisp values after defuzzification and executes it in the SOLAP server. As a result of the query on the SOLAP server, the results matching the searched criteria contain crisp data. We again fuzzify the crisp values in the resulting data with the help of the FM. Here, the fuzzification subcomponent in the FM includes a triangular or trapezoidal membership function for each measurement result. It generates fuzzy class and membership values as output, using the crisp value of input from the relevant membership function. Finally, the results are displayed to the user, including fuzzy terms. For our example, we show the R1 and R4 records in Table 4 as the query result that meets the criteria.

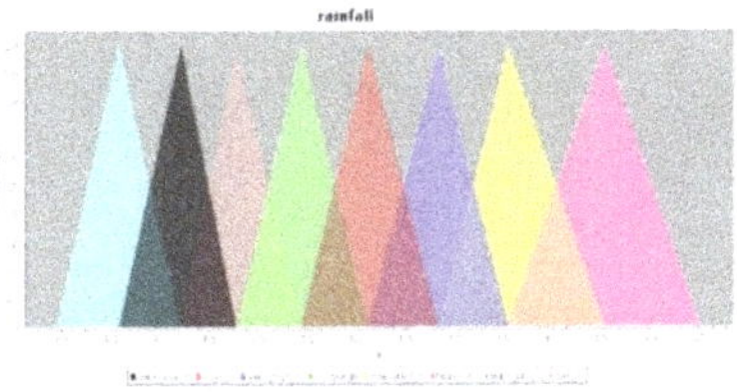


```
FUZZIFY rainfall_in
        TERM nearly_dry_in:= (2.95, 0)  (3.58, 1)  (4.21, 0);
        TERM very_low_in  := (3.58, 0)  (4.21, 1)  (4.77, 0);
        TERM low_in       := (4.21, 0)  (4.77, 1)  (5.43, 0);
        TERM normal_in    := (4.77, 0)  (5.43, 1)  (6.11, 0);
        TERM high_in      := (5.43, 0)  (6.11, 1)  (6.83, 0);
        TERM very_high_in := (6.11, 0)  (6.83, 1)  (7.54, 0);
        TERM overmuch_in  := (6.53, 0)  (7.54, 1)  (8.53, 0);
        TERM heavy_in     := (7.54, 0)  (8.53, 1)  (9.51, 0);
END FUZZIFY
```

Figure 11. Rainfall membership classes.

Suppose we execute this query in a relational database. In that case, we need to thoroughly scan all records, because it is necessary to calculate the rainfall value and find the queried value by grouping based on the city within the station measurement records. The cost of scanning all the data and grouping them is critical; the query execution time is related to the number of records in the database. In the FSOLAP environment, it is not necessary to access all records for the objects that satisfy the query criteria, due to the help of the hierarchical structure. The calculation of the measurements of the cities with which the stations are connected does not imply such a cost. Therefore, the cost of searching rainy stations is limited to the number of stations registered in the database, and the query execution time is less than the relational database query execution time.

Table 4. Sample data for rainfall in database.

ID	Date	City	Crisp Val.	Fuzzy Val.
R1	19 August 2016	Ankara	8.6	heavy (0.7)
R2	19 August 2016	Konya	4.9	low (0.7)
R3	19 August 2016	Adana	4.1	very-low (0.6)
R4	19 August 2016	Rize	8.8	heavy (0.8)

2.5.2. Fuzzy Spatial Query

Fuzzy spatial queries allow the user to interrogate fuzzy spatial objects and their relationships. The QM, the FM, and the SOLAP server components are employed to fetch query results, as shown in Figure 12. The user asks for the objects that have topological relations with the entities under inquiry.

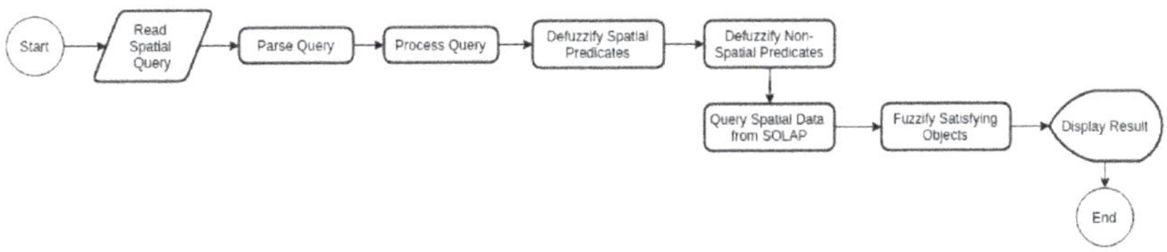

Figure 12. Fuzzy spatial query flow.

Query 2: *Retrieve the appropriate cities for the installation of a solar power plant*

A fuzzy rule definition uses linguistic values, as shown below in the FKB regarding suitable places for solar power plants.

```
if city.sunshine_hour is high and city.position is south
                                then city.solar_power is
                                    high
```

Figure 13 shows how we implemented the MDX query with the parameters entered from the query interface.

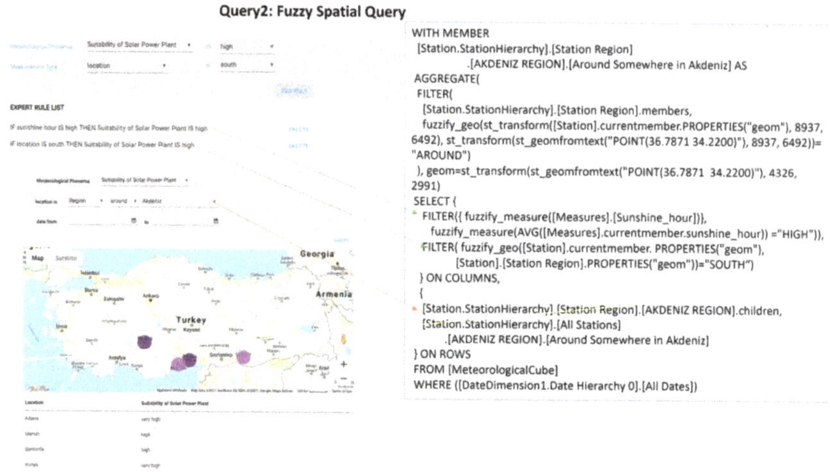

Figure 13. Fuzzy spatial query.

In this query, regions in the south of Turkey with a very high sunshine duration are considered. The intersection of areas with positionally high sunshine hour and south fields are taken into account. We explained the operational structure of the *fuzzify_measure* method in the previous query. Here, the *fuzzify_geo* method is also used. This method is run on the FM and determines the overlap relation between two geometric objects given as parameters. There are as many accesses in the query process as the number of stations in the database. On the other hand, the execution time for the relational database query, given in the following, can be longer due to the averaging of sunshine hour measurements and joining these with the stations.

```
SELECT c.name_1, r.month, r.day, AVG(sunshine_hour)
FROM met_data_rainfall r, tr_city c,
     meteorological_station3 s, tr_region rg
WHERE s.id=r.station_id AND s.city_id=c.gid
  AND rg.id=c.region_id AND c.region_id in(5,7)
GROUP BY c.name_1, r.month, r.day HAVING AVG(sunshine_hour)>7
```

In this query, cities with an average daily sunshine duration of more than seven hours are regarded as having a high sunshine duration. These cities are in the Mediterranean and Southeastern Anatolia regions in the south of the country.

2.5.3. Fuzzy Spatiotemporal Query

In this type of query, the user asks for the fuzzy spatial objects that meet the conditions of the predefined rules within a specified time interval. The rules can be evaluated by an examination of the topological relations between fuzzy regions and fuzzy objects. The query flow is shown in Figure 14.

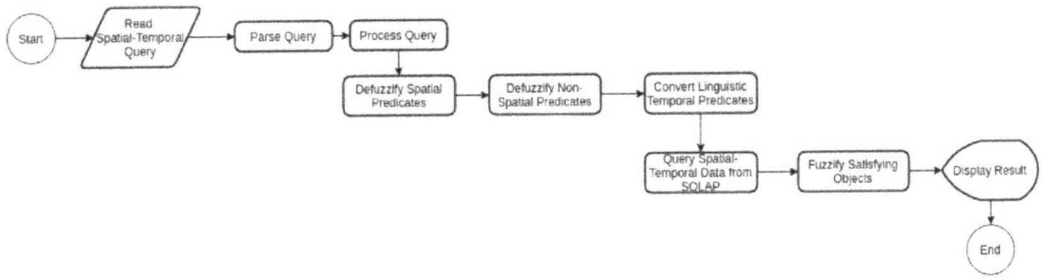

Figure 14. Fuzzy spatiotemporal query flow.

Query 3: *Retrieve locations around Ankara that were at high risk of freezing between 7 January 2012 and 14 January 2012.*

The FKB contains the following fuzzy rule definition that uses linguistic values regarding freezing events.

```
if city.temperature is cold and city.cloudiness is clear
                                   then city.freeze_risk is
                                        high
```

The query syntax's implementation in MDX is represented in Figure 15.

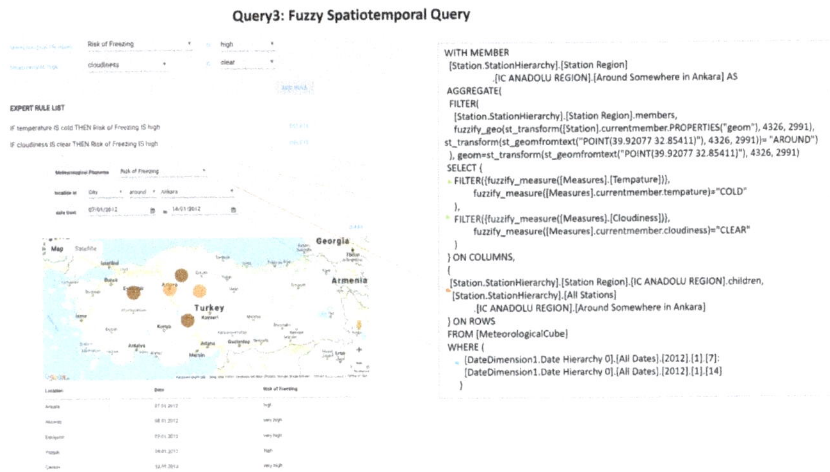

Figure 15. Fuzzy spatiotemporal query.

In addition to the previous query, we can make more specific queries using date attribute conditions. The handling of the fuzzy predicates in the query operation is the same as for the fuzzy spatial query. For the distance attribute, the membership classes in the fuzzy data map are NEAR, CLOSE, and AROUND. We create these fuzzy classes by calculating the paired distances for the geometric data of the stations and applying fuzzy clustering of these values. However, the date predicate greatly reduces the amount of data to be retrieved from the database. As we mentioned earlier, this situation, which requires a full scan of an index-less relational database, is easily handled using the temporal hierarchy in the SOLAP environment. The execution time of the query depends on the number of stations in the database. Relational database systems must be fully searched for temperature and cloudiness between the given dates. In this case, the query execution time is proportional to the number of records and the number of stations in the database.

2.5.4. Fuzzy Spatiotemporal Predictive Query

This type of query asks for fuzzy spatial relations and a specified time with inference. The QM, the FM, the FIS, the FKB, and the SOLAP server components are employed to fetch query results, and the query flow is shown in Figure 16. The QM retrieves the user query, parses it, and sends it to the FM for defuzzification. If the QM detects the inference operand in the query, it sends the conditions to the FKB for inference. When the FKB receives the request from the QM, it determines the fuzzy association rules and sends them to the FIS, and the FIS obtains membership classes/functions from the fuzzy data map subcomponent. The FIS makes predictions with the given parameters and the collected knowledge, and then it sends the inference back to the QM.

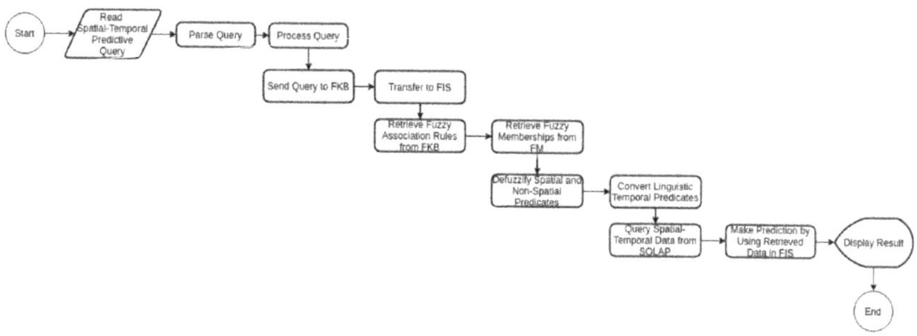

Figure 16. Fuzzy spatiotemporal predictive query flow.

Query 4: *Is there a possibility of a windstorm around Izmir during the last week of December?*

The FKB contains the following rules for meteorological events that occur depending on wind speed.

```
if station.windspeed is high then city.storm_occurrence is
   possible
if station.windspeed is high and actual_pressure is low
                        then city.storm_occurrence is high−
                           possible
```

Unlike other query types, the antecedent part of the association rules is not used in the FKB as a criterion when considering predictive queries. Since the purpose here is to predict the conditions that are the antecedents of the meteorological phenomenon in question, we do not include these fields in the query. Other fuzzy attributes are used as criteria in the MDX query. In addition, the spatial and temporal criteria entered into the interface are used for querying. When the QM detects the *PREDICT* expression in the query, it recognizes that the query requires an inference mechanism. The MDX query constructed with the criteria entered into the meteorological phenomenon query UI is illustrated in Figure 17.

We previously mentioned that the fuzzy association rules which are expert-defined are stored in the FKB. The fuzzy association rules defined for the relevant phenomenon are chosen in the meteorological phenomenon inquiry. The antecedent of each rule is used to look for the fuzzy attribute and membership class found in the consequent part of the fuzzy association rules. In other words, the rules which include these antecedents in the FKB are selected as a consequence of the rules in the fuzzy association rules, and this process is demonstrated in Figure 18.

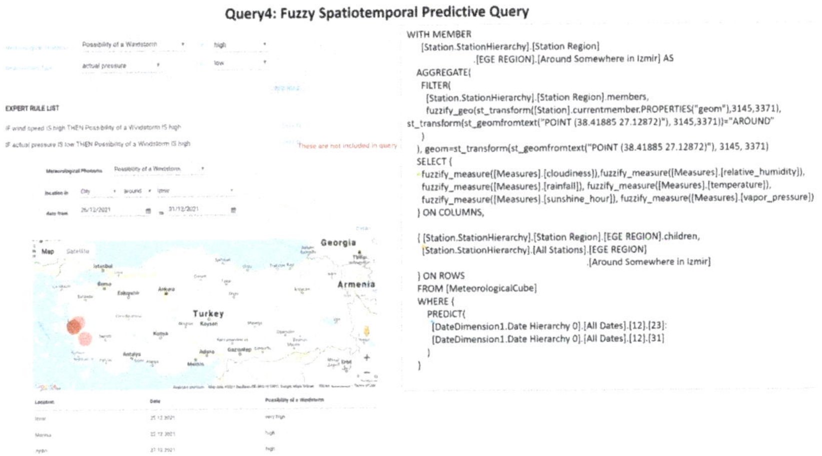

Figure 17. Fuzzy spatiotemporal predictive query.

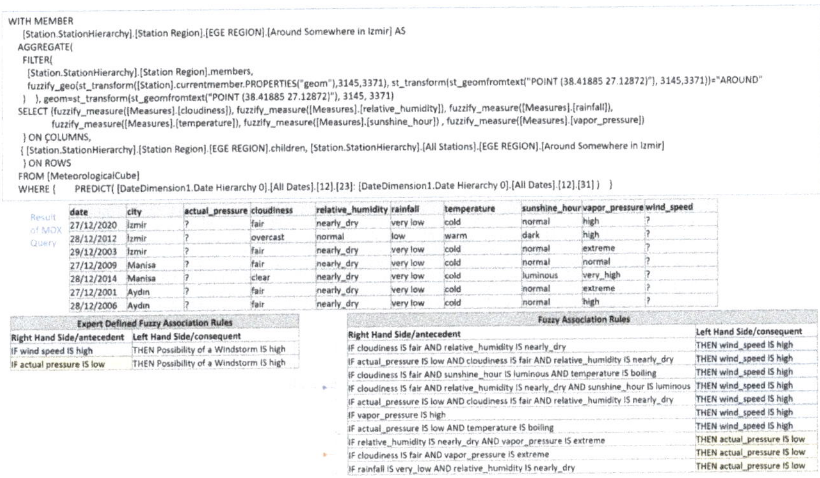

```
WITH MEMBER
  [Station.StationHierarchy].[Station Region].[EGE REGION].[Around Somewhere in Izmir] AS
 AGGREGATE(
  FILTER(
   [Station.StationHierarchy].[Station Region].members,
   fuzzify_geo(st_transform([Station].currentmember.PROPERTIES("geom"),3145,3371), st_transform(st_geomfromtext("POINT (38.41885 27.12872)"), 3145,3371))="AROUND"
  )  ), geom=st_transform(st_geomfromtext("POINT (38.41885 27.12872)"), 3145, 3371)
 SELECT {fuzzify_measure([Measures].[cloudiness]), fuzzify_measure([Measures].[relative_humidity]), fuzzify_measure([Measures].[rainfall]),
        fuzzify_measure([Measures].[temperature]), fuzzify_measure([Measures].[sunshine_hour]) , fuzzify_measure([Measures].[vapor_pressure])
  } ON ÇOLUMNS,
 { [Station.StationHierarchy].[Station Region].[EGE REGION].children, [Station.StationHierarchy].[All Stations].[EGE REGION].[Around Somewhere in Izmir]
  } ON ROWS
 FROM [MeteorologicalCube]
 WHERE (      PREDICT( [DateDimension1.Date Hierarchy 0].[All Dates].[12].[23]: [DateDimension1.Date Hierarchy 0].[All Dates].[12].[31] )   }
```

Result of MDX Query

date	city	actual_pressure	cloudiness	relative_humidity	rainfall	temperature	sunshine_hour	vapor_pressure	wind_speed
27/12/2020	İzmir	?	fair	nearly_dry	very low	cold	normal	high	?
28/12/2012	İzmir	?	overcast	normal	low	warm	dark	high	?
29/12/2003	İzmir	?	fair	nearly_dry	very low	cold	normal	extreme	?
27/12/2009	Manisa	?	fair	nearly_dry	very low	cold	normal	normal	?
28/12/2014	Manisa	?	clear	nearly_dry	very low	cold	luminous	very_high	?
27/12/2001	Aydın	?	fair	nearly_dry	very low	cold	normal	extreme	?
28/12/2006	Aydın	?	fair	nearly_dry	very low	cold	normal	high	?

Expert Defined Fuzzy Association Rules

Right Hand Side/antecedent	Left Hand Side/consequent
IF wind speed IS high	THEN Possibility of a Windstorm IS high
IF actual pressure IS low	THEN Possibility of a Windstorm IS high

Fuzzy Association Rules

Right Hand Side/antecedent	Left Hand Side/consequent
IF cloudiness IS fair AND relative_humidity IS nearly_dry	THEN wind_speed IS high
IF actual_pressure IS low AND cloudiness IS fair AND relative_humidity IS nearly_dry	THEN wind_speed IS high
IF cloudiness IS fair AND sunshine_hour IS luminous AND temperature IS boiling	THEN wind_speed IS high
IF cloudiness IS fair AND relative_humidity IS nearly_dry AND sunshine_hour IS luminous	THEN wind_speed IS high
IF actual_pressure IS low AND cloudiness IS fair AND relative_humidity IS nearly_dry	THEN wind_speed IS high
IF vapor_pressure IS high	THEN wind_speed IS high
IF actual_pressure IS low AND temperature IS boiling	THEN wind_speed IS high
IF relative_humidity IS nearly_dry AND vapor_pressure IS extreme	THEN actual_pressure IS low
IF cloudiness IS fair AND vapor_pressure IS extreme	THEN actual_pressure IS low
IF rainfall IS very_low AND relative_humidity IS nearly_dry	THEN actual_pressure IS low

Figure 18. Fuzzy spatiotemporal predictive query execution: step 1.

We create inferences for each row fetched from the MDX query by running the rules selected from the fuzzy association rule set in the FIS, as shown in Figure 19. The minimum value is calculated by multiplying the results by the weight value of each association rule. The same fuzzy class result is determined by taking the maximum value among the minimum values. If the result value meets the expected criteria, the relevant MDX query result row is marked as satisfied. The results marked as satisfied are shown on the results list and the map.

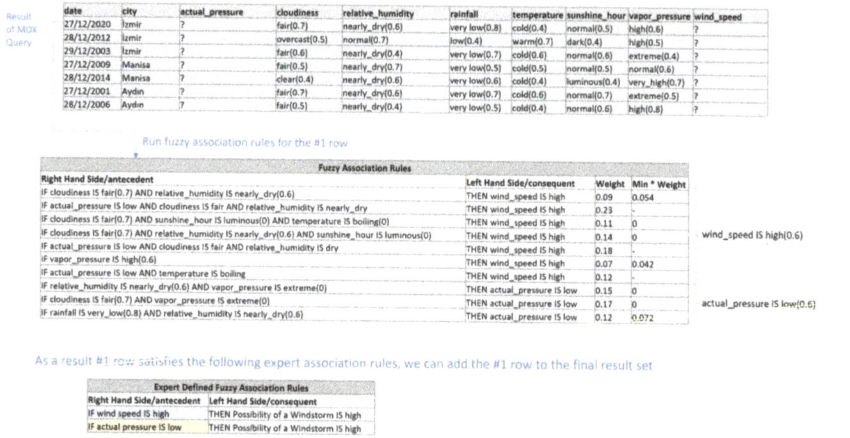

date	city	actual_pressure	cloudiness	relative_humidity	rainfall	temperature	sunshine_hour	vapor_pressure	wind_speed
27/12/2020	İzmir	?	fair(0.7)	nearly_dry(0.6)	very low(0.8)	cold(0.4)	normal(0.5)	high(0.6)	?
28/12/2012	İzmir	?	overcast(0.5)	normal(0.7)	low(0.4)	warm(0.7)	dark(0.4)	high(0.5)	?
29/12/2003	İzmir	?	fair(0.6)	nearly_dry(0.4)	very low(0.7)	cold(0.6)	normal(0.6)	extreme(0.4)	?
27/12/2009	Manisa	?	fair(0.5)	nearly_dry(0.7)	very low(0.5)	cold(0.5)	normal(0.5)	normal(0.6)	?
28/12/2014	Manisa	?	clear(0.4)	nearly_dry(0.6)	very low(0.6)	cold(0.4)	luminous(0.4)	very_high(0.7)	?
27/12/2001	Aydın	?	fair(0.7)	nearly_dry(0.6)	very low(0.7)	cold(0.6)	normal(0.7)	extreme(0.5)	?
28/12/2006	Aydın	?	fair(0.5)	nearly_dry(0.4)	very low(0.5)	cold(0.4)	normal(0.6)	high(0.8)	?

Fuzzy Association Rules			
Right Hand Side/antecedent	Left Hand Side/consequent	Weight	Min * Weight
IF cloudiness IS fair(0.7) AND relative_humidity IS nearly_dry(0.6)	THEN wind_speed IS high	0.09	0.054
IF actual_pressure IS low AND cloudiness IS fair AND relative_humidity IS nearly_dry	THEN wind_speed IS high	0.23	-
IF cloudiness IS fair(0.7) AND sunshine_hour IS luminous(0) AND temperature IS boiling(0)	THEN wind_speed IS high	0.11	0
IF cloudiness IS fair(0.7) AND relative_humidity IS nearly_dry(0.6) AND sunshine_hour IS luminous(0)	THEN wind_speed IS high	0.14	0
IF actual_pressure IS low AND cloudiness IS fair AND relative_humidity IS dry	THEN wind_speed IS high	0.18	-
IF vapor_pressure IS high(0.6)	THEN wind_speed IS high	0.07	0.042
IF actual_pressure IS low AND temperature IS boiling	THEN wind_speed IS high	0.12	-
IF relative_humidity IS nearly_dry(0.6) AND vapor_pressure IS extreme(0)	THEN actual_pressure IS low	0.15	0
IF cloudiness IS fair(0.7) AND vapor_pressure IS extreme(0)	THEN actual_pressure IS low	0.17	0
IF rainfall IS very_low(0.8) AND relative_humidity IS nearly_dry(0.6)	THEN actual_pressure IS low	0.12	0.072

Expert Defined Fuzzy Association Rules	
Right Hand Side/antecedent	Left Hand Side/consequent
IF wind speed IS high	THEN Possibility of a Windstorm IS high
IF actual pressure IS low	THEN Possibility of a Windstorm IS high

Figure 19. Fuzzy spatiotemporal predictive query execution: step 2.

A sample inference is given in Figure 20. In this example, consider a current situation where the relative humidity is 48%, the temperature is +25°, and the cloudiness is 3/8. We want to predict the sunshine hours using this information. The relative humidity of 48% is translated into the linguistic variable value of $\{0.3, 0.7, 0, 0, 0\}$ which can be interpreted as "less, normal". Similarly, linguistic translation can be given as "hot, boiling" for temperature and "partly sunny, partly cloudy" for cloudiness. After all the input variables have been converted to linguistic variable values, the fuzzy inference step can identify the rules that apply to the current situation and can compute the values of the output linguistic variable. As seen in the figure, the five rules of thumb can be translated into a fuzzy rule base using these linguistic terms to describe the meteorological prediction. The rules are selected according to the consequent part. There are three proper rules which have a sunshine hours consequent and can be used for inference. After the rules are executed, the center of gravity method is used to calculate the final predicted value.

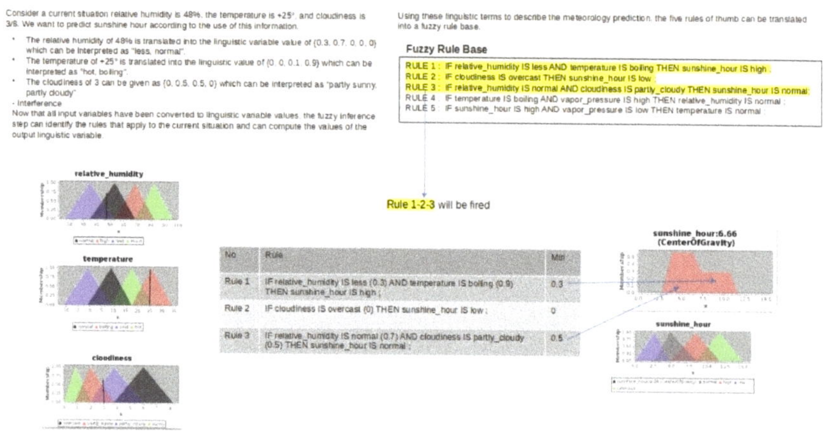

Figure 20. A sample inference.

3. Experimental Results

3.1. Platform

We achieved reasonable performance of the prototype application in the environment and with the specifications, technology, and tools specified below.

- *Application development environment*: Eclipse IDE 2021-03;
- *System*: Windows 10 x64, Intel i5-7200U CPU, 16 GB RAM;
- *Java*: 1.8.0-281, Java HotSpot Client 64-bit Server VM 25.281-b09;
- *SOLAP*: GeoMondrian 1.0 Server;
- *DBMS*: PostgreSQL 13.3 64-bit;
- *FIS*: jFuzzyLogic.jar;
- *Data Size*: approximately 10 GB data consisting of 1161 stations and 15 M records for each measurements (15 M × 10 measurement types).

3.2. Performance Results

We measured the average CPU usage, memory usage, and execution time by running each query type in the fuzzy SOLAP-based framework and the PostgreSQL database. Here, average CPU usage is the average CPU usage rate measured during querying. Similarly, average memory usage is the average memory usage measured in megabytes (MB) during querying. The execution time is the average of the measurements obtained over several query runs.

First, we addressed some of the high-level factors that affect the query performance with regard to CPU usage, memory usage, and execution time. Data size directly affects the performance of the query because the query uses one or more tables with millions of rows or more. Joins are another factor affecting performance; if the query joins two tables, increasing the row count of the result set substantially, the query is likely to be slow. Aggregations also affect performance, as combining multiple rows to produce a result requires more computation than simply retrieving those rows.

In addition to obtaining this information, we also performed the roll-up function provided by SOLAP for aggregating with the UNION operator in relational database queries. In this case, aggregating N dimensions requires N such unions in an SQL query. Another essential issue to consider in terms of query performance is that of cross-tabulations. While SOLAP supports such operations naturally, SQL requires an even more complicated combination of unions and GROUP BY clauses for cross-tabulations. An N-dimensional cross-tabulation requires a 2^N-way union of 2^N different GROUP BY operators to build the underlying representation. In most relational databases, this results in 2^N scans of the data and 2^N sorts or hashes.

The CPU usage for the queries was measured over several query runs, and the average CPU usage for all query types was calculated. The results are given in Table 5.

Table 5. Comparision of average CPU usages between FSOLAP and relational database SQL queries.

	FSOLAP Query Ave. CPU Usage (%)	Relational Database SQL Query Ave. CPU Usage (%)
Query1	29.2	33.7
Query2	30.3	36.6
Query3	30.1	31.3
Query4	30.9	Not Supported

The average CPU usages of the FSOLAP-based query and the relational database query are compared in the column chart shown in Figure 21.

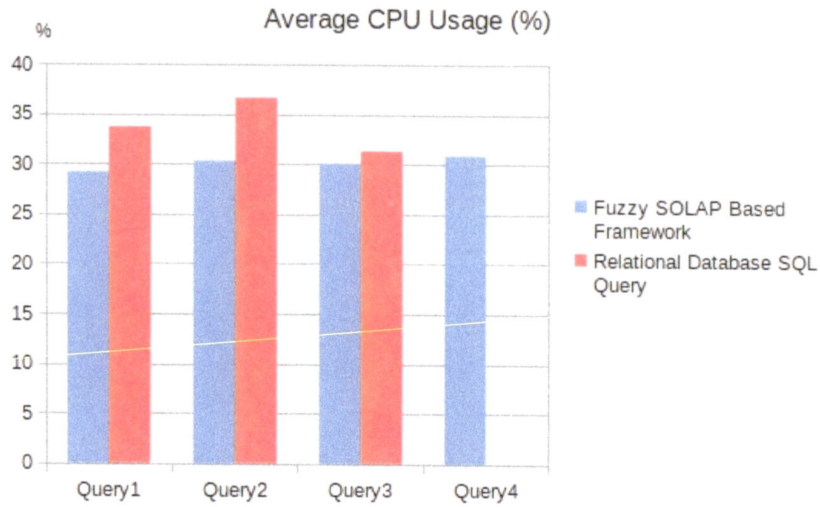

Figure 21. Average CPU usages of FSOLAP and relational database SQL queries.

Similar to the computational power requirement, the measurement results for the average memory usage are given in Table 6.

Table 6. Comparision of average memory usages between FSOLAP and relational database SQL queries.

	FSOLAP Query Ave. Memory Usage (MB)	Relational Database SQL Query Ave. Memory Usage (MB)
Query1	150	278
Query2	228	330
Query3	115	229
Query4	217	Not Supported

The average memory usages of the queries are represented graphically in Figure 22. According to this chart, relational database queries consume more memory than FSOLAP-based queries.

A comparison of the execution times of the queries was used as part of the performance testing, and the results are shown in Table 7.

Table 7. Comparison of average execution times between FSOLAP and relational database SQL queries.

	FSOLAP Query Ave. Execution Time (ms)	Relational Database SQL Query Ave. Execution Time (ms)
Query1	596,480	1,630,362
Query2	257,054	643,642
Query3	18,314	172,303
Query4	183,717	Not Supported

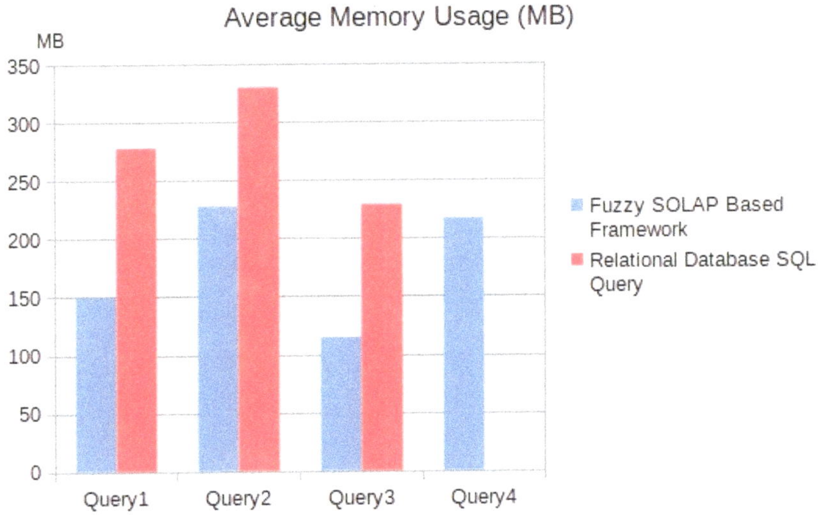

Figure 22. Average memory usage of FSOLAP and relational database SQL queries.

We have shown the time spent between starting the query and finishing the query graphically for each query in Figure 23. The graph shows that relational database queries have a longer execution time.

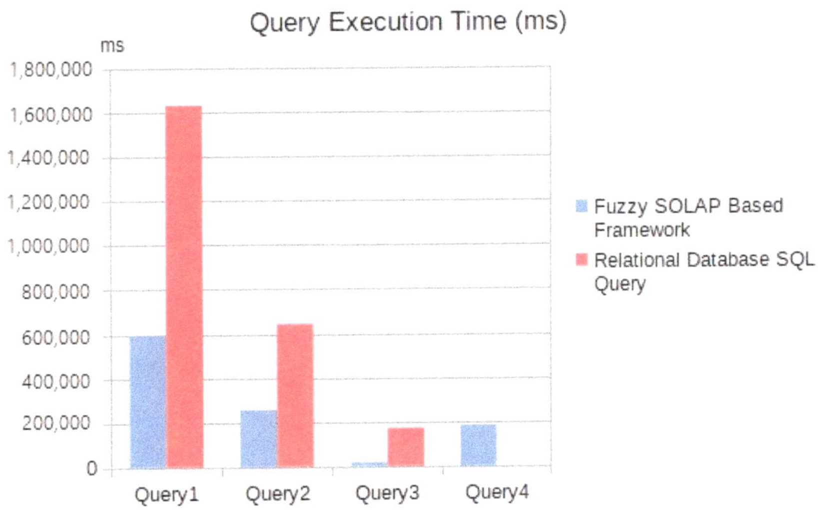

Figure 23. Execution times of FSOLAP and relational database SQL queries.

The implementation of Query 1 in the relational database requires the *having avg* operation as an aggregation for all cities. This requires a great deal of CPU and memory resource usage. Along with these, it also causes a long query time. Query 2 requires *having avg* as an aggregation along with a spatial search. A spatial data search uses index matches with the join operand in the query. This query requires more CPU and memory than other queries, but the query time is comparatively less than Query 1 since the query has a spatial restriction. Query 3, on the other hand, is better in terms of resource usage as it possesses additional time restrictions compared to Query 2, but it also takes less query time. The aggregation process in the queries involves the CPU usage, the union, and the join operands, affecting the memory usage. According to the query criteria, the amount of data in the query process determines the query time. When we evaluate the performance tests in general, we observe that FSOLAP-based query operations require fewer resources and less time than relational database queries. While we obtain adequate CPU and memory usage results, especially in queries containing spatial and temporal criteria, we obtain better results in terms of execution time. In addition, FSOLAP performs well in prediction-type queries, which are not supported for relational database queries.

Based on our experimental analysis and considering all the parameters mentioned, FSOLAP-based querying is preferred over relational database querying, as FSOLAP offers scalability with low resource usage.

4. Discussion

In this paper, we introduced FSOLAP as a new fuzzy SOLAP-based framework to compound the advantages of fuzzy and SOLAP concepts and explained how it supports complex fuzzy spatial queries. We tested the efficiency and effectiveness of FSOLAP in a meteorological application with spatial and temporal hierarchical data, using fuzzy spatial and fuzzy spatiotemporal query types. Moreover, we showed that the fuzzy logic approach is an effective approach for complex applications such as spatiotemporal data with fuzzy spatial queries containing fuzzy terms. In addition, we explained how we handled fuzzy spatiotemporal predictive queries using the inference capability, which has not been previously discussed in the literature. We integrated these queries into FSOLAP with the use of an FIS. It was shown that FSOLAP handles queries effectively and efficiently using fewer resources compared to a relational database system, based on average CPU usage, average memory usage, and average execution time for each type of query. While SOLAP handles hierarchical data naturally, SQL does so with the union operator, which requires high CPU and memory usage as the test results showed. Similarly, SOLAP handles the operation performed by SQL using the group by statement with its core functionality. In extensive performance tests, complex queries structurally containing a group by statement have been shown to require less CPU and memory usage in FSOLAP compared to SQL queries. The average CPU and memory usage of queries during execution is proportionally similar, but the query execution time does not have the same trend. This is because the criteria for query types are determined by the amount of data the query retrieves and processes. As the number of restrictions in query types increased, query execution time decreased inversely.

Related studies on fuzzy SOLAP-based data mining and querying were investigated with regard to whether they have the following concepts or features: fuzziness, OLAP, SOLAP, data mining, inference, temporal querying, fuzzy querying, fuzzy spatial querying, fuzzy predictive querying, high visualization, easy use and performance evaluation. A system known as a fuzzy storage assignment system (FSAS) that provides fuzziness, OLAP, data mining, inference, and fuzzy querying based on fuzzy OLAP was proposed in the study by Lam et al. [15]. Their study was aimed at increasing the availability of decision support data and converting human knowledge into a system for tackling the storage location assignment problem. In another study, David et al. [18] researched fuzzy spatial data warehouses. They proposed a model that supports fuzziness, OLAP, SOLAP, data mining, inference, fuzzy querying, and fuzzy spatial querying. Their work represented a part of the Intelligent Geographical Project (IGP), which integrated fuzzy logic with spatial databases to help in the decision support and OLAP querying processes. Boutkhoum and Hanine [13] also developed software for complex decision-making problems. The software implementation was an integrated decision-making prototype based on an OLAP system and multicriteria analysis (MCA) to generate a hybrid analysis process dealing with complex multicriteria decision-making situations. Their proposal included fuzziness, OLAP, data mining, inference, temporal querying, and fuzzy querying. Ladner et al. [17] studied the use of fuzzy set approaches in spatial data mining to integrate their GIDB geospatial system. They presented an approach to discovering association rules for fuzzy spatial data where they were interested in correlations of spatially related data such as directional or geometric relationships of soil types. They combined and extended techniques developed in spatial and fuzzy data mining to deal with the uncertainty found in typical spatial data, supporting fuzziness, data mining, inference, fuzzy querying, and fuzzy spatial querying. FSOLAP and some related approaches in the literature are compared according to their concepts and characteristics in Table 8.

Table 8. Comparision of FSOLAP and existing approaches.

	FSAS [15]	IGP [18]	OLAP MCA [13]	GIDB [17]	FSOLAP
Fuzziness	✓	✓	✓	✓	✓
OLAP	✓	✓	✓		✓
SOLAP		✓		✓	✓
Data Mining	✓	✓	✓	✓	✓
Inference	✓	✓	✓	✓	✓
Temporal Querying			✓		✓
Fuzzy Querying	✓	✓	✓	✓	✓
Fuzzy Spatial Querying		✓		✓	✓
Fuzzy Predictive Querying					✓
High Visualization	✓		✓		
Performance Evaluation					✓
Easy to Use	✓		✓		

Although the FSOLAP framework brings together the strengths of fuzzy and SOLAP concepts for spatiotemporal applications and offers effective and efficient querying, it has difficulty in defining the expert rules in the representative application domain. As shown in the example queries, the expert-defined rules that the queries refer to must be defined in the system by domain experts. This situation makes it difficult for naïve users to use the framework without the help of a domain expert. Moreover, although FSOLAP provides some visualization, this functionality needs improvement as it is a spatiotemporal application. Future studies aimed at making the framework easy to use can be applied in this context. The realization of these studies would also make it possible to use this framework of analysis and inference in different fields such as agriculture, maritime transport, and others. For example, in the field of agriculture, a future study may develop an early warning system that can alert farmers by mapping the risk of frost.

5. Conclusions

This study proposed a framework based on fuzzy SOLAP (FSOLAP) to analyze fuzzy spatiotemporal data and make predictive analyses of various spatiotemporal events. To achieve this, fuzzy and SOLAP were harmonized to take advantage of the strengths of these two concepts. Moreover, an inference capability was added to the framework to support the predictive type of queries. In summary, some modifications of the SOLAP server and MDX queries were implemented, fuzzification operations were performed, association rules were generated, and pruning and weighting rules were applied to assemble the framework. Then, the performance of the framework was represented by non-spatial, spatial, spatiotemporal, and predictive fuzzy complex queries. We used a case study of a real database involving meteorological objects with specific spatial and temporal attributes. This study showed that the use of fuzzy concepts and SOLAP for spatiotemporal applications was effective and efficient, which was confirmed by both the implementation of query types and performance tests. Features provided by FSOLAP were compared with features in related works, and FSOLAP was shown to have a much broader functionality than the approaches used in similar studies in the literature. Making the framework easy to use for naïve users and enabling it to be utilized in other fields are suggested as avenues for future studies.

The main objective of this paper was to describe a generic fuzzy querying approach to process complex and flexible queries using the FSOLAP framework. We also aimed to manage uncertainty in spatiotemporal database applications when querying the database.

A real-life database that involves meteorological objects with certain spatial and temporal attributes was used as a case study. The proposed mechanism was implemented and several implementation issues that arose when querying the database were discussed.

This study used meteorological aspects and geographic data as spatiotemporal objects. Furthermore, the inference system in the fuzzy SOLAP environment integrated the model with a fuzzy inference system for allowing prediction over spatiotemporal data. As a result, a fuzzy spatiotemporal predictive query could be executed by using the framework.

Modeling and querying spatiotemporal data requires further research in future studies. The model and method presented in this study could be adjusted and/or extended to other fields of application such as agriculture, environment, etc. We implemented some of the fuzzy methods needed in this study, but the set of fuzzy methods should be further extended to different areas. This study implemented a generic fuzzy querying approach to process complex and fuzzy queries using our FSOLAP framework. In this context, the framework supports non-spatial and fuzzy spatial queries as well as fuzzy spatiotemporal query types. The processing of fuzzy aggregation queries and the corresponding algorithms may be studied in future work to explain the involvement of fuzzy spatial hierarchical relationships among members in the computation of the aggregation of numerical measures.

Author Contributions: Conceptualization, Sinan Keskin and Adnan Yazıcı; methodology, Sinan Keskin and Adnan Yazıcı; software, Sinan Keskin; validation, Sinan Keskin and Adnan Yazıcı; formal analysis, Sinan Keskin and Adnan Yazıcı; investigation, Sinan Keskin; resources, Sinan Keskin; data curation, Sinan Keskin; writing—original draft preparation, Sinan Keskin; writing—review and editing, Sinan Keskin and Adnan Yazıcı; visualization, Sinan Keskin; supervision, Adnan Yazıcı. All authors have read and agreed to the published version of the manuscript.

Funding: This research received no external funding.

Institutional Review Board Statement: Not applicable.

Informed Consent Statement: Not applicable.

Data Availability Statement: Not applicable.

Acknowledgments: The meteorological measurement data used in this study were collected to contribute to the Turkish State Meteorological Service. The source code of this study is available at https://github.com/skeskin19/solapfuzzyframework (accessed on 2 March 2022).

Conflicts of Interest: The authors declare no conflict of interest.

Abbreviations

The following abbreviations are used in this manuscript:

CPU	Central Processing Unit
ETL	Extract, Transform, and Load
FCM	Fuzzy c-Means
FIS	Fuzzy Inference System
FM	Fuzzy Module
FKB	Fuzzy Knowledge Base
FP	Frequent Pattern
JDBC	Java Database Connectivity
JSON	JavaScript Object Notation
MBR	Minimum Bounded Rectangle
MDX	Multidimensional Expression
OLAP	Online Analytical Processing
RPF	Rule Power Factor
SOLAP	Spatial Online Analytical Processing

SQL	Structured Query Language
QPr	Query Parser
QPc	Query Processor
QIn	Query Interface
UI	User Interface

References

1. Codd, E.F.; Codd, S.B.; Salley, C.T. *Providing OLAP (On-Line Analytical Processing) to User-Analysts, An IT Mandate*; Arbor Software Corp.: Santa Clara, CA, USA, 1993.
2. Kianmehr, K.; Kaya, M.; ElSheikh, A.M.; Jida, J.; Alhajj, R. Fuzzy association rule mining framework and its application to effective fuzzy associative classification. In *Wiley Interdisciplinary Reviews: Data Mining and Knowledge Discovery*; Wiley: Hoboken, NJ, USA, 2011; pp. 477–495.
3. Rivest, S.; Bédard, Y.; Proulx, M.J.; Nadeau, M. SOLAP: A new type of user interface to support spatio-temporal multidimensional data exploration and analysis. In Proceedings of the 2003 Workshop ISPRS, Quebec, QC, Canada, 2–3 October 2003.
4. Han, J. Towards on-Line Analytical Mining in Large Databases. *ACM Sigmod Rec.* **1998**, *27*, 97–107. [CrossRef]
5. Huang, Y.P.; Kao, L.J.; Sandnes, F.E. Predicting Ocean Salinity and Temperature Variations Using Data Mining and Fuzzy Inference. *Int. J. Fuzzy Syst.* **2007**, *9*, 3–9.
6. Sivaramakrishnan, T.R.; Meganathan, S. Association Rule Mining and Classifier Approach for Quantitative Spot Rainfall Prediction. *J. Theor. Appl. Inf. Technol.* **2011**, *34*, 173–177.
7. Stell, J.G. Part and Complement: Fundamental Concepts in Spatial Relations. In *Annals of Mathematics and Artificial Intelligence*; Kluwer Academic Publishers: Alphen aan den Rijn, The Netherlands, 2004; pp. 1–17.
8. Cheng, T.; Molenaar, M.; Lin, H. Formalizing fuzzy objects from uncertain classification results. *Int. J. Geogr. Inf. Sci.* **2001**, *15*, 27–42. [CrossRef]
9. Fisher, P.; Arnot, C.; Wadsworth, R.; Wellens, J. Detecting change in vague interpretations of landscapes. *Ecol. Inform.* **2006**, *1*, 163–178. [CrossRef]
10. Plewe, B. The Nature of Uncertainty in Historical Geographic Information. *Trans. GIS* **2002**, *6*, 431–456. [CrossRef]
11. Bordogna, G.; Chiesa, S.; Geneletti, D. Linguistic modelling of imperfect spatial information as a basis for simplifying spatial analysis. *Inf. Sci.* **2006**, *176*, 366–389. [CrossRef]
12. Pavan Kumar, K.V.N.N.; Radha Krishna, P.; Kumar De, S. Fuzzy OLAP Cube for Qualitative Analysis. In Proceedings of the 2005 International Conference on Intelligent Sensing and Information Processing, Chennai, India, 4–7 January 2005.
13. Boutkhoum, O.; Hanine, M. An integrated decision-making prototype based on OLAP systems and multicriteria analysis for complex decision-making problems. *Appl. Inform.* **2017**, *4*, 1. [CrossRef]
14. Molina, C.; Prados-Suárez, B.; de Reyes, M.A.P.; Yáñez, M.C.P. Improving the Understandability of OLAP Queries by Semantic Interpretations. In *Flexible Query Answering Systems*; Springer Publishing House: Heidelberg/Berlin, Germany, 2013; pp. 176–185.
15. Lam, C.H.Y.; Chung, S.H.; Lee, C.K.M.; Ho, G.T.S.; Yip, T.K.T. Development of an OLAP Based Fuzzy Logic System for Supporting Put Away Decision. *Int. J. Eng. Bus. Manag.* **2009**, *1*, 1–13. [CrossRef]
16. Duraciova, R.; Chalachanova, J.F. Fuzzy Spatio-Temporal Querying the PostgreSQL-PostGIS Database for Multiple Criteria Decision Making. In *Lecture Notes in Geoinformation and Cartography Dynamics in GIscience*; Springer: Berlin/Heidelberg, Germany, 2017; pp. 81–97.
17. Ladner, R.; Petry, F.E.; Cobb, M.A. Fuzzy Set Approaches to Spatial Data Mining of Association Rules. *Trans. GIS* **2003**, *7*, 123–138. [CrossRef]
18. David, P.; Maria, S.; Ivo, P. Fuzzy Spatial Data Warehouse: A Multidimensional Model. In *Decision Support Systems Advances in*; Devlin, G., Ed.; InTech Publishing House: Rijeka, Croatia, 2010; pp. 57–66.
19. Zheng, K.; Zhou, X.; Fung, P.C.; Xie, K. Spatial Query Processing for Fuzzy Objects. *Vldb. J.* **2012**, *21*, 729–751. [CrossRef]
20. Nurain, N.; Ali, M.E.; Hashem, T.; Tanin, E. Group Nearest Neighbor Queries for Fuzzy Geo-Spatial Objects. In Proceedings of the Second International ACM Workshop on Managing and Mining Enriched Geo-Spatial Data, Melbourne, VIC, Australia, 31 May 2015.
21. Sözer, A.; Oğuztüzün, H.; Petry, F.E. Querying Fuzzy Spatiotemporal Databases: Implementation Issues. In *Uncertainty Approaches for Spatial Data Modeling and Processing Studies in Computational Intelligence*; Springer: Berlin/Heidelberg, Germany, 2010; pp. 97–116.
22. Roman, R.C.; Precup, R.E.; Petriu, E.M. Hybrid data-driven fuzzy active disturbance rejection control for tower crane systems. *Eur. J. Control* **2021**, *58*, 373–387. [CrossRef]
23. Zhu, Z.; Pan, Y.; Zhou, Q.; Lu, C. Event-Triggered Adaptive Fuzzy Control for Stochastic Nonlinear Systems with Unmeasured States and Unknown Backlash-Like Hysteresis. *IEEE Trans. Fuzzy Syst.* **2021**, *29*, 1273–1283. [CrossRef]
24. Yang, H.; Cobb, M.; Shaw, K. A Clips-Based Implementation for Querying Binary Spatial Relationships. In Proceedings of the Joint 9th IFSA World Congress and 20th NAFIPS International Conference, Vancouver, BC, Canada, 25–28 July 2001.
25. Taldmhi, E.; Shima, N.; Kishino, F. An Image Retrieval Method Using Inquires on Spatial Relationships. *J. Inf. Process.* **1992**, *15*, 441–449.

26. Messaoud, R.B.; Boussaid, O.; Rabaseda, S. Mining Association Rules in OLAP Cubes. In Proceedings of the 2006 Innovations in Information Technology, Dubai, United Arab Emirates, 19–21 November 2006.
27. Agrawal, R.; Imieliński, T.; Swami, A. Mining association rules between sets of items in large databases. In Proceedings of the 1993 ACM SIGMOD International Conference on Management of Data—SIGMOD 93, Washington, DC, USA, 26–28 May 1993; pp. 207–216.
28. Schneider, M. A Design of Topological Predicates for Complex Crisp and Fuzzy Regions. In Proceedings of the 20 th International Conference on Conceptual Modeling, Yokohama, Japan, 27–30 November 2001; pp. 103–116.
29. Tang, X.; Fang, Y.; Kainz, W. Fuzzy Topological Relations Between Fuzzy Spatial Objects. In *Fuzzy Systems and Knowledge Discovery Lecture Notes in Computer Science*; Springer: Berlin/Heidelberg, Germany, 2006; pp. 324–333.
30. Zhan, F.B.; Lin, H. Overlay of Two Simple Polygons with Indeterminate Boundaries. *Trans. GIS* **2003**, *7*, 67–81. [CrossRef]
31. Winter, S. Topological Relations between Discrete Regions. In Proceedings of the Fourth Symposium on Large Spatial Databases SSD'95, Portland, ME, USA, 6–9 August 1995; pp. 310–327.
32. Cobb, M.A. Modeling Spatial Relationships within a Fuzzy Framework. *J. Am. Soc. Inf. Sci.* **1998**, *49*, 253–266. [CrossRef]
33. Laurent, A. Querying Fuzzy Multidimensional Databases: Unary Operators and their Properties. *Int. J. Uncertain. Fuzziness Knowl.-Based Syst.* **2003**, *11*, 31–45. [CrossRef]
34. Keskin, S.; Yazici, A.; Oğuztüzün, H. Implementation of X-Tree with 3D Spatial Index and Fuzzy Secondary Index. In Proceedings of the Flexible Query Answering Systems Lecture Notes in Computer Science, Ghent, Belgium, 26–28 October 2011; pp. 72–83.
35. Beckmann, N.; Kriegel, H.; Schneider, R.; Seeger, B. The R*-Tree: An Efficient and Robust Access Method for Points and Rectangles. In Proceedings of the 1990 ACM SIGMOD International Conference on Management of Data—SIGMOD 90, Atlantic City, NJ, USA, 23–26 May 1990.
36. Berchtold, S.; Keim, D.A.; Kriegel, H.-P. The X-tree: An Index Structure for High-Dimensional Data. In Proceedings of the 22th International Conference on Very Large Data Bases (VLDB '96), Mumbai, India, 3–6 September 1996; Morgan Kaufmann Publishers Inc.: San Francisco, CA, USA, 1996; pp. 28–39.
37. Pelleg, D.; Moore, A. X-Means: Extending K-Means with Efficient Estimation of the Number of Clusters. In Proceedings of the 17th International Conference on Machine Learning, San Francisco, CA, USA, 29 June–2 July 2000; pp. 727–734.
38. Dunn, J.C. A Fuzzy Relative of the ISODATA Process and Its Use in Detecting Compact Well-Separated Clusters. *J. Cybern.* **1973**, *3*, 32–57. [CrossRef]
39. Bezdek, J. A convergence theorem for the fuzzy ISODATA clustering algorithms. *IEEE Trans. Pattern Anal. Machine Intell.* **1980**, *PAMI-2*, 1–8. [CrossRef]
40. Soni, H.K.; Sharma, S.; Jain, M. Frequent pattern generation algorithms for Association Rule Mining: Strength and challenges. In Proceedings of the 2016 International Conference on Electrical, Electronics, and Optimization Techniques (ICEEOT), Chennai, India, 3–5 March 2016.
41. Pach, F.P.; Gyenesei, A.; Németh, S.; Orava, P.; Abonyi, J. Fuzzy association rule mining is a historical process for data analysis. *Acta Agrar. Kaposváriensis* **2006**, *10*, 89–107.
42. Ochin, S.; Joshi, N. *Rule Power Factor: A New Interest Measure in Associative Classification, Procedia Computer Science*; Elsevier B.V.: Amsterdam, The Netherlands, 2016; pp. 12–18.
43. Marutho, D.; Handaka, S.H.; Wijaya, E.; Muljono. The Determination of Cluster Number at k-Mean Using Elbow Method and Purity Evaluation on Headline News. In Proceedings of the 2018 International Seminar on Application for Technology of Information and Communication, Semarang, Indonesia, 21–22 September 2018.
44. Rousseeuw, P.J. Silhouettes: A graphical aid to the interpretation and validation of cluster analysis. *J. Comput. Appl. Math.* **1987**, *20*, 53–65. [CrossRef]
45. GeoMondrian SOLAP Server. Available online: http://www.spatialytics.org/blog/geomondrian-1-0-is-available-for-download (accessed on 14 December 2020).
46. Spofford, G.; Harinath, S.; Webb, C.; Huang, D.H.; Civardi, F. *MDX-Solutions*, 2nd ed.; Wiley: Hoboken, NJ, USA, 2006; pp. 1–35.

International Journal of *Geo-Information*

Article

Implicit, Formal, and Powerful Semantics in Geoinformation

Gloria Bordogna , Cristiano Fugazza *, Paolo Tagliolato Acquaviva d'Aragona and Paola Carrara

Institute for Electromagnetic Sensing of the Environment, National Research Council of Italy, via Bassini 15, I20133 Milan, Italy; bordogna.g@irea.cnr.it (G.B.); tagliolato.p@irea.cnr.it (P.T.A.d.); carrara.p@irea.cnr.it (P.C.)
* Correspondence: fugazza.c@irea.cnr.it

Abstract: Distinct, alternative forms of geosemantics, whose classification is often ill-defined, emerge in the management of geospatial information. This paper proposes a workflow to identify patterns in the different practices and methods dealing with geoinformation. From a meta-review of the state of the art in geosemantics, this paper first pinpoints "keywords" representing key concepts, challenges, methods, and technologies. Then, we illustrate several case studies, following the categorization into implicit, formal, and powerful (i.e., soft) semantics depending on the kind of their input. Finally, we associate the case studies with the previously identified keywords and compute their similarities in order to ascertain if distinguishing methodologies, techniques, and challenges can be related to the three distinct forms of semantics. The outcomes of the analysis sheds some light on the diverse methods and technologies that are more suited to model and deal with specific forms of geosemantics.

Keywords: geosemantics; implicit semantics; formal semantics; powerful semantics

Citation: Bordogna, G.; Fugazza, C.; Tagliolato Acquaviva d'Aragona, P.; Carrara, P. Implicit, Formal, and Powerful Semantics in Geoinformation. *ISPRS Int. J. Geo-Inf.* **2021**, *10*, 330. https://doi.org/10.3390/ijgi10050330

Academic Editor: Wolfgang Kainz

Received: 12 March 2021
Accepted: 1 May 2021
Published: 13 May 2021

1. Introduction

Semantics is cornerstone in state-of-the-art data management, notwithstanding the specific domain; without semantics, we would helplessly drown in a deluge of unintelligible Big Data. Let aside the enormous literature on this topic in the field of Linguistics and, even before that, in Philosophy, representing and managing semantics is frequently regarded to as the solution to heterogeneity in data retrieval and exploitation in Computer Science (CS) [1–3]. This paper relates to a specific domain in the landscape of semantics-aware CS, i.e., geospatial information provided in the form of both data and metadata. This is a particularly challenging domain as the non-textual nature of most geospatial data means that the indexing practices of generalist search engines are ineffective; hence the need for semantics representation and management.

Both Sheth et al. [4] and Uschold [5] provide a coarse-grained categorization of semantics; the latter includes the following four categories: (i) implicit semantics, (ii) informally expressed semantics, (iii) formally expressed semantics for human consumption, and (iv) formally expressed semantics for machine processing. In practice, the first three levels fall in the first category defined in [4], which proposes the following classification:

Implicit semantics is the semantics not explicitly represented, i.e., not directly usable by machines to derive new knowledge.

Formal semantics when semantics is represented in some sort of formalism, in order to be machine readable and processable, e.g., in the form of ontologies.

Powerful (soft) semantics when semantics is represented in forms that enable overcoming crisp set-based formalisms, allowing representing degrees of memberships and certainty, e.g., by using fuzzy approaches and contextual time-varying semantics.

In our opinion, the second classification not only includes the first one but, at the same time, empowers the fourth level of the first by offering two distinct representation classifications, opening towards methods that better mimic the human soft and flexible approaches to reasoning and decision making. This is the main motivation for adopting this second classification method, exporting its concepts in the geospatial domain and reflecting them in the forthcoming Sections of this paper. Albeit there is apparently a broad spectrum of technologies that fall under the umbrella of each of these categories (in fact, Almeida et al. [6] elaborate on the notion of *semantic continuum*), we will discuss their common traits.

As regards high level categorization of the forms of semantics, Gärdenfors [7] distinguishes between "symbolic", "associationist", and "conceptual", providing the latter with a spatial characterization. His *cognitive spaces* feature interesting analogies with notions that are typical of the geospatial domain (e.g., spatial intersection). Still, non-symbolic approaches are mostly contained in the category of implicit semantics according to the classification by Sheth.

The ultimate purpose of this work is to provide the reader with awareness of directions on the main issues, challenges, and possible solutions to address the different categories of semantics defined by Sheth in the domain of geoinformation. In a nutshell, this paper outlines which technologies are more appropriate to consider when tackling a given research problem. The importance of this topic for the geospatial community is attested by the increasing relevance of semantics as the "glue" between heterogeneous thematic domains and also across their individual workflows. On the one hand, inter-disciplinary interoperability requires mapping of the individual terminologies used for annotating data. On the other, effective discovery and provision of geospatial data requires fine-grained characterization of resources (i.e., semantic metadata) not only for data, but also for services, APIs, instruments, data providers, etc.

The hypothesis behind our work is that Sheth's three forms of semantics are also reflected in the geosemantics context. The objective of our paper is then to identify technologies, methodologies, challenges, and solutions that are distinctive for the implicit, the formal, and the powerful geosemantics in order to orient the reader in problem solving. To achieve this, by analyzing recent reviews and editorial papers on geosemantics, we first mine which are the main technologies, methodologies, research challenges, and solutions presented by the authors, regarding them as keywords (Section 2.3).

Successively, we perform a two-step analysis by first discussing selected case studies involving the management of implicit, formal, and powerful geosemantics. The choice of the case studies has been performed by taking into account both their belonging to one of the semantic categories of Sheth (depending on the characteristics of their inputs) and the variety and representativeness of application domains as outlined in [8]. Specifically, the varied and most representative applications to which geomatics can be put include urban planning, disaster management, assessment of biodiversity, and land administration. We then associate the keywords with the case studies and assess whether Sheth's categories are characterized by distinguishing keywords, i.e., specific methods, technologies and solutions, thus allowing for a more distinctive clustering of the keywords with respect to what emerged from the metareview in Section 2.3.

A contribution of this paper is also the methodological workflow we followed in order to characterize the forms of semantics in geoinformation with their preferred/elective approaches.

2. Materials and Methods

This Section is organized as follows: Section 2.1 details our aim and the workflow we followed to confirm our hypothesis. Section 2.2 explains the categorization of semantics in the main reference work [4] inspiring this paper; then, Section 2.3 presents a meta-analysis of the literature on geosemantics as discussed in recent surveys and review papers. Sections 2.4–2.6 present the case studies we selected according to the criteria expressed above.

2.1. Workflow

In this work, we aim at investigating whether the three forms of semantics by Sheth et al. [4] can be related to distinguishing methodologies, techniques, and knowledge sources among those found in the literature on geospatial information. This is by no means a foregone conclusion and these distinguishing methodologies may not be the same as in other contexts. In fact, the geospatial domain sometimes diverges from current trends because of its specificities (e.g., proposing service-oriented architectures as opposed to resource-oriented ones).

To this aim, we define the workflow whose main phases are depicted in Figure 1: Top-left, a meta-review of recent surveys of papers illustrating applications of geospatial information management is performed (Section 2.3). The meta-review allows for identifying topics, research challenges, and solutions; these are considered to be keywords and represented on the right side of Figure 2. On the top-right hand side, assuming as starting point of our analysis the aforementioned three forms of semantics (whose definitions are clarified in Section 2.2), we select and analyze several case studies, categorizing them according to these three forms of semantics on the basis of the characteristics of their inputs (Sections 2.4–2.6).

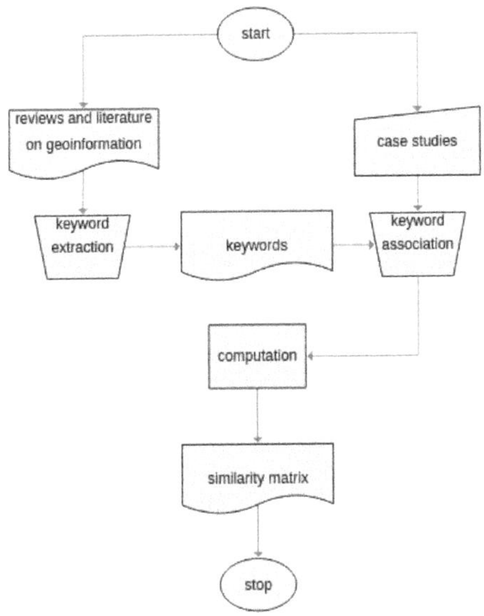

Figure 1. Depiction of the workflow followed.

Finally, in order to substantiate the hypothesis behind this work—that Sheth's categories are also reflected in the geosemantics context—the results yielded by the two previous independent phases are cross-referenced in order to compute a similarity matrix on the basis of the keywords associated with the case studies. Specifically, this is achieved by verifying that the intra-similarities (similarity degrees between pairs of case studies belonging to the same form of geosemantics) are greater than the inter-similarity degrees between pairs of study cases classified as different forms of geosemantics. The greater the intra-similarity with respect to the inter-similarity, the more distinctive the methods and technologies characterizing the three forms of geosemantics.

2.2. Three Shades of Semantics

Looking at geosemantics through the lenses proposed by Sheth allows for categorizing in a minimal set of classes the broad (and ever-growing) landscape of topics (comprising both methods and technologies) that populate this domain. Otherwise, the implications of information source heterogeneity (as far as genre and nature are concerned), information multidimensionality, and domain knowledge dependency easily yield a multiplicity of classes that configures a semantic continuum *à la* Almeida [6]. Since data source heterogeneity, cross-domain interaction, data/process imperfection, and big data volumes are common traits in the geospatial domain, distinguishing between *implicit*, *formal*, and *powerful* semantics allows us to divide the presented case studies in three categories with a clear solution of continuity.

Implicit semantics refers to the kind that is implicit in data and that is not represented explicitly in any machine-processable syntax. It is typically related to concepts and relationships between them that are not represented in a formal way but are embedded in multimedia documents, i.e., their "meaning is conveyed based on a shared understanding derived from human consensus" [5]. These can be natural language documents, multi-spectral images, time series of measurements, video frame sequences, audio recordings, undocumented tabular data, etc. The main objective of extracting implicit semantics is to cope with the inherent ambiguity characterizing it. In fact, terms in texts, visual aspects in images, etc., can mean different things depending on both context and knowledge of people [9]. It should be noted that *implicit* does not mean missing a knowledge-based underpinning but that the latter does not (or cannot) be given a formal representation, such as in the assessment by a domain expert.

In more general terms, we can state that semantics that are represented in some well-formed syntax (governed by syntax rules) is referred to as formal semantics. In 2001, Berners-Lee et al. [10] stated that "The Semantic Web is an extension of the current web in which information is given well-defined meaning, better enabling computers and people to work in cooperation". As such, the Semantic Web (SW) is the most apparent embodiment of semantics in the field of Internet-mediated contents and applications. Here, the inflection we give to this term is that of *formal* semantics, specifically those provided by decidable fragments of First-Order Logic (FOL) [11]. In fact, "formal" is the category name that Sheth et al. give to this kind of semantics [4], analogous to "formal semantics for machine-processing" in Uschold's categorization [5]. Explicit representations of formal semantics include knowledge graphs, ontologies, and the like.

Finally, Sheth et al. introduce the concept of *powerful* semantics, intended as formal semantics which is empowered with the ability to represent not only precise and well-defined concepts and relationships, but also imprecise and uncertain concepts and gradual relationships, whose meaning can be subjective, vague, and variable depending on several contextual conditions [12]. The ability of formal frameworks to represent and manage powerful semantics is indeed aimed at performing approximate and qualitative reasoning in order to discover implicit concepts and relationships, possibly uncertain and imprecise too, although accurate enough to be useful to solve some needed task.

2.3. A Meta-Analysis Perspective

Timothy Tambassi, in his Preface to the book "The Philosophy of GIS" [13], pointed out that the literature on GIS is heterogeneous and scattered, primarily because of the multiple branches of knowledge that use, manage, and create geographic information. This is also true for geosemantics, whose literature configures a conceptual 'forest' of issues, topics, technologies, methodologies, challenges, and solutions where it is easy to loose orientation. To frame approaches in the field of geosemantics, we have taken into account some stimulating overview papers on this subject which appeared in the last decade and tried to examine and categorize the topics, research challenges, and solutions described. It is a meta-analysis exercise that considered the papers described in the following.

Kokla et al. [14] offers a comprehensive review of the contributions that represent a progress in geospatial semantics since 2015; it focuses around two main topics, i.e., information modeling (ontologies and their development) and (latent) knowledge elicitation (from unstructured or semi-structured content, based in particular on textual contents). This paper reviews more than 150 works; among them are papers that present categorizations of methods and approaches to geosemantics, such as [15–19]. Other cited contributions report on the efforts for describing the methods at hand: [20–32]. Furthermore, in this review the reader can find many works that exemplify the former within a great number of applications; among these [33–55].

Hu [56] provides an overview and a review of important contributions dealing with six major research areas in geospatial semantics, i.e., "semantic interoperability and ontologies" [16,24,38,57–70] , "digital gazetteers" [71–86] , "Geographic Information Retrieval" [32,87–106], "geospatial Semantic Web and Linked Data" [43,107–114], "place semantics" [47,115–121], "cognitive geographic concepts and qualitative reasoning" [70,119–124].

Janowicz et al. [125] is a rich overview of the geosemantics landscape focusing on some selected topics that the authors deem of particular interest; the contributions reviewed are organized according to these. With respect to the question on what kinds of Geospatial Classes should be distinguished, they cite [16–18,63,65,66,68,126–129]; instead, the question on how to reference Geospatial Phenomena is supported by [113,123,130–132]. Discovering events and accounting for geographic change are faced and fostered in [133–136], Handling places and moving object trajectories is dealt with in [70,77,90,133,137–143]. The following papers are cited with reference to comparison, alignment, and translation of Geospatial Classes [15,69,90,144–149]. Finally, the issues raised by processing, publishing, and retrieving geodata are tackled by in [150–156].

The approach changes in [157]: Rather than reviewing papers dealing with projects and issues related to geosemantics, it reviews ideas rooted in cognitive science and linguistics for sketching their application to semantics of geographic information. It discusses notions from 1990 to 2010 and shows why and how these ideas have been productive for dealing with semantics.

We also considered a couple of papers that are not strictly reviews but, in our opinion, are worth being included as they offer a landscape of trends and contributions in geosemantics. Janowicz et al. [112], an editorial paper on the Semantic Web, outlines the research field of geospatial semantics, highlights major research directions and trends, and takes a glance at future challenges. Another editorial paper [158], considers VGI (Voluntary Geographic Information) and claims that geospatial Linked Data and Knowledge Graphs, when used for implementing intelligent data search, can result in precise data-sharing services.

The less recent work we considered is [159], where the author observes that the main approaches to overcome semantic heterogeneity rely on ontologies that, having a priori definitions, are decontextualized. On the contrary, he affirms that semantics reconciliation needs to take into account context-based meanings. Since "meaning and context are dynamically emergent from activity and interaction, determined in the moment and in the doing". He further highlights the limitations of representational approaches. In fact, the latter assume that context is stable, delimited information that can be known and encoded in just another information layer or another ontology in an information system. These are the reasons why this work encourages non-representational modelling formalisms to cope with semantic interoperability in sharing and integrating geographic information.

By analyzing the above overviews, we have extracted a list of terms that the authors pinpointed as topics of interest, research challenges, or solutions, which we regard as keywords. The correspondence between the keywords and the respective originating reviews can be found in supplementary material. The keywords are listed on the right side of the diagram in Figure 2. The list is wide enough to suggest how large is the playground offered by geosemantics.

Still, this list may be biased, being based on authors' views and reviews of a rapidly evolving literature, and some terms can have overlapping meanings. For instance, more recent reviews, such as Kokla et al. [14], produced an increase in this term list, due to the emergence of mobile and social applications, IoT, AI, etc. in the last five years. These research fields introduced novel concepts, such as lightweight ontologies. This increase is also due to the paradigm shift, dating back in 2012 [20], from the general-purpose Web to communities and their specific perspectives, pushed in turn by the movement of Critical GIS [160]. With reference to the notion of Digital Earth, in [161] the authors solicited "a network of theories that fosters interoperability without giving up on semantic heterogeneity". As such, it is possible that more recent works may further populate the list in Figure 2.

In the papers we examined, the authors suggested a grouping of these keywords according to some categories, listed on the left side of the diagram. Some keywords can be related to multiple categories as they can be good suggestions in diverse application scenarios. As an example, term "gazetteers" has been presented in some works as dealing with either "geospatial Semantic Web" or "elicitation of semantic information"; "domain ontologies" have been used in works coping with "geo-semantics formalization" and "semantic interoperability". On the other hand, there are categories that can be tackled with multiple strategies; for example, geosemantics issues falling under category "cognitive geographic concepts" have been dealt with in projects on either "events-change discovery", "place-based GIS", or "qualitative reasoning".

Figure 2 makes it apparent that the categories on the left are not associated with distinguishing topics and solutions on the right, i.e., the reviews did not succeed in letting patterns emerge in the geosemantics "forest", thus making order in the diverse practices.

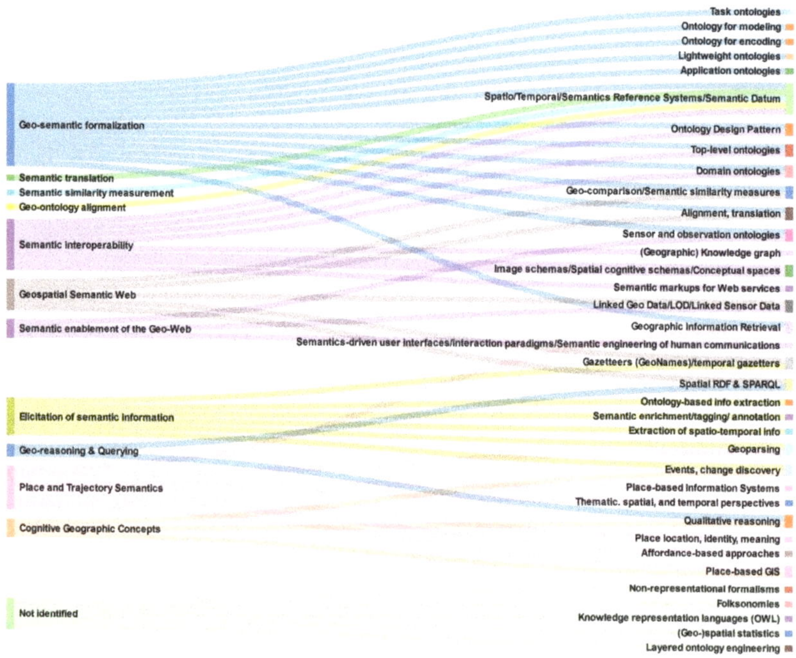

Figure 2. Diagram connecting keywords in geosemantics (**right**) and their categorization (**left**), as found in the reviews taken into consideration.

2.4. Implicit Geosemantics

In [14], the extraction of implicit *geo*semantics is named "elicitation of semantic information". Under this interpretation, the term is used in a broader sense to encompass processes aimed to make latent knowledge explicit from unstructured or semi-structured contents. These processes focus on eliciting a structured representation of information in various forms, such as semantic metadata, links to ontology concepts, collections of topics, geotagged maps and images, etc. Sources of implicit geosemantics are multimedia documents, in the form of unstructured and semi-structured textual documents, pictures taken from cameras, images from remote sensing, audio and video files. In most cases, metadata are available but are generally insufficient to representing and understanding the contents.

Typically, unstructured texts, posts in social networks, and news streams may refer to geographic names into their contents to describe events, points of interest (POIs), and places. The discipline that extracts geographic contents from unstructured and semistructured texts in order to index them and enable the evaluation of both content and spatial queries is Geographic Information Retrieval (GIR) [87]. Images are another potential source of geosemantic information. Photos may depict geographic places without explicitly mentioning their name or geolocation. With regard to video files, we can consider TV news reporting events relative to specific geographic areas. Finally, remote sensing images may contain representations of the status of the environment with respect to the occurrence of geo-temporal phenomena and events going on in a given area. The segmentation of images in order to extract geographic footprints of places and events can be performed by applying spatio-temporal analysis. The latter is primarily based on (i) domain experts' knowledge; (ii) statistical and machine learning approaches, or (iii) hybrid approaches combining the previous two [162].

Some important challenges of implicit geosemantics extraction within multimedia documents are related to three main objectives:

(i) reconciling the place and space conceptualizations of geosemantics: while the "platial" (based on place) perspective is usually defined within texts by textual place names, linguistic descriptions, and the semantic relationships between places, the spatial perspective typical of georeferenced maps explicitly represents the geometries by their coordinates, distances, topology, and directions, but mostly lack descriptions of their meanings. This reconciliation from platial to spatial and vice-versa requires modeling uncertainty of the recognition process;
(ii) increasing human perception of the semantics of geoinformation by considering users' spatial, temporal, and content needs and preferences. This amounts to identifying and summarizing geographic contents on the basis of distinct spatial, temporal, and content granularities;
(iii) enhancing interoperability of the geoinformation semantics representation in order to be able to re-use it within different contexts and applications. This is achieved by adopting standards and domain/task/application ontologies.

Basically, artificial intelligence approaches comprising different methodologies (such as soft computing, clustering, genetic algorithms, geostatistic analysis, neural networks, support vector machines, and the like) are applied to extract implicit semantics from multimedia documents. Knowledge bases are used to support the analysis: These may take the form of gazetteers, DBpedia (https://wiki.dbpedia.org/ accessed on 1 April 2021), generic and domain thesauri such as WordNet (https://wordnet.princeton.edu/ accessed on 1 April 2021), geo ontologies, and thematic geospatial information. In the following, we present some case studies focused on to the above challenges which consider different genres of geographic contents (basically, objects, events and moving objects' trajectories) within distinct categories of multimedia documents (textual documents and social media posts).

A synoptic view of the four case studies dealing with an implicit form of semantics is reported in Table 1: Besides the identifier of the case study, its acronym, and a brief description, the table reports the type of input, the method it applies, the type of generated output, and its potential use. It can be noticed that the type of input is either unstructured textual documents or social media documents, a kind of data that typically contain the implicit form of semantics. It can be also noticed that the outputs contain more explicit geosemantics, constituted by geofootprints of documents, spatio-temporal clusters of events, trajectories, and georeferenced placenames.

As for the application domains that are covered by the case studies, that in Section 2.4.1 is related to retrieval of georeferenced information, providing urban planners with effective means for mining knowledge of territorial resources. The case study in

Section 2.4.2 performs trajectory mining to support mobility planning for tourists. The case study in Section 2.4.4 sows that disaster management can be fostered by timely event detection and, finally, the case study in Section 2.4.4 is about geo-gazetteer creation from VGI, in support of land administration.

2.4.1. From "Place" to "Space": Representing Uncertainty of Geoinformation within Texts to Support Geographic Information Retrieval

In [163], a GIR system was proposed that allows for extracting implicit geosemantics within contents of textual documents through the identification of *fuzzy geographic footprints*, i.e., the distinct locations on Earth referred to by documents.

The GIR model applies soft computing methods; specifically, the evaluation of multiple bipolar criteria [164,165] aggregated based on a p-norm operator [166] to extract the fuzzy footprints of documents representing their geographic focus. In a nutshell, some criteria have a positive influence on the selection of geographic names within the text as footprints of a document (for example, when the initial characters of the term is a capital letter, when the term occurrence is close to positive anchor terms such as "street", "city", "nation", etc.). Others have a negative influence (for example when the term is preceded by negative anchor terms such as "Sir", "Mr", "Mrs", etc.).

The prototypical system, has the classic structure of an Information Retrieval System (IRS) [163], consisting of two main components: the *Indexing Module* and the *Retrieval Module*. The Indexing Module has two main sub-modules: the *Full-Text Indexing* and the *GeoIndexing* sub-modules. The former performs full text indexing of the documents to represent their significant contents, and generates the textual inverted index to enable content based searches. Instead, the GeoIndexing sub-module identifies the fuzzy footprints of documents by the support of a knowledge base that comprises both a geo-ontology and a rule-base that encodes the heuristic knowledge required to cope with geo/non-geo ambiguities during geoparsing, and with geo/geo ambiguities during geocoding. An example of geo/non-geo ambiguity is the case of a place name having also a non geographic meaning such as "Nice" (France), "Crema, Brindisi" (Italy), and "Of" (Turkey). Instead, geo/geo ambiguities are due to distinct locations on Earth having the same place name, such as Rome, Paris, London, etc.). The disambiguation rules take into account both the geographic context, based on the shared assumption that "close places are more closely related than far places", and the textual context, based on the consideration that distinct geographic names appearing close in text are also closely related in geographic space. This way, place names within documents are associated with a fuzzy footprint in the geographic space, thus reconciling the two conceptualizations of geosemantics and enabling both content and spatial searchers.

Table 1. Dimensions of case studies (Implicit geosemantics).

Subsection	Name	Description	Input Data	Method Classification	Output Data	Potential Use
Section 2.4.1	GeoFinder	GIR extracting and modeling uncertainty of geofootprints in textual documents	textual documents from etherogeneous sources (tested on a collection of articles about Energy power stations and on CLEF2008 collection)	fuzzy computational intelligence	uncertain geofootprints of textual documents, i.e., fuzzy sets of geographic points with uncertainty degrees in [0,1] associated with the index terms of textual documents	Performing spatial queries (expressing metrical and topological conditions) in combination with content-based queries on textual collections
Section 2.4.2	Events spatio-temporal footprint	footprint detection of events' popularity (from Twitter)	semi-structured text (Twitter) messages with explicit geotags	unsupervised learning	clusters of punctual geo-temporal footprints of an event or topic identified by a set of keywords in a given time-lapse	geo temporal analysis of events reported in social networks
Section 2.4.3	Tour miner	mining popular tourists' tours (from Twitter)	semi-structured text (Twitter) messages with explicit geotags	knowledge-based semi-supervised learning	popular tours identified by a hierarchy of clusters containing sets of "close" paths (a path being an ordered lists of geographic entity names)	geo temporal analysis of tourists' mobility based on social network messages
Section 2.4.4	Eliciting Geographic Gazetteer	Extracting place names and their footprints from social networks	images, captions and metadata from Flickr	unsupervised learning	geographic gazetteer of place names	updating or creating geographic gazetteer

2.4.2. Detecting Periodic/Episodic Events from Social Networks with Desired Spatio-Temporal Granularity

The paper by [167] proposes an approach to discover events of interest from social media by modeling the distinct spatio-temporal granularity. The main characteristic of this study is flexibility in detecting events characterized by either an hypothetical periodic or episodic timestamp, thus allowing confirming a priori knowledge of their possible geotemporal regularities. Given a set of sources of spatio-temporal information, such as Twitter, the methodology first performs a focused crawling of the selected social media contents to collect candidate messages related to an event of interest; successively, the collected messages are analyzed by means of an original, density-based spatio-temporal clustering algorithm. The latter is defined by extending the DBSCAN algorithm to group messages densely located in the spatio-temporal domain. Its output is a set of spatio-temporal clusters with arbitrary shapes: these identify the areas on Earth where an event matching the keywords (i.e., the parameters used to filter the messages) occurred within a given time span, possibly with a given periodicity.

The exploration is interactive and multi-granular, allowing analysts to customize not only the topics of interest, i.e., the category, but also the time period and the spatial density so as to fit different spatio-temporal scales. One can specify (i) a set of keywords of interest to filter the messages about an event or a topic (e.g., traffic jam, hurricane, landslide, football match), (ii) the desired granularity of the time period of analysis (such as each day, month, year) and (iii) the desired spatial granularity needed to form a cluster, defined by spatio-temporal density of messages. Each cluster generated by the algorithm can be identified by the list of the most representative keywords that were found in the messages of the cluster, thus representing the cluster's semantics. The use of thesauri [168] helps identify the more general terms expressing the meaning of the specific terms found in individual messages of the cluster. As far as the representation of the geographic footprint of each cluster is concerned, a convex hull can be computed from the geographic coordinates of the messages in each cluster to obtain a polygon representation of the geo-footprint.

2.4.3. Discovering and Summarizing Moving Object Trajectories from Twitter

The work described in [169] proposes an approach to identify, track, and analyze popular tours of tourists visiting a Region Of Interest (ROI) based on the Tweets they publish.

The solution is constituted by two main suites of tools: the FollowMe suite for tourist identification and tracking and the TripsAnalysis suite for popular tour mining.

The FollowMe suite allows users to submit spatial queries to the Twitter API to find *hang* tweets, i.e., tweets posted in the area of the monitored airports. For each user identified by means of hang tweets, the FollowMe suite queries (through the Twitter API) his/her timeline, i.e., the history of tweets posted by the user, to get tweets tracked.

Given a ROI, trips that occur in the ROI are reconstructed and extracted by querying hang tweets and tracked tweets previously stored in the local data base. Reconstructed trips are represented by a list of geographic coordinates, ordered according to message creation time and are exported through the web service interface.

The Trip Analysis Suite performs the activities of knowledge discovery on trips collected by the FollowMe Suite. A knowledge-based trajectory clustering method allows analyzing trips based on customizable semantics. The analyst can specify both the desired granularity and semantics of the analysis by providing a vector layer of geographic slots (geo-slots) of interest. These are drawn from external interoperable sources that the algorithm exploits to conflate the trips' points to ease their grouping. For example, it is possible to conflate and then analyse trips with respect to the visited municipalities, regions, countries, city's neighborhoods, ZIP codes, etc. This way, the algorithm first geo-partitions the trips represented based on the ordered sequence of geographic coordinates into a conflated trip representation consisting of an ordered sequence of geo-slot identifiers, i.e., a string. This way, different geo-slots partitions provide different interpretations, scales, and semantics of the analysis.

The conflated trips can be easily clustered using a complete-link hierarchical trajectory clustering algorithm using a string-similarity matching. Matching is applied to the concatenated identifiers of the geo-slots in the conflated trips' representation. Finally, popular tours can be identified by selecting a partition of the clusters' hierarchy by specifying either a threshold on the minimum desired inter-similarity of conflated trips within a popular tour, or a minimum number of trips that a popular tour must contain.

2.4.4. Creation of Geographic Gazetteers by Volunteered Geographic Information Analysis

Constructing geographic gazetteers is very costly in terms of human effort and, once created, they need to be constantly updated. The work [81] proposes to exploit data science for the extraction of semantic information on toponyms, places, and POIs from big geoinformation created by volunteers on the Web, specifically from geotagged Flickr pictures. The aim is to enrich and update current gazetteers by automatically creating digital gazeteers of georeferenced place names such as "city center", "shopping district", and POIs associated with keywords and geofootprints. The ultimate purpose is to support diverse applications, such as geographic information retrieval (GIR), digital library services, and systems using spatio-temporal knowledge. The geographic footprints are extracted from the GPS locations of Flicker pictures while place descriptions are distilled from their tags. Close GPS locations associated with similar textual descriptions created by distinct volunteers are assumed as identifying the same place. These locations are generally not perfectly matching but usually have a cluster structure in space. This suggested the authors to use a distance-decaying function to measure the membership of candidate point locations assigned to a place so as to present an intuitive user reputation model for trust evaluation.

2.5. *Formal Geosemantics*

The reason the use case in Section 2.5.1 is exemplar to the transition from implicit to formal geosemantics is twofold. On the one hand, it upholds ontologies as the formalization means, offering less constrained expressiveness to the modeling of geospatial entities; on the other, it tackles a research issue, that of next generation maps, that has roots in cartography and, as such, is typically bound to the interpretation of implicit information mediated by the domain expertise of end users. Most applications of semantics to geospatial information use "lowercase" semantics, such as that of SKOS vocabularies [170] which are not harnessing full expressiveness of ontology languages; others mistake RDF encoding for semantics. Instead, it is important to keep in mind that far more expressive modeling criteria (ontology languages) and inference tools (reasoners) exist. Section 2.5.1 provides both a conceptual model for geo-entities and an exemplar implementation.

Discovery, in the sense of "retrieval of geospatial information", is largely dependent on metadata. In turn, semantic characterization of metadata is regarded to as the primary means to achieve interoperability [171] in a domain that is otherwise fraught with heterogeneities [14,56]. Unleashing this potential typically amounts to relating metadata items to entities in the Web of Data (the Linked Open Data Cloud: https://lod-cloud.net/ accessed on 1 April 2021), such as terms from SKOS vocabularies, people and organizations in FOAF representations [172], etc. Whereas this step may not be strictly necessary for semantics-aware discovery [173], leveraging on these categories of data structures can easily yield semantics-aware resource descriptions. The advantages of this practice are manifold. On the one hand, these data structures may greatly improve user experience in metadata production. On the other hand, traditional metadata can be enriched in order to enable smarter discovery criteria. This is the focus of Section 2.5.2.

Let aside the aforementioned virtuous data structures, there is a large corpus of web-accessible data structures that does not take advantage of ontologies expressed in OWL/OWL2, such as those mentioned above, or schema languages compatible with them (e.g., RDF Schema). As an example, consider the Microdata that is typically embedded in web pages or the XML/JSON data structures that are often used in the enactment of APIs. Section 2.5.3 proposes creation of "semantic twins" of JSON data structures to allow for transparently accessing heterogeneous data sources. It should be noted that although we already considered the JSON format in the previous Section, in this context the semantics underlying the JSON data (its implicit schema) is made explicit by the mapping to RDF, assuming an interpretation. Some of the (augmneted) information contained in the RDF data structures could be fed back to the original JSON ones so as to realize a JSON-LD [174] representation of resources.

Finally, Section 2.5.4 describes a model for *semantic mediation* with the aim of improving geospatial discovery, e.g., by exploiting the smarter metadata originating from creation methodologies akin to those presented in Sections 2.5.2 and 2.5.3. In fact, it is apparent that discovery constitutes a "crucial first step" in the enactment of Spatial Data Infrastructures (SDIs) and nevertheless is "mostly neglected and approached following old paradigms" [112]. Beside harnessing the richer information entailed by semantic characterization of metadata, another key objective of this practice is to implement geospatial data management as a machine-processable API, thus fostering FAIR access to geospatial resources [175]. The rationale for this is that it makes little sense to strive for semantic characterization of metadata and not accomplish the last mile toward their full exploitation by automated agents. The synoptic view of the case studies analysed in relation to formal geosemantics is reported in Table 2.

2.5.1. Holistic Map Representation with Geographic Scenarios

The work in [176] illustrates *Geographic Scenarios* [177], a notion developed on the basis of General System Theory [178] integrating spatial, process, and relational information related to geographical elements and georeferenced events. In contrast with reductionist approaches (such as those dividing geo-entities into themes), Geographic Scenarios propose a holistic view that should be better suited to represent hierarchical connections among geo-entities. Moreover, by favoring space over time, state-of-the-art GIS may fall sort of portraying dynamic relationships and causalities.

Basing the conceptual framework of Geographic Scenarios on an ontology allows for expressing multi-hierarchy categorizations and fuzzy boundaries, portraying diverse and complex entities at different scales and dimensions. Geo-characterization is the process by means of which scenarios as well as their individual components are assigned properties and relationships not only on the basis of traditional notions, such as regionalization and classification, but also according to ecology and human-orientation (that are often regarded to as mere thematic dimensions). Events are made first-class citizens in the ontological modeling of geographic scenarios, thus allowing attribution of dynamic relationships between geo-entities.

From a technical viewpoint, the realization that is presented combines relational data with ontology classes and properties by applying SWRL rules [179]; the resulting information is stored in a graph database for querying. Whereas the proposed example does not fully demonstrate the augmented capabilities of geographic scenarios, modularity of the possible semantic underpinning (the ontology) and the scalable solution for storage (a graph database) suggest more extensive implementations.

Table 2. Dimensions of case studies (Formal geosemantics).

Subsection	Name	Description	Input Data	Method Classification	Output Data	Potential Use
Section 2.5.1	Holistic map representation with Geographic Scenarios	Ontology-based ingestion of geo-entities into graph databases.	relational databases	rule-based system	graph databases	express diacronic relations and causalities with respect to traditional GIS
Section 2.5.2	EDI/Liftboy	semantic augmentation of geospatial metadata	structured text (metadata documents, XML Schema-based)	information retrieval in a graph database	metadata enriched (aka annotated) with links to RDF entities (often defined by authoritative sources) in semantic-web (RDF) resources (organized in graphs)	semantic discovery (e.g., multilingualism, semantic expansion); disambiguation and preservation of information meaning (in the future and with respect to different audiences)
Section 2.5.3	Semantic twins	augmentation of geospatial metadata based on heterogeneous sources	structured and semi-structured data, JSON- and HTML-based)	information retrieval in a graph database	metadata enriched (i.e., annotated) with links to json entities (exploiting semantic twins that grant consistency of metadata items)	same as previous
Section 2.5.4	Semantic mediation for FAIR access to resources	machine-actionable search of geospatial resources [once geospatial service interfaces (e.g., standard CSW) has been extended to semantic machine actionable API]	REST service interface (API) definitions enriched with semantics	information retrieval based on a task-ontology	machine-actionable semantic augmentation of REST API definitions expressed in Hydra	enablement of semantic agent

2.5.2. Ex-Ante and Ex-Post Semantic Characterization of Metadata

In the last decade, our work group has been tasked with the development of the SDI for a national flagship project on marine research. The key approaches were (i) creation of a decentralized network of nodes providing data [180] and (ii) the extensive use of semantics-aware technologies in metadata management [181]. The latter entailed development of a metadata editor that could easily adapt to the ever-changing landscape of metadata formats and profiles [182].

Since no tool in the state-of-the-art allowed for this degree of flexibility, we decided to develop EDI, a brand new metadata editor [183]. Beside allowing for an extremely user-friendly interface for metadata provision, the tool allows for both compliance with any XML or text-based metadata format as well as pluggability of heterogeneous RDF-based resources (made available as SPARQL [184] endpoints) as the reference data sources for providing auto-completion functionalities. This feature allows for the integration of a broad range of third-party data structures (e.g., code lists, controlled vocabularies, gazetteers, and registries) in the Web of Data.

Field values can also be generated on demand, can duplicate the content of another field, and even use generic XPath functions in order to mix-and-match values taken from the output XML document. Finally, this output document can be fed into an arbitrary chain of XSLT transformations (e.g., to generate a text-based output, such as JSON). All these functionalities are governed by a *template*, expressed in XML, that regulates production of the output document, defines the external data sources to be accessed via SPARQL, etc. Please refer to [185] for a comprehensive description of the template language.

Addressing semantic augmentation of metadata at editing-time (i.e., ex-ante) leaves an enormous amount of resource descriptions not featuring this important characteristic. As a consequence of this, important capabilities enabled by semantically enriched metadata (e.g., multilingualism, query expansion) could not be implemented by geoportals in discovery workflows. Then, we started working on offline, ex-post semantic lift of metadata records and realized it was possible to employ templates the other way around to search traditional XML metadata for correspondences in RDF data sources. The resulting application, named Liftboy, is described in [186] and made available on GitHub (https://github.com/IREA-CNR-MI/liftboy-python accessed on 1 April 2021) in its newer, improved implementation.

As a final note, we want to stress the importance of semantic characterization of metadata. Typically, this is seen as a solution to semantic heterogeneity and an opportunity for applying query expansion in information retrieval (in [186] the authors provide examples for both of these). In our opinion, semantic metadata can serve a higher purpose, that of "normalizing" resource description by conflation into a kind of pointer instead of repeatedly duplicating metadata property values (such as keywords, names, e-mail addresses of people, etc.) that frequently lead to inconsistencies, a practice we named *metadata delegation* [187]. It would be easier if all references to a keyword provided by a well-known controlled vocabulary were tagged with a unique identifier for that term (the URI of a `skos:Concept` [170]), if all references to a researcher pointed to her FOAF record [172], creating a web of decentralized metadata.

2.5.3. Exploiting Non-Rdf Data Structures for Semantic Metadata Creation

This case study builds on a software named SPARQL-Generate (https://ci.mines-stetienne.fr/sparql-generate/ accessed on 1 April 2021) [188] that extends the syntax of SPARQL 1.1 [189] with constructs that allow for extracting data from heterogeneous data structures and generating RDF descriptions. The application to the geospatial domain we describe is production of metadata for *samples* (also called *specimens*) in the International Geo Sample Number (IGSN) format [190]. The target data structures are the entities made available by the European Long Term Ecological Research Network (eLTER) in its Sites and Data Registry (DEIMS-SDR) [191,192] (specifically, the entities representing *activities*, *sites*, and *sensors*).

We wanted to build on EDI, the metadata editor presented in the previous Section, but the originating sources are in JSON format and thus could not be directly integrated in the autocompletion functionalities provided by the former. We then decided to create RDF descriptions as signpost for the aforementioned entities and relate samples to them by plugging-in these RDF "semantic twins" in a custom EDI template. Then, the metadata maintainer can access the HTML5 interface generated by the EDI client and select the entities in the originating data structures via the many widgets made available by the software, drawing information from external data structures.

2.5.4. Semantic Mediation for FAIR Access to Resources

This case study considers the articulation of geospatial discovery as a web API in order to make catalogs accessible by automated agents. One may argue that the Catalogue Service for the Web (CSW) by OGC [193] serves this purpose and, of course, when the automated agent knows where the endpoint is and which protocol to use, resource harvesting and search are straightforward. Still, when the agent only knows the homepage of the data provider and no information on the protocol applying, these operations may get difficult to achieve.

The problem (and the link to the subject of this paper, i.e., semantics) is that the Web, as experienced by human agents, is unlike web APIs in that there is a *semantic gap* to be bridged [194] before machines can fully participate. Overcoming this gap requires internalizing the key principles of REST (REpresentational State Transfer) as expressed by Roy Fielding in his Ph.D. dissertation [195]; specifically:

1. identification of resources
2. self-descriptive messages
3. hypermedia as the engine of application state

Please refer to Chapter 5 of the dissertation for an explanation of these. The attentive reader may already have spotted how the breadth of this research topic can be extended so as to encompass FAIR (Findable, Accessible, Interoperable, and Reusable) practices [175].

Since their inception, the FAIR principles have been deeply rooted in the notion of machine-actionability. Among the technologies for a machine-actionable Web, it is generally acknowledged that, despite the apparent differences, there is a broad overlapping between REST principles and FAIR practices (FORCE11 Guiding Principles for Findable, Accessible, Interoperable and Re-usable Data Publishing: https://www.force11.org/fairprinciples 1 April 2021). In fact [196], the machine-actionable behaviors of REST match the requirements of (at least) the first three letters in "FAIR", as both recur to specification of semantics for their enactment and both rely on resolvable identifiers.

In order to achieve machine-actionability for geospatial services, the European Plate Observing System research infrastructure [197,198] exploits Hydra [199], an RDF vocabulary that is capable of expressing the mechanics of APIs in a way that is both intelligible to automated agents and also semantically rich. Please refer to the Hydra Core Vocabulary (https://www.hydra-cg.com/spec/latest/core 1 April 2021) for a more thorough descriptions of the features of this formalism.

The potential of this characterization of APIs is apparent. As an example, search for processing services matching a given set of parameters, such as the Normalized Difference Vegetation Index (NDVI) for a specific bounding box can greatly take advantage of semantics-aware service description [200]. Moreover, automated workflow composition on the basis of more precisely defined inputs and outputs can be easier than with other technologies [201].

2.6. *Powerful Geosemantics*

There are concepts and relationships in the real world that are intrinsically imprecise and fuzzy, due to their gradual nature. This characteristic is particularly evident in the geographic context, in which natural entities and spatio-temporal phenomena are characterized by blurred and time-varying contours. For instance, it is impossible to encode in a classic ontology based on OWL vague concepts like "most streets in Naples center are very narrow", which involve some fuzziness for which a crisp definition does not make sense. What is the size of a street that makes it *"narrow"*? This is a matter of degrees depending on a subjective interpretation and, certainly, there is not a crisp transition between a street being *large* and *narrow* that may be agreed upon by all observers. The term *most* means that there are exceptions, i.e., a few streets are large, but its hard to quantify a crisp percentage. Furthermore there may be cases in which one needs to define a fuzzy concept hierarchy, a fuzzy taxonomy, in which a class is a specialization to a degree of several super classes such as "In Italy churches, beside being (1) places of worship, are often (0.8) historical buildings". Furthermore, it may be necessary to define fuzzy relationships between concepts such as in "bell towers are very close to churches".

Another possible source of imperfection occurs when an ontology is used for quality assurance to tag observations such as in Citizen Science (CS) projects. Such projects are at present a common practice to collect geospatial data in many domains such as natural sciences by involving volunteers to create georeferenced observations of objects of interest. A volunteer may be not completely sure about his/her observation, which is the case of epistemic uncertainty. This may happen because (s)he does not have adequate knowledge of the problem or because of deficiencies in the means of observation. This may also happen when the domain knowledge is precise.

Finally, there are more complex situations that may involve both ill-defined knowledge and epistemic uncertainty [202].

To cope with the above issues, powerful semantics approaches are needed which "extend" classic ontologies with the ability to represent and manage uncertainty and imprecision: To this end, the literature proposes soft ontologies [12]. In particular, there are three main groups defined on the basis of the probabilistic, the fuzzy, and the possibilistic or evidential frameworks. They have been adopted for extending propositional logic with probability, possibility, belief, or truth of a statement.

Fuzzy ontologies have been defined to model ill-defined knowledge with several purposes, depending on the kind of imperfection they need to represent and manage in the application [202]. Although a standard representation of a fuzzy ontology is still to come, a lot of researches have fuzzified the existing Description Logics (DL) and have defined fuzzy DL reasoners. The most up-to-date and complete fuzzyDL ontology reasoner has been proposed in [203].

To model epistemic uncertainty, fuzzy ontologies have been defined within a possibilistic framework that deals with certainty and possibility degrees of truth thus modeling the epistemic uncertainty characterizing experts' subjective knowledge and the evaluation of the certainty of this knowledge. To this end, several possibilistic DL reasoners have been defined [204], which allow for representing and reasoning on uncertain statements such as "It is possible that this town is an Historic Area". To this end, each concept, relation, and axiom is associated with a real value u in (0, 1] representing its certainty level.

Nevertheless, fuzzy ontologies do not allow to model the time varying nature of concepts and their context-dependent meaning. Specifically, most geographic concepts are represented by prototypes that vary with time: The prototypical modern city to an Italian person has changed during centuries, and it different for Chinese people. Fuzzy set theory cannot completely model how humans use concepts, in particular the fact that their meaning is influenced by context and states that vary with human knowledge in time. To this end, the framework known as state-context-property (SCOP) based on quantum mechanics [205] has been defined to map elements taken from operational foundations of quantum mechanics (like states, measurements, and observables) onto concepts and contexts.

In the following Subsections, we recap three case studies exploiting powerful semantics. Their synoptic view is reported in Table 3. They have been selected as representative of distinct application domains such as the creation of biodiversity observations (Section 2.6.1), remote sensing to aid disaster management (Section 2.6.2), and dynamic urban planning (Section 2.6.3). The first two of them exploit a fuzzy ontology enconding epistemic uncertainty of volunteers when creating georeferenced observations (i.e., VGI) and the vague and incomplete knowledge of experts when interpreting a phenomenon from remote sensing evidence, respectively. By representing epistemic uncertainty and vagueness of knowledge, it is possible to model the distinct quality of the results of a decision process.

The last case study illustrates the application of the SCOP framework to model retrieval of maps within a GIR with increasing precision, achieved by exploiting the varying states of knowledge of user needs.

2.6.1. A Fuzzy Ontology to Support Volunteered Geographic Information Creation and Search

Within the *Space4agri* [206] project, agronomists surveyed agronomic fields by tagging the observed crops and their phenological growth stages based on an agronomic ontology [207]. In this process, texts or pictures were added to report a difficulty or doubt of the agronomists when selecting a phenological growth stage from the ontology. This is due to different reasons:

- doubt in interpreting the meaning of the descriptions in the ontology;
- difficulty to distinguish the characteristic aspects of a phenological stage in the observed crop sample, because of a deficiency of the observation means (e.g., a far point of view);
- hesitancy to select a unique growth stage for several observed crop samples close in space within the same parcel, because of variability of their characteristics.

This suggested the need for extending the classic ontology-based reasoning by representing the epistemic uncertainty of the agronomists in creating VGI items (i.e., when selecting tags from the ontology [207]). Specifically, volunteers can create georeferenced annotations of crops they are observing in situ with the support of a fuzzy ontology. They are bound to select linguistic predicates, possibly fuzzy, to tag the observed crops and with each selected predicate they can associate a degree d in [0,1] representing the overall deficiency of their observation. This way, they can represent epistemic uncertainty due to both limitations of the means of observation (e.g., a far point of view, low resolution of the means of observation) and difficulty of precisely quantifying some properties of the observed crops. The linguistic predicates such as "crop has large leaf", "crop has long stamen", "crop has many branches" describe possibly fuzzy properties of the distinct kinds of crops: For example, a rice crop during its germination can appear with "elongated and thin branches" and "very small seeds". The semantics of these linguistic predicates can be defined by level-1 fuzzy sets (whose membership degrees are numeric in the range [0,1]). The fuzzy ontology can then explicitly represent linguistic concepts in both symbolic form (encoded by the linguistic terms "large", "long", "many") and quantitative form. The latter is expressed by the membership functions defined on the numeric domains of the properties: For example, "large" is defined with a membership function on numeric values in cm. In the fuzzy ontology, compatibility between linguistic predicates is represented by Level-2 fuzzy relations, i.e., fuzzy sets on multidimensional basic domains whose membership degrees are not numbers but linguistic values. Fuzzy relations between linguistic predicates are used to perform approximate reasoning in the fuzzy ontology to automatically classify the crops, possibly into distinct types with different membership degrees. The defect degrees are interpreted as minimum thresholds, i.e., uncertainty levels, on the compatibility degrees between the linguistic predicates so that the final membership to a type of crop is modified by epistemic uncertainty. When formulating queries to the database of georeferenced crop observations, for example, requesting to map "rice crop fields", the stored observations can be mapped Onto different shades of color depending on their membership degrees to type "rice crop", thus accounting for both fuzziness and observation uncertainty.

Table 3. Dimensions of case studies (Powerful geosemantics).

Subsection	Name	Description	Input Data	Method Classification	Output Data	Potential Use
Section 2.6.1	Fuzzy ontology supporting VGI	VGI quality assurance and assessment by modeling both imprecision/vagueness of domain knowledge and uncertainty of volunteer's perceptions	VGI items created by selecting linguistic predicates from a fuzzy ontology and by associating uncertainty of observations	fuzzy computational intelligence	VGI quality assessment based on qualitative reasoning (level-based uncertainty reasoning)	quality assurance and assessment; ontology enrichment
Section 2.6.2	Environmental status indicator mapping	fuzzy classification of standing water from remote sensing images	remote sensing images and in situ observations plus incomplete (fuzzy) ontology (contributing factors and their (soft) constraints to derive partial evidence of watered water)	fuzzy computational intelligence+ machine learning	identification of watered areas and the fuzzy ontology enrichment	monitoring water bodies; ontology enrichment
Section 2.6.3	Modeling user interaction in GIR	modeling user intention and concepts' status	State-Context-Property representation of concepts and user queries	context-sensitive measurement of conceptual distance	identification of collapsed states representing answers to the user intention	modeling evolving and context dependent geographic concepts

2.6.2. Fuzzy Ontology to Support Remote Sensing Image Interpretation

In remote sensing, Geographic Object-Based Image Analysis (GEOBIA) groups techniques aiming at segmenting and classifying objects and phenomena (represented by groups of pixels sharing common properties) in satellite images based on image analysis procedures that rely on a priori expert knowledge [35]. In recent years, application of ontologies enconding experts' knowledge is emerging [14]. Ontologies are used to associate some perceived concepts with their data representation [35]. A widely applied approach to detect the geographic footprint of environmental phenomena is to compute spectral indexes (SI) maps. SI values integrate reflectance measurements at different wavelengths into a synthetic feature that can highlight some perceived aspects of the phenomenon in each pixel. SI maps are then segmented to identify target phenomena, such as vegetation presence and vigor (biomass presence, Leaf Area Index, Chlorophyll content, etc.), bare soil condition, and soil properties composition, burned areas, water presence, and so on. The segmentation consists of thresholding the pixel SI values by different thresholds specified in the ontology to define the different environmental phenomena.

Nevertheless, using the same ontology to segment a given phenomenon such as "green areas" in a new image may cause inaccuracies with many omissions and commission errors, since the value of the threshold must be tuned depending on several factors, such as the context and observation conditions. In fact, accurate calibration is needed to set a proper threshold for each study area. Thus uncertainty and imprecision must be represented since the kind of knowledge is perceptual by very nature [35]. These are the reasons why powerful semantics approaches are appealing. In fact, these techniques allow for explicit representation of perceptual characteristics of phenomena in images by means of fuzzy ontologies. Thus, they can cope with the limitations of both traditional GEOBIA solutions using ontologies and machine learning techniques requiring huge amounts of training data often unavailable.

In [162], an approach based on powerful semantics was proposed to map standing water areas from optical multispectral remote sensing images. Ill-defined knowledge of experts on the perceptual characteristics of standing water within optical images is represented by defining fuzzy sets on spectral indexes identified as features. The membership functions of these fuzzy sets relax the crisp segmentation thresholds defined in the vast literature on standing water mapping so as to tolerate imprecision and uncertainty. A fuzzy ontology is thus defined describing standing water in terms of fuzzy sets on spectral indexes. For each spatial unit with given values of spectral indexes, partial evidence degrees of standing water are computed by evaluating the membership degrees to the fuzzy sets in the fuzzy ontology. Finally, the partial evidence degrees in each spatial unit are combined by applying a fuzzy aggregation operator, learnt by a shallow machine learning algorithm trained on a small reference data set. Beside not requiring big training data, the approach offers the advantage of explicating the criteria used to map standing water, allowing discovering how many spectral indexes, which of them, and to which extent they contributed to map standing water in each spatial unit. The fuzzy ontology with new fuzzy relationships between fuzzy concepts.

2.6.3. State-Context-Property Framework to Model Human Interaction within a Geographic Information Retrieval System

According to [208], human-computer interaction is based on the exchange of words (or graphical tokens on maps) which are interpreted in the context of the conversation. The words used may originally have a broad meaning; through conversation the context becomes more precise and the concepts obtain more specific meanings. The authors present a proof of concept that shows the selection of several predetermined map types (e.g., street map, political map, map for hiking, ski routes) in a GIR by formalizing their approach in SCOP [205]. Specifically, SCOP is applied to predict an answer to the question: "Which map is appropriate for a given context?" where the context is declined as the intended purpose of the user.

A concept and a context serve as input parameters to the inference model that calculates the collapsed state and returns it. In this collapsed state, probability values for prototypes of the concept can be calculated. A use case is illustrated, in which a user states to a GIR query interface that she needs a map, without stating the kind of map. So far the concept "map" is in ground state, where all maps have some non-zero probability to be relevant. The user then states the intended usage that is to go on a bicycling trip. Now the state of the concept "map" collapses into a bicycling map. The user interaction may continue to indicate the region where the trip is planned, and this new information further restricts the map to an area. The application of SCOP is still at its early stage; it needs further developments and investigations to be practically applied, but its potential is great as far as prototypical modeling of contexts and states is concerned.

3. Results and Discussion

To organize the material, we started from the notion of semantics as a function that maps the world of syntax onto the world of meaning, in analogy with the studies on denotational semantics [209]. Once put on these lenses, we analyzed the presented case studies considering the original information they dealt with (syntactic objects with a certain amount of semantics), the meaning that is extracted and formalized (the new semantic objects), and the techniques that are applied to map the former onto the latter (the incremental semantic mapping function). This analysis of the case studies is presented in Table 4, where each row resumes one of them.

The first two columns identify the case study by indicating the corresponding subsection and a short name. In the columns that follow, one can find information about the mapping of the input information onto the new semantic objects: Specifically, column 3 contains the description of the input information pertaining the case study; column 4 provides the incremental semantic function that is used to map the original information with partial semantics onto the output with augmented meaning; finally, column 5 indicates the final information, i.e., the semantic domain of the case study. Column 6 indicates the delta between the input and output information; finally, column 7 enumerates the keywords, among those on the right side of Figure 2, that can be related to the case study: The more relevant keywords are in bold font and are assigned a weight w = 2 in the analysis that follows.

Here, complexity degree is intended as the level at which semantics is made explicit in either the input or the output data structures considered by the specific case study. Specifically, the complexity degree is an integer in the range 1–7, following the principle of indiscernibility of Miller [210]. The general criterion for attributing this value is that complexity lower than 4 accounts for objects presenting scarce or no machine understandable information about their meaning; values between 4 and 7 indicate that meaning is more and more machine understandable and processable. For instance, the most simple case is that of unstructured text (complexity = 1), such as in case study 3.1 where input is constituted by free text keywords. The degree increases when more information is added such as in case study 3.2 and 3.3 (complexity = 2) where input is enriched both by the presence of structure (JSON documents) and by geographic coordinates. When the previous information is further augmented, complexity increases (complexity = 3) such as in the output data of case study 3.1 where uncertainty degrees are added. The next step in explication of semantics may involve schema information or categorization of data (complexity = 4). Then, when relationships among the entities (topological, order, metric, broader/narrower) are taken into account, complexity increases to 5. Complexity is 6 when vague and uncertain concepts and relationships are represented. Finally, when information can be generated by approximate reasoning or has fully reached semantic interoperability, complexity is 7.

Table 4. Dimensions of case studies.

Subsection	Name	Complexity Degree of Input Data Explicit Semantics	Incremental Semantic Function (Methods, Techniques)	Complexity Degree of Output Data Explicit Semantics	Added Semantic Value	Keywords in Geosemantics (cfr.Sankey Diagram in Figure2)
Section 2.4.1	GeoFinder	1	euristic rules and explicit geographic information in gazetteer	3	2	**Geographic Information Retrieval**; semantic enrichment/tagging/annotation; **gazetteer**; **geoparsing**; **geonames**; **place-based information systems**; geo-comparison; extraction of spatio-temporal information
Section 2.4.2	Events spatio-temporal footprint	2	clustering	4	2	semantic enrichment/tagging/annotation; **geo-comparison**; **extraction of spatio-temporal information**; **events, change discovery**; thematic/spatial and temporal perspective
Section 2.4.3	Tour miner	2	knowledge-based clustering	5	3	**geo-comparison**; **folksonomies**; semantic enrichment/tagging/annotation; **extraction of spatio-temporal information**; **events, change discovery**; thematic/spatial and temporal perspective; **(Geographic) Knowledge graph**; geospatial statistics
Section 2.4.4	Eliciting Geographic Gazetteer	2	statistic analysis and clustering	4	2	**Geographic Information Retrieval**; semantic enrichment/tagging/annotation; **gazetteer**; **place-based information systems**; **folksonomies**; geo-comparison; extraction of spatio-temporal information; Place location, identity, meaning; geospatial statistics; top-level ontologies
Section 2.5.1	Holistic map representation with Geographic Scenarios	4	SWRL rules combining relational data and ontologies	5	1	**Ontology for modeling**; Domain ontologies; **Ontology-based info extraction**; Semantic enrichment/tagging/ annotation; Thematic, spatial, and temporal perspectives; **Knowledge representation languages (OWL)**
Section 2.5.2	EDI/Liftboy	5	entity annotation. Specific metadata profile specifying where and how to find the Semantic Web (RDF) resources	7	2	Domain ontologies; **Linked Geo Data/LOD/Linked Sensor Data**; Gazetteers (GeoNames)/temporal gazetters; Spatial RDF and SPARQL; Ontology-based info extraction; **Semantic enrichment/tagging/ annotation**; Knowledge representation languages (OWL); **Semantics-driven user interfaces/interaction paradigms/Semantic engineering of human communications**
Section 2.5.3	Semantic twins	4	same as example EDI/Liftboy, but to grant consistency of metadata items the semantic twins of JSON entities is exploited	7	3	Domain ontologies; **Linked Geo Data/LOD/Linked Sensor Data**; Gazetteers (GeoNames)/temporal gazetters; **Spatial RDF and SPARQL**; Ontology-based info extraction; **Semantic enrichment/tagging/ annotation**; Knowledge representation languages (OWL)
Section 2.5.4	Semantic mediation for FAIR access to resources	5	information retrieval based on an application ontology	7	2	Ontology for encoding; Application ontologies; Geographic Information Retrieval; Linked Geo Data/LOD/Linked Sensor Data; **Semantic enrichment/tagging/ annotation**; Knowledge representation languages (OWL); Semantics-driven user interfaces/interaction paradigms/Semantic engineering of human communications; **Semantic markups for Web services**
Section 2.6.1	Fuzzy ontology supporting VGI	6	fuzzy rule based inference engine	7	1	**task-ontology**; **ontology for modeling**; **ontology for encoding**; sensor and observation ontology; **qualitative reasoning**; **application ontologies**; ontology design pattern; ontology-based information extraction; semantics-driven user interfaces
Section 2.6.2	Environmental status indicator mapping	5	incomplete (fuzzy) ontology + machine learning exploiting in situ classified data	7	2	task ontology; **ontology for modeling**; **ontology for encoding**; **qualitative reasoning**; **application ontologies**; ontology design pattern; ontology-based information extraction
Section 2.6.3	Modeling user interaction in GIR	5	inference in SCOP framework	7	2	**conceptual space**; **semantic engineering of human communications**; **qualitative reasoning**; Non-representational formalisms

The four case studies presented in Section 2.4 share the same type of input geoinformation, which is essentially not explicit, being dispersed within unstructured and loosely structured texts. In the output geoinformation of these case studies, semantics is made explicit but not always in a standard, interoperable format; because of this, it may be difficult or even impossible to reuse the results in different contexts.

The first case study in Section 2.5 portrays a model exploiting semantics at its full potential, via ontologies. The second applies to semi-structured geoinformation in the form of metadata, possibly compliant with OGC standards. The third case study involves structured (JSON) and semi-structured (HTML) information that lacks the relations between the entities involved (e.g., between descriptions of sensors and the corresponding points of contact) and, in general, can not be easily reused in a Web of Data context. Finally, the fourth case study applies to unstructured information intended to the human agent (i.e., the specification of computer interaction protocols). For each of these, the output is information that can be shared and reused in an interoperable way by enabling querying and retrieval in a Linked Data perspective. The first two case studies in Section 2.6 involve explicit and rich geoinformation in the form of soft ontologies, while the last case study uses the SCOP formalism. All these case studies enable qualitative and approximate reasoning to deduce novel geoinformation automatically.

A preliminary observation that can be made is that complexity of the inputs is lower for the case studies in Section 2.4, medium for those in Section 2.5, and maximum for the case studies in Section 2.6. The same for the outputs. More insights come from cross-referencing of the case studies and the keywords listed on the right of Figure 2, yielding the representation in Figure 3. This last figure illustrates the weighted associations between case studies and keywords: Case studies within the same Section (i.e., associated with the same form of geosemantics) are characterized by shades of the same color (yellow for implicit, blue for formal, and grey for powerful geosemantics). On the x axis, the length of the bar represents the different importance of the method/technique in the case study while the pair hue-color uniquely identifies both the case study and its belonging semantic category. It can be visually noticed that the case studies classified in the same form of geosemantics are mostly associated with distinctive keywords. For example, the case studies in Section 2.6 (powerful geosemantics) are associated with "Non-representational formalisms", "Task ontologies", and "Qualitative reasoning". Nevertheless, some keywords (e.g., "Semantic enrichment/tagging/annotation") are associated with case studies classified in "adjacent" forms of semantics.

To confirm the conjecture suggested by Figure 3, i.e., that the three geosemantics forms are good categorizations for the keywords, we also computed the similarity measure known as Jaccard coefficient between any pair of case studies on the basis of the aforementioned weighted keywords, as shown in Figure 4. The figure clearly shows that the intra-similarities (regarding pairs of case studies belonging to the same form of geosemantics, grouped within the colored rectangles) are greater than the inter-similarity degrees between pairs of case studies classified as different forms of geosemantics (i.e., appearing outside the colored rectangles).

It can be noticed that all case studies have greatest intra-similarity with another case study of the same geosemantics form. Only case studies in the yellow group share some inter-similarity with those of the blue group, which is anyway an order of magnitude lower than the intra-similarity. Specifically, as far as the case studies dealing with the implicit form of geosemantics are concerned, their overall intra-similarity, computed as percentage of shared keywords among all the case studies of the same category, reaches 54.3%, while their overall inter-similarity with any other case study of the others two categories is only 1.7%; as far as the case studies dealing with the explicit form are concerned, they have an overall intra-similarity of 58% and an overall inter-similarity of 2.6%; finally the case studies dealing with the powerful form have overall intra-similarity of 37% and an overall inter-similarity of only 0.9%. These findings confirm our hypothesis that the three forms of semantics are characterized by distinguishing techniques, methods, and knowledge sources in the geospatial domain.

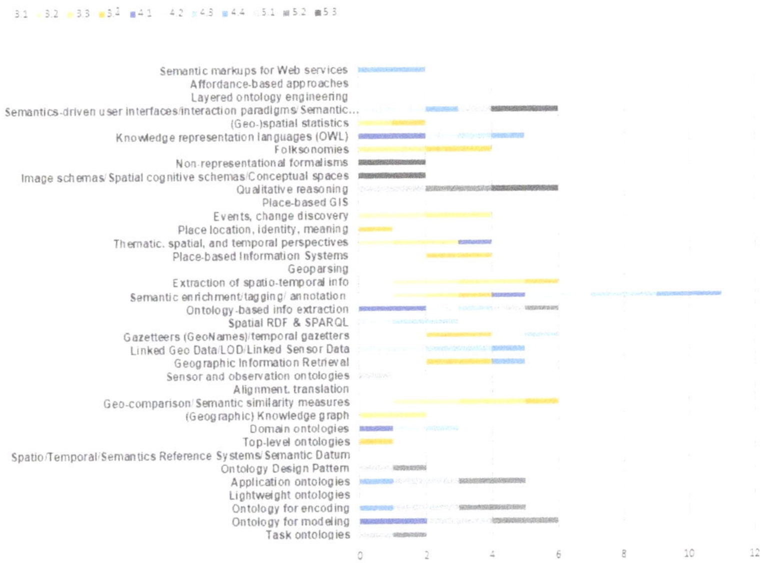

Figure 3. Case studies and the keywords representing their main activities and technologies.

Similarity matrix between case studies (based on shared keywords)	3.1	3.2	3.3	3.4	4.1	4.2	4.3	4.4	5.1	5.2	5.3
3.1	1.00	0.19	0.14	0.13	0.03	0.03	0.03	0.02	0.00	0.00	0.00
3.2		1.00	0.57	0.14	0.03	0.03	0.03	0.03	0.00	0.00	0.00
3.3			1.00	0.27	0.03	0.03	0.03	0.03	0.00	0.00	0.00
3.4				1.00	0.05	0.04	0.03	0.03	0.00	0.00	0.00
4.1					1.00	0.25	0.24	0.09	0.00	0.00	0.00
4.2						1.00	0.75	0.24	0.00	0.00	0.00
4.3							1.00	0.25	0.00	0.00	0.00
4.4								1.00	0.15	0.10	0.00
5.1									1.00	0.11	0.23
5.2										1.00	0.12
5.3											1.00

Figure 4. Jaccard similarity between study cases represented as fuzzy sets of keywords.

Besides revealing the distinguishing features of the geosemantics forms, we also found in this analysis that case studies related to implicit and formal semantics have many activities in common, identified by the shared keywords "Thematic spatial and temporal perspectives", "semantic enrichment/tagging/annotation", "Gazetteers (GeoNames)/temporal gazetteers", and "Geographic Information Retrieval". Formal and powerful semantics share "Semantics-driven user interfaces/interaction paradigms/...", "ontology based information extraction", "Application ontologes", "Ontology for encoding", and "Ontology for modeling". This means that there is not a clear-cut partition between the forms of semantics. This shows that a "semantic continuum" is present, gently blending the groups, moving from implicit to powerful semantics. Conversely, the approaches related to powerful and implicit semantics share no keywords. These findings reveal that the ordering of categories introduced by Sheth [4] also seems to emerge from our analysis even in the context of geographic information.

Figure 5 provides an even more synoptic view on the relations between the keywords and the three forms of semantics, complementing Figure 2 with the findings described in this section. In fact, the figure clearly visualizes that, once the keywords are grouped according to the forms of semantics that are associated with the case studies presented in this paper, they are much more clustered. This means that patterns emerge in the geosemantics "forest", thus making order among the diverse practices.

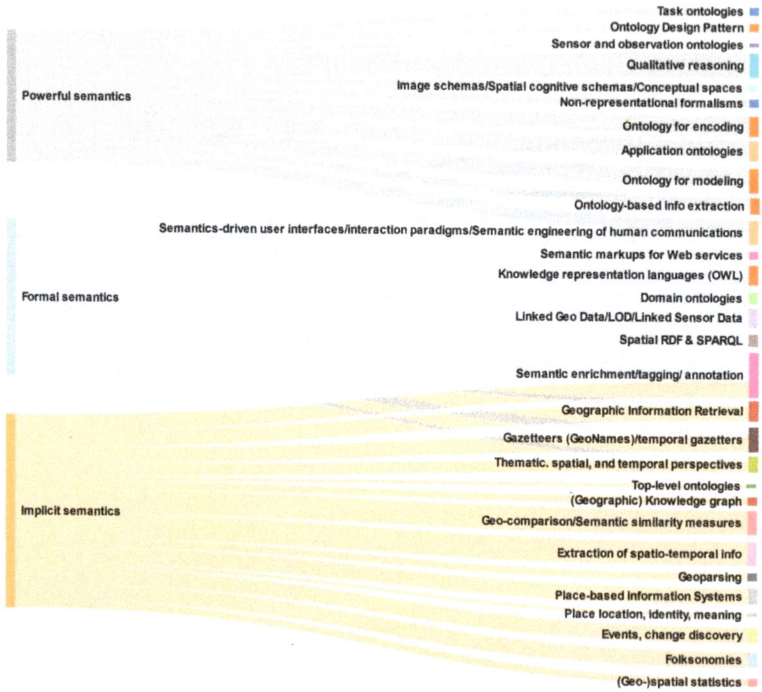

Figure 5. Comparison between the grouping of keywords in Figure 2 (on the right-hand side) and the grouping induced by the three forms of geosemantics (via the case studies) makes it apparent their greater distinguishing power.

Of course, this analysis can be enriched both by extending the meta-review to encompass more methodologies, techniques, and knowledge bases and by analyzing other case studies in the literature. Nevertheless we think that this contribution has the merit of setting a methodological workflow to characterize the forms of semantics in geoinformation and their preferred/elective approaches.

4. Conclusions

This paper applied the categories of semantics defined by Sheth to the domain of geoinformation in order to orient the reader in problem solving. We first analyzed recent reviews and editorial papers on geosemantics, mining which are the main technologies, methodologies, research challenges, and solutions presented by the authors. Then, we discussed selected case studies for the implicit, formal, and powerful geosemantics, respectively. The two-step analysis culminates with cross-referencing these two sources in order to confirm that the three forms of geosemantics are characterized by distinguishing techniques, methods, and knowledge sources.

The subsistence of this conjecture is attested by the Jaccard distances computed between members of the same/different categories of semantics (see Figure 4). This can also be visually assessed by looking at Figures 3 and 5. In the latter, it is also apparent that there are fringe keywords associated with "adjacent" categories (i.e., categories with similar semantics explicitation degrees). This paper contributes to structuring the approaches to semantics in geoinformation, partitioning the semantic continuum suggested in [6] in discrete, distinguishing techniques and methods.

Further insight may come from categorizing in the three forms of semantics the papers considered in the meta-review (Section 2.3) according to the associated keywords. Future work will also investigate scaling-up of the workflow by applying content representation methods used in information retrieval. In fact, these can automatically identify the keywords from the text of the reviewed literature.

Supplementary Materials: The following are available online at https://www.mdpi.com/article/10.3390/ijgi10050330/s1.

Author Contributions: All authors equally contributed to the writing and revising of this paper. All authors have read and agreed to the published version of the manuscript.

Funding: This research received no external funding.

Conflicts of Interest: The authors declare no conflict of interest.

Abbreviations

The following abbreviations are used in this manuscript:

CS	Citizen Science
CSW	Catalogue Service for the Web
eLTER	European Long-Term Ecological Research
FAIR	Findable, Accessible, Interoperable, and Reusable
FOAF	Friend Of A Friend
FOL	First-Order Logic
GEOBIA	GEographic Object-Based Image Analysis
GIR	Geographic Information Retrieval
IGSN	International Geo-Sampling Number
IRS	Information Retrieval System
NDVI	Normalized Difference Vegetation Index
NER	Named Entity Recognition
PID	Persistent IDentifier
POI	Point Of Interest
REST	REpresentational State Transfer
ROI	Region Of Interest
SEM	Simple Event Model
SI	Spectral Indexes
WKT	Well-Known Text

References

1. Guha, R.; McCool, R.; Miller, E. Semantic search. In Proceedings of the WWW '03, 12th international conference on World Wide Web, Budapest, Hungary, 20–24 May 2003; ACM Press: New York, NY, USA, 2003; pp. 700–709.
2. Duval, E.; Hodgins, W. Metadata principles and practicalities. *D-Lib Mag.* **2002**, *8*, 1–10. [CrossRef]
3. Nogueras-Iso, J.; Muro-Medrano, P.; Zarazaga-Soria, F. *Geographic Information Metadata for Spatial Data Infrastructures: Resources, Interoperability and Information Retrieval*; Springer: Berlin/Heidelberg, Germany, 2005.
4. Sheth, A.; Ramakrishnan, C.; Thomas, C. Semantics for the Semantic Web: The Implicit, the Formal and the Powerful. *Int. J. Semant. Web Inf. Syst.* **2005**, *1*, 1–18. [CrossRef]
5. Uschold, M. Where are the semantics in the semantic web. *AI Mag.* **2003**, *24*, 25.
6. Almeida, M.; Rocha Souza, R.; Fonseca, F. Semantics in the Semantic Web: A Critical Evaluation. *Knowl. Organ. J.* **2011**, *38*, 187–203. [CrossRef]
7. Gärdenfors, P. How to make the Semantic Web more semantic. In *Formal Ontology in Information Systems*; Varzi, A., Vieu, L., Eds.; IOS Press: Amsterdam, The Netherlands, 2004; pp. 19–36.
8. Lemmens, M. Geo-information, Technologies, Applications and the Environment. In *Geotechnologies and the Environment (GEOTECH)*; Springer: Berlin/Heidelberg, Germany, 2013; Volume 5, p. 349.

9. Ait-Ameur, Y.; Gibson, J.; Mery, D. On Implicit and Explicit Semantics: Integration Issues in Proof-Based Development of Systems. In Proceedings of the 6th International on Symposium Leveraging Applications of Formal Methods, Verification and Validation—ISoLA 2014, Corfu, Greece, 8–11 October 2014; Springer: Berlin/Heidelberg, Germany, 2014; pp. 604–618.
10. Berners-Lee, T.; Hendler, J.; Lassila, O. The Semantic Web. *Sci. Am.* **2001**, *284*, 34–43. [CrossRef]
11. Baader, F.; Calvanese, D.; Mcguinness, D.; Nardi, D.; Patel-Schneider, P. *The Description Logic Handbook: Theory, Implementation, and Applications*; Cambridge University Press: Cambridge, UK, 2007.
12. Caglioni, M.; Fusco, G. Formal Ontologies and Uncertainty. In geographical knowledge. *TeMA J. Land Use Mobility Environ.* **2014**, 187–198. [CrossRef]
13. Tambassi, T. (Ed.) *The Philosophy of GIS*; Springer International Publishing: Cham, Switzerland, 2019. [CrossRef]
14. Kokla, M.; Guilbert, E. A Review of Geospatial Semantic Information Modeling and Elicitation Approaches. *ISPRS Int. J. Geo-Inf.* **2020**, *9*, 146. [CrossRef]
15. Kuhn, W. Geospatial semantics: Why, of what, and how? In *Journal on Data Semantics III*; Springer: Berlin/Heidelberg, Germany, 2005; pp. 1–24.
16. Janowicz, K. Observation-Driven Geo-Ontology Engineering. *Trans. GIS* **2012**, *16*, 351–374. [CrossRef]
17. Kuhn, W. Semantic engineering. In *Research Trends in Geographic Information Science*; Springer: Berlin/Heidelberg, Germany, 2009; pp. 63–76.
18. Kuhn, W. Core concepts of spatial information for transdisciplinary research. *Int. J. Geogr. Inf. Sci.* **2012**, *26*, 2267–2276. [CrossRef]
19. Hu, Y. 1.07–Geospatial Semantics. In *Comprehensive Geographic Information Systems*; Huang, B., Ed.; Elsevier: Oxford, UK, 2018; pp. 80–94. [CrossRef]
20. Janowicz, K.; Hitzler, P. The Digital Earth as knowledge engine. *Semant. Web J.* **2012**, *3*, 213–221. [CrossRef]
21. Shadbolt, N.; Smart, P. *Knowledge Elicitation: Methods, Tools and Techniques*; CRC Press: Boca Raton, FL, USA, 2015.
22. Kavouras, M.; Kokla, M. *Theories of Geographic Concepts: Ontological Approaches to Semantic Integration*; CRC Press: Boca Raton, FL, USA, 2007.
23. Casati, R.; Smith, B.; Varzi, A.C. *Ontological Tools for Geographic Representation*; IOS Press: Amsterdam, The Netherlands, 1998.
24. Hu, Y.; Janowicz, K. Enriching top-down geo-ontologies using bottom-up knowledge mined from linked data. *Adv. Geogr. Inf. Sci. Past Next Twenty Years* **2016**, *183–198*.
25. Hong, J.H.; Kuo, C. A semi-automatic lightweight ontology bridging for the semantic integration of cross-domain geospatial information. *Int. J. Geogr. Inf. Sci.* **2015**, *29*, 2223–2247. [CrossRef]
26. Gandon, F.L. A survey of the first 20 years of research on semantic Web and linked data. *Ingénierie des Systèmes d Inf.* **2018**, *23*, 11–38. [CrossRef]
27. Giunchiglia, F.; Zaihrayeu, I. Lightweight Ontologies. In *Encyclopedia of Database Systems*; Springer: New York, NY, USA, 2009.
28. Gangemi, A.; Presutti, V. Ontology design patterns. In *Handbook on Ontologies*; Springer: Berlin/Heidelberg, Germany, 2009; pp. 221–243.
29. Martinez-Rodriguez, J.L.; Hogan, A.; Lopez-Arevalo, I. Information extraction meets the semantic web: A survey. *Semant. Web* **2020**, 11, 1–81. [CrossRef]
30. Allahyari, M.; Pouriyeh, S.; Assefi, M.; Safaei, S.; Trippe, E.D.; Gutierrez, J.B.; Kochut, K. A Brief Survey of Text Mining: Classification, Clustering and Extraction Techniques. *arXiv* **2017**, arXiv:1707.02919.
31. Monteiro, B.R.; Davis, C.A., Jr.; Fonseca, F. A survey on the geographic scope of textual documents. *Comput. Geosci.* **2016**, *96*, 23–34. [CrossRef]
32. Purves, R.S.; Clough, P.; Jones, C.B.; Hall, M.H.; Murdock, V. Geographic information retrieval: Progress and challenges in spatial search of text. *Found. Trends Inf. Retr.* **2018**, *12*, 164–318. [CrossRef]
33. Rajbhandari, S.; Aryal, J.; Osborn, J.; Musk, R.; Lucieer, A. Benchmarking the applicability of ontology in geographic object-based image analysis. *ISPRS Int. J. Geo-Inf.* **2017**, *6*, 386. [CrossRef]
34. Guilbert, É.; Moulin, B. Towards a Common Framework for the Identification of Landforms on Terrain Models. *ISPRS Int. J. Geo Inf.* **2017**, *6*, 12. [CrossRef]
35. Arvor, D.; Belgiu, M.; Falomir, Z.; Mougenot, I.; Durieux, L. Ontologies to interpret remote sensing images: Why do we need them? *GISci. Remote Sens.* **2019**, *56*, 911–939. [CrossRef]
36. Ballatore, A. Prolegomena for an Ontology of Place. In *Advancing Geographic Information Science*; GSDI Association Press: Needham, MA, USA, 2016.
37. Garbacz, P.; Lawrynowicz, A.; Szady, B. Identity Criteria for Localities. In Proceedings of the FOIS, Cape Town, South Africa, 19–21 September 2018.
38. Krisnadhi, A.; Hu, Y.; Janowicz, K.; Hitzler, P.; Arko, R.; Carbotte, S.; Chandler, C.; Cheatham, M.; Fils, D.; Finin, T.; et al. The GeoLink modular oceanography ontology. In Proceedings of the International Semantic Web Conference, Bethlehem, PA, USA, 11–15 October 2015; Springer: Berlin/Heidelberg, Germany, 2015; pp. 301–309.
39. Gould, N.; Mackaness, W. From taxonomies to ontologies: formalizing generalization knowledge for on-demand mapping. *Cartogr. Geogr. Inf. Sci.* **2016**, *43*, 208–222. [CrossRef]
40. Yan, J.; Guilbert, É.; Saux, E. An ontology-driven multi-agent system for nautical chart generalization. *Cartogr. Geogr. Inf. Sci.* **2017**, *44*, 201–215. [CrossRef]
41. Varanka, D.; Usery, E.L. The map as knowledge base. *Int. J. Cartogr.* **2018**, *4*, 201–223. [CrossRef]

42. Janowicz, K.; Haller, A.; Cox, S.J.; Le Phuoc, D.; Lefrançois, M. SOSA: A lightweight ontology for sensors, observations, samples, and actuators. *J. Web Semant.* **2019**, *56*, 1–10. [CrossRef]
43. Auer, S.; Lehmann, J.; Hellmann, S. LinkedGeoData: Adding a Spatial Dimension to the Web of Data. In Proceedings of the International Semantic Web Conference, Chantilly, VA, USA, 25–29 October 2009.
44. Vatant, B.; Wick, M. GeoNames Ontology. 2012. Available online: http://www.geonames.org/ontology/ (accessed on 1 April 2021)
45. J. Paul Getty Trust [Los Angeles, CA]. Getty Thesaurus of Geographic Names. [Software, E-Resource] Retrieved from the Library of Congress. 1999. Available online: https://lccn.loc.gov/99483604 (accessed on 1 April 2021).
46. Derungs, C.; Purves, R. Mining nearness relations from an n-grams Web corpus in geographical space. *Spat. Cogn. Comput.* **2016**, *16*, 301–322. [CrossRef]
47. Ballatore, A. Extracting Place Emotions from Travel Blogs. In Proceedings of the AGILE, Helsinki, Finland, 25–29 May 2015.
48. Strapparava, C.; Valitutti, A. WordNet Affect: An Affective Extension of WordNet. In Proceedings of the LREC, Lisbon, Portugal, 26–28 May 2004.
49. Derungs, C.; Purves, R. From text to landscape: locating, identifying and mapping the use of landscape features in a Swiss Alpine corpus. *Int. J. Geogr. Inf. Sci.* **2014**, *28*, 1272–1293. [CrossRef]
50. Wartmann, F.; Acheson, E.; Purves, R. Describing and comparing landscapes using tags, texts, and free lists: an interdisciplinary approach. *Int. J. Geogr. Inf. Sci.* **2018**, *32*, 1572–1592. [CrossRef]
51. Ilarri, S.; Stojanovic, D.; Ray, C. Semantic management of moving objects: A vision towards smart mobility. *Expert Syst. Appl.* **2015**, *42*, 1418–1435. [CrossRef]
52. Fileto, R.; May, C.; Renso, C.; Pelekis, N.; Klein, D.; Theodoridis, Y. The Baquara2 knowledge-based framework for semantic enrichment and analysis of movement data. *Data Knowl. Eng.* **2015**, *98*, 104–122. [CrossRef]
53. Han, L.; Kashyap, A.L.; Finin, T.W.; Mayfield, J.; Weese, J. UMBC-EBIQUITY-CORE: Semantic Textual Similarity Systems. In Proceedings of the *SEM@NAACL-HLT, Atlanta, GA, USA, 13–14 June 2013.
54. Jiang, Y.; Li, Y.; Yang, C.; Liu, K.; Armstrong, E.; Huang, T.; Moroni, D.; Finch, C. A comprehensive methodology for discovering semantic relationships among geospatial vocabularies using oceanographic data discovery as an example. *Int. J. Geogr. Inf. Sci.* **2017**, *31*, 2310–2328. [CrossRef]
55. Li, W.; Bhatia, V.; Cao, K. Intelligent polar cyberinfrastructure: Enabling semantic search in geospatial metadata catalogue to support polar data discovery. *Earth Sci. Inform.* **2015**, *8*, 111–123. [CrossRef]
56. Hu, Y. Geospatial Semantics. *Compr. Geogr. Inf. Syst.* **2018**, 80–94. [CrossRef]
57. Hakimpour, F.; Timpf, S. Using Ontologies for resolution of Semantic Heterogeneity in GIS. In Proceedings of the 4th AGILE Conference on Geographic Information Science, Brno, Czech Republic, 19–21 April 2001.
58. Fallahi, G.; Frank, A.; Mesgari, M.; Rajabifard, A. An ontological structure for semantic interoperability of GIS and environmental modeling. *Int. J. Appl. Earth Obs. Geoinf.* **2008**, *10*, 342–357. [CrossRef]
59. Fonseca, F.; Câmara, G.; Monteiro, A. A Framework for Measuring the Interoperability of Geo-Ontologies. *Spat. Cogn. Comput.* **2006**, *6*, 309–331. [CrossRef]
60. Brodaric, B. The design of GSC FieldLog: ontology-based software for computer aided geological field mapping. *Comput. Geosci.* **2004**, *30*, 5–20. [CrossRef]
61. Schuurman, N. Social Dimensions of Object Definition in GIS. In *Re-Presenting GIS*; Fisher, P., Unwin, D., Eds.; John Wiley & Sons: Hoboken, NJ, USA, 2005.
62. Baglioni, M.; Masserotti, M.V.; Renso, C.; Spinsanti, L. Building Geospatial Ontologies from Geographical Databases. In Proceedings of the GeoS, Mexico City, Mexico, 29–30 November 2007.
63. Scheider, S. Grounding Geographic Information in Perceptual Operations. In *Frontiers in Artificial Intelligence and Applications*; IOS Press: Amsterdam, The Netherlands, 2012.
64. Worboys, M.; Stewart, K. From Objects to Events: GEM, the Geospatial Event Model. In Proceedings of the GIScience, Adelphi, MD, USA, 20–23 October 2004.
65. Raskin, R.G.; Pan, M.J. Knowledge representation in the semantic web for Earth and environmental terminology (SWEET). *Comput. Geosci.* **2005**, *31*, 1119–1125. [CrossRef]
66. Couclelis, H. Ontologies of geographic information. *Int. J. Geogr. Inf. Sci.* **2010**, *24*, 1785–1809. [CrossRef]
67. Hu, Y.; Janowicz, K.; Carral, D.; Scheider, S.; Kuhn, W.; Berg-Cross, G.; Hitzler, P.; Dean, M.; Kolas, D. A Geo-ontology Design Pattern for Semantic Trajectories. In Proceedings of the COSIT, Scarborough, UK, 2–6 September 2013.
68. Carral, D.; Scheider, S.; Janowicz, K.; Vardeman, C.; Krisnadhi, A.A.; Hitzler, P. An Ontology Design Pattern for Cartographic Map Scaling. In Proceedings of the ESWC, Montpellier, France, 26–30 May 2013.
69. Cruz, I.; Sunna, W.; Makar, N.; Bathala, S. A visual tool for ontology alignment to enable geospatial interoperability. *J. Vis. Lang. Comput.* **2007**, *18*, 230–254. [CrossRef]
70. Goodchild, M. *Formalizing Place in Geographic Information Systems*; Springer: New York, NY, USA, 2011.
71. Goodchild, M.; Hill, L. Introduction to digital gazetteer research. *Int. J. Geogr. Inf. Sci.* **2008**, *22*, 1039–1044. [CrossRef]
72. Alani, H.; Jones, C.; Tudhope, D. Voronoi-based region approximation for geographical information retrieval with gazetteers. *Int. J. Geogr. Inf. Sci.* **2001**, *15*, 287–306. [CrossRef]

73. Rice, M.T.; Aburizaiza, A.O.; Jacobson, R.; Shore, B.M.; Paez, F.I. Supporting Accessibility for Blind and Vision-impaired People With a Localized Gazetteer and Open Source Geotechnology. *Trans. GIS* **2012**, *16*, 177–190. [CrossRef]
74. Schlieder, C.; Vögele, T.; Visser, U. Qualitative Spatial Representation for Information Retrieval by Gazetteers. In Proceedings of the COSIT, Morro Bay, CA, USA, 19–23 September 2001.
75. Janowicz, K.; Keßler, C. The role of ontology in improving gazetteer interaction. *Int. J. Geogr. Inf. Sci.* **2008**, *22*, 1129–1157. [CrossRef]
76. Davies, C.; Holt, I.; Green, J.; Harding, J.; Diamond, L. User Needs and Implications for Modelling Vague Named Places. *Spat. Cogn. Comput.* **2009**, *9*, 174–194. [CrossRef]
77. Hollenstein, L.; Purves, R. Exploring place through user-generated content: Using Flickr tags to describe city cores. *J. Spat. Inf. Sci.* **2010**, *1*, 21–48.
78. Grothe, C.; Schaab, J. Automated Footprint Generation from Geotags with Kernel Density Estimation and Support Vector Machines. *Spat. Cogn. Comput.* **2009**, *9*, 195–211. [CrossRef]
79. Keßler, C.; Maué, P.; Heuer, J.T.; Bartoschek, T. Bottom-up gazetteers: Learning from the implicit semantics of geotags. In Proceedings of the International Conference on GeoSpatial Sematics, Mexico City, Mexico, 3–4 December 2009; Springer: Berlin/Heidelberg, Germany, 2009; pp. 83–102.
80. Li, L.; Goodchild, M.F. Constructing places from spatial footprints. In Proceedings of the 1st ACM SIGSPATIAL International Workshop on Crowdsourced and Volunteered Geographic Information, Redondo Beach, CA, USA, 7–9 November 2012; pp. 15–21.
81. Gao, S.; Li, L.; Li, W.; Janowicz, K.; Zhang, Y. Constructing gazetteers from volunteered Big Geo-Data based on Hadoop. *Comput. Environ. Urban Syst.* **2017**, *61*, 172–186. [CrossRef]
82. Uryupina, O. Semi-supervised learning of geographical gazetteers from the internet. In Proceedings of the HLT-NAACL 2003, Edmonton, AB, Canada, 27 May–1 June 2003.
83. Zhu, R.; Hu, Y.; Janowicz, K.; McKenzie, G. Spatial signatures for geographic feature types: examining gazetteer ontologies using spatial statistics. *Trans. GIS* **2016**, *20*, 333–355. [CrossRef]
84. Samal, A.; Seth, S.; Cueto, K. A feature-based approach to conflation of geospatial sources. *Int. J. Geogr. Inf. Sci.* **2004**, *18*, 459–489. [CrossRef]
85. Sehgal, V.; Getoor, L.; Viechnicki, P. Entity resolution in geospatial data integration. In Proceedings of the GIS '06, Arlington, VA, USA, 10–11 November 2006.
86. Hastings, J. Automated conflation of digital gazetteer data. *Int. J. Geogr. Inf. Sci.* **2008**, *22*, 1109–1127. [CrossRef]
87. Jones, C.; Purves, R. Geographical information retrieval. *Int. J. Geogr. Inf. Sci.* **2008**, *22*, 219–228. [CrossRef]
88. Jones, R.; Zhang, W.; Rey, B.; Jhala, P.; Stipp, E. Geographic intention and modification in web search. *Int. J. Geogr. Inf. Sci.* **2008**, *22*, 229–246. [CrossRef]
89. Sanderson, M.; Kohler, J. Analyzing geographic queries. In Proceedings of the SIGIR Workshop on Geographic Information Retrieval, Sheffield, UK, 25–29 July 2004.
90. Janowicz, K.; Raubal, M.; Kuhn, W. The semantics of similarity in geographic information retrieval. *J. Spat. Inf. Sci.* **2011**, *2*, 29–57. [CrossRef]
91. Hu, Y.; Janowicz, K.; Prasad, S. Improving wikipedia-based place name disambiguation in short texts using structured data from DBpedia. In Proceedings of the GIR '14, Dallas, TX, USA, 4 November 2014.
92. Cucerzan, S. Large-Scale Named Entity Disambiguation Based on Wikipedia Data. In Proceedings of the EMNLP-CoNLL, Prague, Czech Republic, 28–30 June 2007.
93. Overell, S.; Rüger, S. Using co-occurrence models for placename disambiguation. *Int. J. Geogr. Inf. Sci.* **2008**, *22*, 265–287. [CrossRef]
94. Leidner, J.L. *Toponym Resolution in Text: Annotation, Evaluation and Applications of Spatial Grounding of Place Names*; Universal-Publishers: Irvine, CA, USA, 2008.
95. Ju, Y.; Adams, B.; Janowicz, K.; Hu, Y.; Yan, B.; McKenzie, G. Things and Strings: Improving Place Name Disambiguation from Short Texts by Combining Entity Co-Occurrence with Topic Modeling. In Proceedings of the EKAW, Bologna, Italy, 19–23 November 2016.
96. Gelernter, J.; Balaji, S. An algorithm for local geoparsing of microtext. *GeoInformatica* **2013**, *17*, 635–667. [CrossRef]
97. Vasardani, M.; Winter, S.; Richter, K.F. Locating place names from place descriptions. *Int. J. Geogr. Inf. Sci.* **2013**, *27*, 2509–2532. [CrossRef]
98. Hu, Y.; Janowicz, K.; Prasad, S.; Gao, S. Metadata Topic Harmonization and Semantic Search for Linked-Data-Driven Geoportals: A Case Study Using ArcGIS Online. *Trans. GIS* **2015**, *19*, 398–416. [CrossRef]
99. Li, W.; Goodchild, M.F.; Raskin, R. Towards geospatial semantic search: exploiting latent semantic relations in geospatial data. *Int. J. Digit. Earth* **2014**, *7*, 17–37. [CrossRef]
100. Amitay, E.; Har'El, N.; Sivan, R.; Soffer, A. Web-a-where: geotagging web content. In Proceedings of the SIGIR '04, Sheffield, UK, 25–29 July 2004.
101. Silva, M.; Martins, B.; Chaves, M.; Afonso, A.; Cardoso, N. Adding geographic scopes to web resources. *Comput. Environ. Urban Syst.* **2006**, *30*, 378–399. [CrossRef]
102. Wang, C.; Xie, X.; Wang, L.; Lu, Y.; Ma, W. Detecting geographic locations from web resources. In Proceedings of the GIR '05, Bremen, Germany, 4 November 2005.

103. Frontiera, P.; Larson, R.; Radke, J. A comparison of geometric approaches to assessing spatial similarity for GIR. *Int. J. Geogr. Inf. Sci.* **2008**, *22*, 337–360. [CrossRef]
104. Jones, C.; Purves, R.; Ruas, A.; Sanderson, M.; Sester, M.; Kreveld, M.V.; Weibel, R. Spatial information retrieval and geographical ontologies an overview of the SPIRIT project. In Proceedings of the SIGIR '02, Tampere, Finland, 11–15 August 2002.
105. Keßler, C.; Janowicz, K.; Bishr, M. An agenda for the next generation gazetteer: Geographic information contribution and retrieval. In Proceedings of the 17th ACM SIGSPATIAL International Conference on Advances in Geographic Information Systems, Seattle, WA, USA, 3–6 November 2009; pp. 91–100.
106. Gey, F.; Larson, R.; Sanderson, M.; Joho, H.; Clough, P.D. GeoCLEF: the CLEF 2005 Cross-Language Geographic Information Retrieval Track. In Proceedings of the CLEF, Vienna, Austria, 21–23 September 2005.
107. Egenhofer, M. Toward the semantic geospatial web. In Proceedings of the GIS '02, McLean, VA, USA, 8–9 November 2002.
108. Hart, G.; Dolbear, C. *Linked Data: A Geographic Perspective*; CRC Press: Boca Raton, FL, USA, 2013.
109. Kuhn, W.; Kauppinen, T.; Janowicz, K. Linked data-a paradigm shift for geographic information science. In Proceedings of the International Conference on Geographic Information Science, Vienna, Austria, 24–26 September 2014; Springer: Berlin/Heidelberg, Germany, 2014; pp. 173–186.
110. Goodwin, J.; Dolbear, C.; Hart, G. Geographical Linked Data: The Administrative Geography of Great Britain on the Semantic Web. *Trans. Gis* **2008**, *12*, 19–30. [CrossRef]
111. Patroumpas, K.; Alexakis, M.; Giannopoulos, G.; Athanasiou, S. TripleGeo: An ETL Tool for Transforming Geospatial Data into RDF Triples. In Proceedings of the Edbt/Icdt Workshops, Citeseer, Athens, Greece, 24–28 March 2014; pp. 275–278.
112. Janowicz, K.; Scheider, S.; Pehle, T.; Hart, G. Geospatial semantics and linked spatiotemporal data–Past, present, and future. *Semant. Web* **2012**, *3*, 321–332. [CrossRef]
113. Battle, R.; Kolas, D. GeoSPARQL: Enabling a Geospatial Semantic Web. *Semantic Web J.* **2011**, *3*, 355–370. [CrossRef]
114. Athanasis, N.; Kalabokidis, K.; Vaitis, M.; Soulakellis, N. Towards a semantics-based approach in the development of geographic portals. *Comput. Geosci.* **2009**, *35*, 301–308. [CrossRef]
115. Purves, R.; Edwardes, A.; Wood, J. Describing place through user generated content. *First Monday* **2011**, *16*. [CrossRef]
116. Manning, C.D.; Surdeanu, M.; Bauer, J.; Finkel, J.R.; Bethard, S.; McClosky, D. The Stanford CoreNLP natural language processing toolkit. In Proceedings of the 52nd Annual Meeting of the Association for Computational Linguistics: System Demonstrations, Baltimore, MD, USA, 23–24 June 2014; pp. 55–60.
117. Adams, B.; McKenzie, G.; Gahegan, M. Frankenplace: Interactive Thematic Mapping for Ad Hoc Exploratory Search. In Proceedings of the 24th International Conference on World Wide Web, WWW '15, Florence, Italy, 18–22 May 2015; International World Wide Web Conferences Steering Committee: Geneva, Switzerland, 2015; pp. 12–22. [CrossRef]
118. Kim, J.; Vasardani, M.; Winter, S. Similarity matching for integrating spatial information extracted from place descriptions. *Int. J. Geogr. Inf. Sci.* **2017**, *31*, 56–80. [CrossRef]
119. Ye, M.; Shou, D.; Lee, W.; Yin, P.; Janowicz, K. On the semantic annotation of places in location-based social networks. In Proceedings of the KDD, San Diego, CA, USA, 21–24 August 2011.
120. Rattenbury, T.; Naaman, M. Methods for extracting place semantics from Flickr tags. *ACM Trans. Web (TWEB)* **2009**, *3*, 1–30. [CrossRef]
121. Hu, Y.; Gao, S.; Janowicz, K.; Yu, B.; Li, W.; Prasad, S. Extracting and understanding urban areas of interest using geotagged photos. *Comput. Environ. Urban Syst.* **2015**, *54*, 240–254. [CrossRef]
122. Klippel, A.; Tappe, H.; Kulik, L.; Lee, P.U. Wayfinding choremes—A language for modeling conceptual route knowledge. *J. Vis. Lang. Comput.* **2005**, *16*, 311–329. [CrossRef]
123. Renz, J.; Nebel, B. Qualitative Spatial Reasoning Using Constraint Calculi. In *Handbook of Spatial Logics*; Springer: Berlin/Heidelberg, Germany, 2007.
124. Gao, S.; Janowicz, K.; McKenzie, G.; Li, L. Towards Platial Joins and Buffers in Place-Based GIS. In Proceedings of the COMP '13, London, UK, 12–13 March 2013.
125. Janowicz, K.; Scheider, S.; Adams, B. A Geo-semantics Flyby. In Proceedings of the Reasoning Web. Semantic Technologies for Intelligent Data Access, Reasoning Web 2013, Mannheim, Germany, 30 July–2 August 2013; Rudolph, S., Gottlob, G., Horrocks, I., van Harmelen, F., Eds.; Springer: Berlin/Heidelberg, Germany, 2013; Volume 8067, pp. 230–250. [CrossRef]
126. Bateman, J.; Farrar, S. Towards a generic foundation for spatial ontology. In *Proceedings of the Third International Conference on Formal Ontology in Information Systems*; IOS Press: Amsterdam, The Netherlands, 2004.
127. Bittner, T.; Donnelly, M.; Smith, B. A spatio-temporal ontology for geographic information integration. *Int. J. Geogr. Inf. Sci.* **2009**, *23*, 765–798. [CrossRef]
128. Brodaric, B.; Gahegan, M. Experiments to Examine the Situated Nature of Geoscientific Concepts. *Spat. Cogn. Comput.* **2007**, *7*, 61–95. [CrossRef]
129. Bennett, B.; Mallenby, D.; Third, A. An Ontology for Grounding Vague Geographic Terms. In Proceedings of the FOIS, Saarbrücken, Germany, 31 October–3 November 2008.
130. Chrisman, N. *Exploring Geographic Information Systems*; Wiley: Hoboken, NJ, USA, 2001.
131. Kuhn, W. Semantic reference systems. *Int. J. Geogr. Inf. Sci.* **2003**, *17*, 405–409. [CrossRef]
132. Mark, D.M.; Smith, B.; Egenhofer, M.; Hirtle, S. Ontological foundations for geographic information science. *Res. Challenges Geogr. Inf. Sci.* **2004**, 335–350.

133. Kauppinen, T.; Hyvönen, E. Modeling and Reasoning About Changes in Ontology Time Series. In *Ontologies: A Handbook of Principles, Concepts and Applications in Information Systems*; Sharman, R., Kishore, R., Ramesh, R., Eds.; Springer: Boston, MA, USA, 2007; pp. 319–338. [CrossRef]
134. Christakos, G.; Bogaert, P.; Serre, M. *Temporal GIS: Advanced Functions for Field-Based Applications*; Springer: Berlin/Heidelberg, Germany, 2002.
135. Galton, A.; Mizoguchi, R. The water falls but the waterfall does not fall: New perspectives on objects, processes and events. *Appl. Ontol.* **2009**, *4*, 71–107. [CrossRef]
136. Hage, W.V.; Malaisé, V.; Segers, R.; Hollink, L.; Schreiber, G. Design and use of the Simple Event Model (SEM). *J. Web Semant.* **2011**, *9*, 128–136. [CrossRef]
137. Montello, D.R.; Goodchild, M.F.; Gottsegen, J.; Fohl, P. Where's downtown?: Behavioral methods for determining referents of vague spatial queries. *Spat. Cogn. Comput.* **2003**, *3*, 185–204. [CrossRef]
138. Abdelmoty, A.I.; Smart, P.; Jones, C.B. Building Place Ontologies for the Semantic Web: Issues and Approaches. In Proceedings of the 4th ACM Workshop on Geographical Information Retrieval, GIR'07, Lisbon, Portugal, 9 November 2007; Association for Computing Machinery: New York, NY, USA, 2007; pp. 7–12. [CrossRef]
139. Lutz, M.; Klien, E. Ontology-based retrieval of geographic information. *Int. J. Geogr. Inf. Sci.* **2006**, *20*, 233–260. [CrossRef]
140. Jones, C.; Alani, H.; Tudhope, D. Geographical Information Retrieval with Ontologies of Place. In Proceedings of the COSIT, Morro Bay, CA, USA, 19–23 September 2001.
141. Jordan, T.; Raubal, M.; Gartrell, B.; Egenhofer, M. An Affordance-Based Model of Place in GIS. 1999. Available online: https://citeseerx.ist.psu.edu/viewdoc/download?doi=10.1.1.4.8628&rep=rep1&type=pdf (accessed on 1 April 2021)
142. Alazzawi, A.; Abdelmoty, A.; Jones, C. What can I do there? Towards the automatic discovery of place-related services and activities. *Int. J. Geogr. Inf. Sci.* **2012**, *26*, 345–364. [CrossRef]
143. Ying, J.; Lee, W.; Weng, T.C.; Tseng, V.S. Semantic trajectory mining for location prediction. In Proceedings of the 19th ACM SIGSPATIAL International Conference on Advances in Geographic Information Systems, Chicago, IL, USA, 1–4 November 2011.
144. Harvey, F.; Kuhn, W.; Pundt, H.; Bishr, Y.; Riedemann, C. Semantic interoperability: A central issue for sharing geographic information. *Ann. Reg. Sci.* **1999**, *33*, 213–232. [CrossRef]
145. Dou, D.; McDermott, D.; Qi, P. Ontology Translation on the Semantic Web. *J. Data Semant.* **2005**, *2*, 35–57.
146. Raubal, M. Formalizing conceptual spaces. Formal ontology in information systems. In Proceedings of the Third International Conference (FOIS 2004), Torino, Italy, 4–6 November 2004; Volume 114, pp. 153–164.
147. Rodriguez, M.A.; Egenhofer, M. Comparing geospatial entity classes: an asymmetric and context-dependent similarity measure. *Int. J. Geogr. Inf. Sci.* **2004**, *18*, 229–256. [CrossRef]
148. Schwering, A.; Raubal, M. Measuring Semantic Similarity Between Geospatial Conceptual Regions. In Proceedings of the GeoS, Mexico City, Mexico, 29–30 November 2005.
149. Li, B.; Fonseca, F. TDD: A comprehensive model for qualitative spatial similarity assessment. *Spat. Cogn. Comput.* **2006**, *6*, 31–62. [CrossRef]
150. Martin, D.; Burstein, M.; McDermott, D.; McIlraith, S.; Paolucci, M.; Sycara, K.; McGuinness, D.L.; Sirin, E.; Srinivasan, N. Bringing semantics to web services with OWL-S. *World Wide Web* **2007**, *10*, 243–277. [CrossRef]
151. Fensel, D.; Bussler, C. The Web Service Modeling Framework WSMF. *Electron. Commer. Res. Appl.* **2002**, *1*, 113–137. [CrossRef]
152. Vaccari, L.; Shvaiko, P.; Marchese, M. A geo-service semantic integration in Spatial Data Infrastructures. *Int. J. Spat. Data Infrastruct. Res.* **2009**, *4*, 24–51.
153. Lemmens, R.; Wytzisk, A.; de By, R.; Granell, C.; Gould, M.; Van Oosterom, P. Integrating semantic and syntactic descriptions to chain geographic services. *IEEE Internet Comput.* **2006**, *10*, 42–52. [CrossRef]
154. Lutz, M. Ontology-based descriptions for semantic discovery and composition of geoprocessing services. *Geoinformatica* **2007**, *11*, 1–36. [CrossRef]
155. Janowicz, K.; Schade, S.; Bröring, A.; Keßler, C.; Maué, P.; Stasch, C. Semantic Enablement for Spatial Data Infrastructures. *Trans. GIS* **2010**, *14*, 111–129. [CrossRef]
156. Klien, E. A Rule-Based Strategy for the Semantic Annotation of Geodata. *Trans. GIS* **2007**, *11*, 437–452. [CrossRef]
157. Kuhn, W. Cognitive and Linguistic Ideas in Geographic Information Semantics. In *Cognitive and Linguistic Aspects of Geographic Space*; Lecture Notes in Geoinformation and Cartography; Springer: Berlin/Heidelberg, Germany, 2013; pp. 159–174. [CrossRef]
158. Zhu, Y. Geospatial semantics, ontology and knowledge graphs for big Earth data. *Big Earth Data* **2019**, *3*, 187–190. [CrossRef]
159. Di Donato, P. Geospatial Semantics: A Critical Review. In *Computational Science and Its Applications—ICCSA 2010*; Taniar, D., Gervasi, O., Murgante, B., Pardede, E., Apduhan, B.O., Eds., Springer: Berlin/Heidelberg, Germany, 2010; Volume 6016. [CrossRef]
160. Schuurman, N. Formalization Matters: Critical GIS and Ontology Research. *Ann. Assoc. Am. Geogr.* **2006**, *96*, 726–739. [CrossRef]
161. Wang, C.; Kantor, C.M.; Mitchell, J.T.; Bacastow, T.S. Digital Earth Education. In *Manual of Digital Earth*; Springer Nature: Berlin/Heidelberg, Germany, 2020; pp. 755–783.
162. Goffi, A.; Bordogna, G.; Stroppiana, D.; Boschetti, M.; Brivio, P.A. Knowledge and Data-Driven Mapping of Environmental Status Indicators from Remote Sensing and VGI. *Remote Sens.* **2020**, *12*, 495. [CrossRef]
163. Bordogna, G.; Ghisalberti, G.; Psaila, G. Geographic information retrieval: Modeling uncertainty of user's context. *Fuzzy Sets Syst. FSS* **2012**, *196*. [CrossRef]

164. Zadrozny, S.; Kacprzyk, J. Bipolar Queries Using Various Interpretations of Logical Connectives. In Proceedings of the Foundations of Fuzzy Logic and Soft Computing, 12th International Fuzzy Systems Association World Congress, IFSA 2007, Cancun, Mexico, 18–21 June 2007; Melin, P., Castillo, O., Aguilar, L.T., Kacprzyk, J., Pedrycz, W., Eds.; Springer: Berlin/Heidelberg, Germany, 2007; Volume 4529, pp. 181–190. [CrossRef]
165. Dubois, D.; Prade, H. An introduction to bipolar representations of information and preference. *Int. J. Intell. Syst.* **2008**, *23*, 866–877. [CrossRef]
166. Dujmović, J.; Larsen, H. Generalized conjunction/disjunction. *Int. J. Approx. Reason.* **2007**, *46*, 423–446. [CrossRef]
167. Arcaini, P.; Bordogna, G.; Ienco, D.; Sterlacchini, S. User-driven geo-temporal density-based exploration of periodic and not periodic events reported in social networks. *Inf. Sci.* **2016**, *340–341*, 122–143. [CrossRef]
168. Fellbaum, C. (Ed.) *WordNet: An Electronic Lexical Database*; Language, Speech, and Communication; MIT Press: Cambridge, MA, USA, 1998.
169. Psaila, G.; Toccu, M.; Bordogna, G.; Frigerio, L.; Cuzzocrea, A. An Interoperable Open Data Framework for Discovering Popular Tours Based on Geo-Tagged Tweets. *Int. J. Intell. Inf. Database Syst.* **2017**, *10*, 1. [CrossRef]
170. Miles, A.; Bechhofer, S. SKOS Simple Knowledge Organization System Reference. In Proceedings of the W3C Recommendation, W3C, Maputo, Mozambique, 1–2 April 2009.
171. Perego, A.; Fugazza, C.; Vaccari, L.; Lutz, M.; Smits, P.; Kanellopoulos, I.; Schade, S. Harmonization and Interoperability of EU Environmental Information and Services. *IEEE Intell. Syst.* **2012**, *27*, 33–39. [CrossRef]
172. Brickley, D.; Miller, L. The Friend Of A Friend (FOAF) Vocabulary Specification. 2007. Available online: http://xmlns.com/foaf/spec/ (accessed on 1 April 2021)
173. Santoro, M.; Mazzetti, P.; Nativi, S.; Fugazza, C.; Granell, C.; Díaz, L. Methodologies for augmented discovery of geospatial resources. In *Discovery of Geospatial Resources: Methodologies, Technologies, and Emergent Applications*; Díaz, L., Granell, C., Huerta, J., Eds.; IGI Global: Hershey, PA, USA, 2012; Chapter 9, pp. 172–203.
174. Sporny, M.; Longley, D.; Kellogg, G.; Lanthaler, M.; Champin, P.A.; Lindström, N. JSON-LD 1.1 A JSON-Based Serialization for Linked Data. Ph.D. Thesis, W3C Recommendation, W3C, Cambridge, MA, USA, 2020.
175. Wilkinson, M.D.; Dumontier, M.; Aalbersberg, I.J.; Appleton, G.; Axton, M.; Baak, A.; Blomberg, N.; Boiten, J.W.; da Silva Santos, L.B.; Bourne, P.E.; et al. The FAIR Guiding Principles for scientific data management and stewardship. *Sci. Data* **2016**, *3*, 1–9. [CrossRef]
176. Huang, Y.; Yuan, M.; Sheng, Y.; Min, X.; Cao, Y. Using Geographic Ontologies and Geo-Characterization to Represent Geographic Scenarios. *ISPRS Int. J. Geo-Inf.* **2019**, *8*, 566. [CrossRef]
177. Lü, G.; Batty, M.; Strobl, J.; Lin, H.; Zhu, A.X.; Chen, M. Reflections and speculations on the progress in Geographic Information Systems (GIS): A geographic perspective. *Int. J. Geogr. Inf. Sci.* **2019**, *33*, 346–367. [CrossRef]
178. Bertalanffy, L.V. *General Systems Theory: Foundations, Development, Applications / by Ludwig von Bertalanffy*, rev. ed.; Braziller: New York, NY, USA, 1968.
179. Horrocks, I.; Patel-Schneider, P.F.; Boley, H.; Tabet, S.; Grosof, B.; Dean, M. SWRL: A Semantic Web Rule Language Combining OWL and RuleML. *W3C Memb. Submiss.* **2004**, *21*, 1–31.
180. Fugazza, C.; Menegon, S.; Pepe, M.; Oggioni, A.; Carrara, P. The RITMARE Starter Kit—Bottom-up Capacity Building for Geospatial Data Providers. In Proceedings of the 9th International Conference on Software Paradigm Trends (ICSOFT), INSTICC, Vienna, Austria, 29–31 August 2014; SciTePress: Setúbal, Portugal, 2014; Volume 1, pp. 169–176. [CrossRef]
181. Fugazza, C.; Basoni, A.; Menegon, S.; Oggioni, A.; Pavesi, F.; Pepe, M.; Sarretta, A.; Carrara, P. RITMARE: Semantics-aware Harmonisation of Data in Italian Marine Research. *Procedia Comput. Sci.* **2014**, *33*, 261–265. [CrossRef]
182. Tagliolato, P.; Fugazza, C.; Oggioni, A.; Carrara, P. Semantic Profiles for Easing SensorML Description: Review and Proposal. *ISPRS Int. J. Geo-Inf.* **2019**, *8*, 340. [CrossRef]
183. Pavesi, F.; Basoni, A.; Fugazza, C.; Menegon, S.; Oggioni, A.; Pepe, M.; Tagliolato, P.; Carrara, P. EDI—A Template-Driven Metadata Editor for Research Data. *JORS J. Open Res. Softw.* **2016**, *4*, e40. [CrossRef]
184. SPARQL Working Group. SPARQL 1.1 Query Language. W3C Recommendation 21 March 2013, World Wide Web Consortium. 2013. Available online: http://www.w3.org/TR/sparql11-query/ (accessed on 1 April 2021).
185. Fugazza, C.; Pepe, M.; Oggioni, A.; Tagliolato, P.; Pavesi, F.; Carrara, P. Describing Geospatial Assets in the Web of Data: A Metadata Management Scenario. *ISPRS Int. J. Geo-Inf.* **2016**, *5*, 229. [CrossRef]
186. Fugazza, C.; Tagliolato, P.; Frigerio, L.; Carrara, P. Web-Scale Normalization of Geospatial Metadata Based on Semantics-Aware Data Sources. *ISPRS Int. J. Geo-Inf.* **2017**, *6*, 354. [CrossRef]
187. Fugazza, C.; Pepe, M.; Oggioni, A.; Tagliolato, P.; Carrara, P. Raising Semantics-Awareness in Geospatial Metadata Management. *ISPRS Int. J. Geo-Inf.* **2018**, *7*, 370. [CrossRef]
188. Lefrançois, M.; Zimmermann, A.; Bakerally, N. A SPARQL extension for generating RDF from heterogeneous formats. In Proceedings of the Extended Semantic Web Conference (ESWC'17), Portorož, Slovenia, 28 May–1 June 2017.
189. Prud'hommeaux, E.; Harris, S.; Seaborne, A. SPARQL 1.1 Query Language; Technical Report W3C. 2013. Available online: https://www.w3.org/TR/sparql11-query/ (accessed on 1 April 2021).
190. Lehnert, K. IGSN: International Geo Sample Number. Unambiguous Citation of Physical Samples. 2015. Available online: https://zenodo.org/record/31788 (accessed on 1 April 2021).

191. Wohner, C.; Peterseil, J.; Poursanidis, D.; Kliment, T.; Wilson, M.; Mirtl, M.; Chrysoulakis, N. DEIMS-SDR—A web portal to document research sites and their associated data. *Ecol. Inform.* **2019**, *51*, 15–24. [CrossRef]
192. Zilioli, M.; Oggioni, A.; Tagliolato, P.; Pugnetti, A.; Carrara, P. Feeding Essential Biodiversity Variables (EBVs): Actual and potential contributions from LTER-Italy. *Nat. Conserv.* **2019**, *34*, 477–503. [CrossRef]
193. Percivall, G. OGC's Open Standards for Geospatial Interoperability. In *Encyclopedia of GIS*; Springer International Publishing: Cham, Switzerland, 2017; pp. 1466–1473. [CrossRef]
194. Richardson, L.; Amundsen, M.; Ruby, S. *RESTful Web APIs*; O'Reilly Media, Inc.: Newton, MA, USA, 2013.
195. Fielding, R.T. REST: Architectural Styles and the Design of Network-based Software Architectures. Ph.D. Thesis, University of California, Irvine, CA, USA, 2000.
196. Wilkinson, M.; Verborgh, R.; Bonino da Silva Santos, L.O.; Clark, T.; Swertz, M.; Kelpin, F.; Gray, A.; Schultes, E.; van Mulligen, E.M.; Ciccarese, P.; et al. Interoperability and FAIRness through a novel combination of Web technologies. *PeerJ Comput. Sci.* **2016**. [CrossRef]
197. Trani, L.; Atkinson, M.; Bailo, D.; Paciello, R.; Filgueira, R. Establishing Core Concepts for Information-Powered Collaborations. *Future Gener. Comput. Syst.* **2018**, *89*, 421–437. [CrossRef]
198. Trani, L.; Paciello, R.; Sbarra, M.; Ulbricht, D. Representing Core Concepts for solid-Earth sciences with DCAT—The EPOS-DCAT Application Profile. In Proceedings of the EGU General Assembly Conference Abstracts, Vienna, Austria, 8–13 April 2018; p. 9797.
199. Lanthaler, M.; Gütl, C. Hydra: A Vocabulary for Hypermedia-Driven Web APIs. In Proceedings of the WWW2013 Workshop on Linked Data on the Web, Rio de Janeiro, Brazil, 14 May 2013; Volume 996, ISSN 1613-0073.
200. Lanucara, S.; Fugazza, C.; Tagliolato, P.; Oggioni, A. Information Systems for Precision Agriculture: Monitoring Computation of Prescription Maps. *ERCIM News* **2018**, *113*, 24–25.
201. Ventura, D.; Verborgh, R.; Catania, V.; Mannens, E. Autonomous Composition and Execution of REST APIs for Smart Sensors. In Proceedings of the Joint Proceedings of the 1st Joint International Workshop on Semantic Sensor Networks and Terra Cognita and the 4th International Workshop on Ordering and Reasoning, CEUR Workshop Proceedings, Bethlehem, PA, USA, 11–12 October 2015; Volume 1488, pp. 25–30.
202. Cross, V.; Chen, S. *Fuzzy Information Processing*; Communications in Computer and Information Science; Chapter Fuzzy Ontologies: The State of the Art revised; Springer: Berlin/Heidelberg, Germany, 2018.
203. Bobillo, F.; Straccia, U. The fuzzy ontology reasoner fuzzyDL. *Knowl.-Based Syst.* **2016**, *95*, 12–34. [CrossRef]
204. Safia, B.-B.; Aicha, M. Poss-OWL 2: Possibilistic Extension of OWL 2 for an uncertain geographic ontology. In Proceedings of the 18th International Conference on Knowledge-Based and Intelligent Information & Engineering Systems—KES2014, Gdynia, Poland, 16–18 September 2014; Elsevier Procedia: Amsterdam, The Netherlands, 2014; pp. 407–416.
205. Aerts, D.; Gabora, L. A theory of concepts and their combinations II: A Hilbert space representation. *Kybernetes* **2005**, *34*, 192–221. [CrossRef]
206. Bordogna, G.; Kliment, T.; Frigerio, L.; Brivio, P.A.; Crema, A.; Stroppiana, D.; Boschetti, M.; Sterlacchini, S. A Spatial Data Infrastructure Integrating Multisource Heterogeneous Geospatial Data and Time Series: A Study Case in Agriculture. *ISPRS Int. J. Geo-Inf.* **2016**, *5*, 73. [CrossRef]
207. Bordogna, G.; Frigerio, L.; Kliment, T.; Brivio, P.; Hossard, L.; Manfron, G.; Sterlacchini, S. Contextualized VGI Creation and Management to Cope with Uncertainty and Imprecision. *Int. J. Geo-Inf.* **2016**, *5*, 234–253. [CrossRef]
208. Hahn, J.; Frank, A.U. Select the appropriate map depending on context in a hilbert space model (scop). In *International Symposium on Quantum Interaction; Lecture Notes in Computer Science*; Springer: Berlin/Heidelberg, Germany, 2013; Volume 8369. [CrossRef]
209. Slonneger, K.; Kurtz, B. *Formal Syntax and Semantics of Programming Languages: A Laboratory Based Approach*, 1st ed.; Addison-Wesley Longman Publishing Co., Inc.: Boston, MA, USA, 1995.
210. Miller, G. *The Magical Number Seven, Plus Or Minus Two: Some Limits on Our Capacity for Processing Information*; Bobbs-Merrill Reprint Series in the Social Sciences; P-241, College Division of Bobbs-Merrill Company: Indianapolis, IN, USA, 1975.

MDPI
St. Alban-Anlage 66
4052 Basel
Switzerland
Tel. +41 61 683 77 34
Fax +41 61 302 89 18
www.mdpi.com

ISPRS International Journal of Geo-Information Editorial Office
E-mail: ijgi@mdpi.com
www.mdpi.com/journal/ijgi

www.ingramcontent.com/pod-product-compliance
Lightning Source LLC
Chambersburg PA
CBHW040802250926
51850CB00007BA/1060
9783036563879